The Organization and Management of Construction

SHAPING THEORY AND PRACTICE

VOLUME ONE

Managing the Construction Enterprise

CIB W65

International Symposium for

The Organization and Management of Construction

SHAPING THEORY AND PRACTICE

VOLUME ONE

Managing the Construction Enterprise

Edited by

D.A. Langford and A. Retik

UNIVERSITY OF STRATHCLYDE
200 YEARS OF USEFUL LEARNING

GLASGOW CALEDONIAN UNIVERSITY

Routledge
Taylor & Francis Group
LONDON AND NEW YORK

First published 1996 by E & FN SPON

This edition published 2014 by Routledge
2 Park Square, Milton Park, Abingdon, Oxon OX14 4RN
52 Vanderbilt Avenue, New York, NY 10017, USA

First issued in paperback 2020

Routledge is an imprint of the Taylor & Francis Group, an informa business

A catalogue record for this book is available from the British Library

Publisher's Note
This book has been produced from camera ready copy provided by the individual contributors in order to make the book available for the Conference.

ISBN 13: 978-0-367-57957-9 (pbk)
ISBN 13: 978-0-419-22230-9 (hbk)

CONTENTS

FOREWORD

What makes top performing companies different...?

> They are better organized to meet the needs of customers so that they are either more innovative in anticipating customer needs, more reliable in meeting customer expectations, better able to deliver their product or service more cheaply, or some combination of the above.
>
> Robert Waterman, from *Frontiers of Excellence*

The 1990s has seen the construction industry in many parts of the world subject to the worst effects of global recession. Demand for its products has fallen dramatically, nowhere more so than in the UK. Not only has the industry had to cope with this collapse of its market, it has also been beset by internal strife; the fragmented structure and self-protectionist attitude of individual groups being reminiscent of many western manufacturing industries of the 1960s and 1970s. During the boom construction period of the 1980s, there was strife between professions: architects belittling surveyors, surveyors criticizing engineers; and also, disputes between contractors; management contractors criticizing performance from trade contractors and they, in turn, complaining of late, or non-payment. In such circumstances in time of growth, it comes as no surprise to see that matters could only get worse in times of recession.

In the period of intense competition for a drastically reduced workload, in the past 5–7 years, internal strife has increased. Internal competition has also intensified, to the point of professionals and contractors submitting uneconomic, or even suicidal, bids for available work. Whilst this internal conflict has grown more intense and may seem to be 'life-threatening' to the construction sector, the industry has, by and large, ignored the greatest threat to its existence: the changing nature and demands of its customers. The lessons which may be learned from other industries which have, in the past, made the same fundamental error, appear to have been missed by the construction industry. For example, whilst internal strife in the UK automotive industry of the 1960s and 1970s may appear to have been the main cause of its 'terminal illness', the true cause of its ultimate failure lay, not in internal conflict, but in parallel failure to see the changes which were taking place in society and in competitive markets, at a global level. The industry failed to see the underlying strengths behind the superficial weaknesses of the early products of its new global competitors, in addressing the growth of consumer choice, and customer-based product development. It failed, also, to address the emergence of the 'informed client'; the consumer with more choice of products, a more enquiring and discerning mind in selection of products, and with more power to influence the development of the product to meet his/her particular needs.

There have been some attempts to address areas of perceived conflict and problems within the construction sector, as in the Latham Report, a critical study of internal politics, interdisciplinary and contractual relationships in the UK. In attempting to reduce internal conflict, Sir Michael Latham proposes that there should be fundamental change to the nature of the construction contract, moving from a predisposition for adversarial relationships to support for team building and negotiation. The Latham proposals appear bold and innovative in the context of the construction industry, stating that customers should expect to see savings of up to 30 per cent in construction costs by the year 2000. Such 'ambitious' targets are, however, decried by some, who state that such ridiculously high levels of saving are unachievable. If, however, such 'miserably small' levels of improvement were to have been proposed to Motorola's management in the 1980s, when they were introducing their 'Six-Sigma' programme of quality improvement, the proposers would have been sent back to the drawing board. In quality improvement, Motorola set initial targets of a tenfold improvement over 5 years, to be followed by a further tenfold improvement on the level achieved within the following period! Cynics said that these targets could not be met – yet they were, and were surpassed.

The difference in target levels for improvement may be traced to their origins, in addressing either an internal, industry-based, or external, market-based agenda. It is the difference between continuing to adopt a 'supply-side' dominated approach and moving to a 'demand-led' approach to construction. Setting ambitious targets for improvement in the construction industry cannot be based solely upon study of the internal market, even if the study is of the best performance in the world in the field. Many of the customers will be basing their appraisal of construction sector quality and productivity, not upon other construction providers but, upon other suppliers of major capital equipment. Whilst the Latham Report on the UK construction sector acknowledges the importance of clients, it is still primarily concerned with the internal problems of the industry, rather than the changing nature of the market into which the products are delivered.

If an airline board is to assess the success, or otherwise, of the building procurement process, why should they not compare it with aircraft procurement? If they do, how will the product stand up in terms of quality of manufacture, predictability of capital and revenue cost, performance and reliability etc.? How will the members of the board assess its 'added value' to the company? If these board members are making their annual round of decisions on areas of capital investment, will they be judging their investment in buildings, one against the other, or in terms of likely return to the business – building versus aircraft? It is in this scenario, of looking outward to the global market for products and services, seeking out the best performers in both areas and benchmarking current internal performance against these, that the construction industry will find the real measure of its current level of performance. Only when this has been done can the industry determine the targets which it must meet, in terms of increased productivity, quality and performance, in order to meet the expectations of its customers.

Such change will require revolutionary, not evolutionary, change to the management structure and systems of the construction sector, and fundamental change to the relationships of the internal players. If comparison is made with ‘best practice’ in aeronautical or marine construction, one might predict that the future for building construction lies in the delivery of ‘one-stop shopping’ products, constructed at lower cost, to tighter timescales and to far higher standards of quality than are currently dreamt of.

George M. Cairns

University of Strathclyde
April 1996

INTRODUCTION TO VOLUME ONE

Managing the Construction Enterprise contains papers concerned with three major sub-themes:

- The firm and its business environment
- The firm, its markets and competitors
- The firm and its resources

The papers collected together in 'The firm and its business environment', have been grouped into two sets, those dealing with strategic change in the construction environment and those dealing with strategic management. The first set consider long-term issues influencing managerial activity at the apex of construction firms. Issues such as managing change, industry-based scenario plans and the relationship between firms and environmental movements are placed together to give a forward look to the issues and controversies of the future. In the second part of this sub-theme papers which address the theme of contractors' strategic response to their business environment are presented. International companies are included in this sub-theme.

'The firm, its markets and competitors' is again broken into two sets of papers. One set is contained in an area which deals with marketing in the construction industry. The second explores research in the relatively recent area of partnering.

The third sub-theme focuses upon the resources that firms use to achieve strategic plans and market ambitions. One set of papers uncovers research in the field of human resource management and how the energies of people in construction enterprises can be engaged to realise corporate plans. These papers speak of developing people, equal opportunities, creating learning organizations. The second set tackles the question of building in and managing quality.

The structure of the volumes

To organize the range of topics covered by the W65 Commission the papers have been structured into three volumes. Each volume focuses upon a particular theme. These themes are:

- Managing the Construction Enterprise
- Managing the Project
- Managing Information

Within each theme there are sub-themes which collect together papers around a specialist area of interest and then within each sub-theme there is a further selection to collect together papers on similar topics.

This approach is intended to assist the reader in identifying and consolidating areas of special interest and to link together subjects appearing in each volume.

Maintaining standards

In line with the growing confidence and maturity of the discipline, the papers contained within the volumes have been subjected to exacting standards of refereeing. Of the 241 abstracts received, 183 were selected for development into full papers. All of the full papers received were then refereed and authors invited to incorporate the referees' comments into their papers. At the time of sending the material to the publishers 161 papers had been accepted. The standards achieved ensure that the papers presented in the volumes are of a standard expected in refereed journal articles.

Acknowledgments

We would like to record our special thanks to members of the organizing committee, the scientific committee and the conference organizers, Meeting Makers, especially Susan Miller and Elaine Bone. Particular thanks are due to Edith Henry of E & FN Spon for the painstaking attention to the detail associated with presenting the papers for the 8th International Symposium of the W65 Commission on Organization and Management of Construction.

Dave Langford
Arkady Retik
Glasgow, June 1996

Part I
The firm and its business environment

1 The firm and strategic change

A FRAMEWORK FOR MANAGING STRATEGIC CHANGE IN THE CONSTRUCTION INDUSTRY

Ben Obinero Uwakweh
Construction Innovation Center, University of Cincinnati, Cincinnati, Ohio, USA

Abstract

The management of change is a dynamic and on-going process. Organizations undergo changes but there is no one defined approach in the management of change. What is proposed in this paper is a model that may be used in the managerial process. This model is based on Technology, Political and Cultural factors. It is suggested that all these factors must be managed for the change process to be effective. The concept of Performing Arts is also used to illustrate that the management process must be continuous and any change in one factor will trigger a response in another area.
Keywords: Change, Political, Technical, Cultural, Organization, Strategic.
Sommaire

La gestion du changement est un processus dynamique et permanent. Toute organisation est soumise à des changements; or, il n'existe pas une façon unique de faire face à ces changements. Le présent papier propose un modèle applicable au processus de gestion du changement. Ce modèle est construit sur des facteurs technologiques, politiques et culturels, dont il est suggéré ici que seule la gestion tryptique aboutit à un processus de changement efficace. Le concept d'"arts performants" est utilisé pour illustrer la nécessaire continuité du processus de gestion et la répercussion immédiate de tout changement survenant dans un facteur d'un domaine donné sur un autre facteur d'un autre domaine.

1 Introduction

In the light of the technological innovations in the world today, organizations will also need to undergo changes to remain competitive and profitable in the market place. For example, with advancements in technology, it is possible for firms in Europe to respond to an invitation to bid for a project in the United States or Africa within a day. The implication is that organizations which are not responsive to changes in the environment will miss business opportunities. In recent times, the Construction Industry is undergoing changes in the way it conducts business. In the past, the relationship between Constructors and Designers was very adversarial thus resulting in disputes. However, the emphasis in the industry now is on teams because owners of construction are now

The Organization and Management of Construction: Shaping theory and practice (Volume One). Edited by D.A. Langford and A. Retik. Published in 1996 by E & FN Spon. ISBN 0 419 22230 8.

taking active role in the project delivery process. Owners are now concerned with not only completing their project on time, but are also concerned with the constructors safety records, quality of projects and the companies management capabilities. This demand by owners are encouraging the various parties involved in the construction process to examine their business practices and management styles.

The objective of this paper is therefore to present a framework for managing change in the industry. This framework is based on the literature in organizational development and research and would be used to illustrate how a midwestern firm in the United States is managing its change.

2 The Need For A Model For Managing Change

There are three dominant views that have guided thinking about organizations and the practice of change. The first tradition views organization and change from a technical perspective and hence prescribe change strategies based on empiricism and self interest (Arygris & Schon, 1978). The focus is upon the acquisition and application of knowledge useful for effective performance of organizational tasks. The technical view is instrumental and rational and the focus is upon the acquisition and application of the knowledge useful for effective performance of organizational tasks.

A second tradition views organizations as political entities which can only be changed by the exercise of power by the dominant group over those with less power or by bargaining among powerful groups. The third tradition, however, views organizations as cultural systems of values with shared symbols and shared cognitive schemes which tie people together and form a common organizational culture. Change comes about by altering the norms and cognitive schemes of the members of the organization.

Practicing managers, students of organizations, and experts on change tend to think in terms of only one of the above traditions. The result of this thinking often leads to unanticipated negative consequences. For example, management scientists and field engineers frequently view work and organization design as an engineering or technical problem. This can lead to problems because the tendency in this approach is to design work such that it will no longer be challenging to workers and hence lead to such unanticipated behaviors as absenteeisms and low motivation. On the other hand, a purely political orientation to organization change is likely to be dysfunctional. It can lead to low levels of trust, cynicism, and a view that all interactions are either win or lose situation. The cultural orientation when overemphasized is also not effective. This is because when based solely on trust, and collaboration, it avoids the problem of power politics of change.

Because the use of any one of these traditions provides results or effects that tend to be dysfunctional, a more comprehensive view is needed. This view acknowledges all three approaches and also recognizes that organizations must make continuous adjustment in order to manage change. Further, this comprehensive view of managing change likens the process to performing arts because there are characteristics exhibited in the management of performing arts that managers perform. These three properties are; particularity, variety and contextuality (Vaill, 1989).

Particularity aims at explaining the uniqueness of any organization. This quality makes it particularly difficult to generalize any situation. The term, variety, refers to the basic heterogeneity of human systems. Each personnel will respond differently to a given stimulus. Thus, the manager must constantly aim at having all the personnel to work as a system and consider how they will respond to each situation. Contextuality, on the other hand, focuses on the culture of the organization. Hence in contextuality we emphasize the fact that individuals have the potential to perform at a much greater potential provided they get adequate support from the environment.

3 The Proposed Model: Technical-Political-Cultural (TPC) Model.

The model that is proposed acknowledges that technical, political and cultural issues facing organizations change continuously and as a result organizations make adjustments in order to resolve these conflicts as they arise.

3.1 Technical problem

All organizations have production problem. This is in the context of environmental threats and opportunities, social, financial and technical resources that must be arranged to produce some desired output. In order to solve this problem, Construction firms are constantly seeking for competitive edge as they try to acquire projects. The leverage may be in financial engineering or their past experience in the engineering of the project. Construction firms may also resort to goal setting, strategy formulation, or designing management systems.

3.2 Political problem

All organizations face the problem of allocating power and resources. The uses to which the organization will be put as well as who will reap the benefits of the organization must be determined. Discussion around these issues are reflected in compensation programs, career decisions, budget decisions, and internal power structure of the organization. Unlike the technical problem where we have the formal tools such as strategic planning and organizational design, the concepts of political allocation problem are less formal and less obvious. In construction firms, as in most firms, there are constant discussions among employees on who gets what position, or who is the next in line to the President or Chief Executive Officer.

3.3 Cultural Problem

Organizations are held together by their culture. This consists of values, beliefs, objectives and the interpretation shared by organizational members. Some of the important issues facing management in cultural problems is to determine what values should be shared, what objectives are worth striving for, what beliefs should the employees be committed to and what interpretations of past events would be most beneficial to the firm. Once these decisions are made, then the top leadership must communicate these in a memorable and believable fashion. The idea is to ensure that this will not be instantly forgotten and be discussed as an organizational propaganda.

The model that is being proposed draws heavily from the work done by Noel Tichy (1982). In describing these three factors that must be managed, he describes it as a strategic rope with three strands. Figure 1. presents the strategic rope.

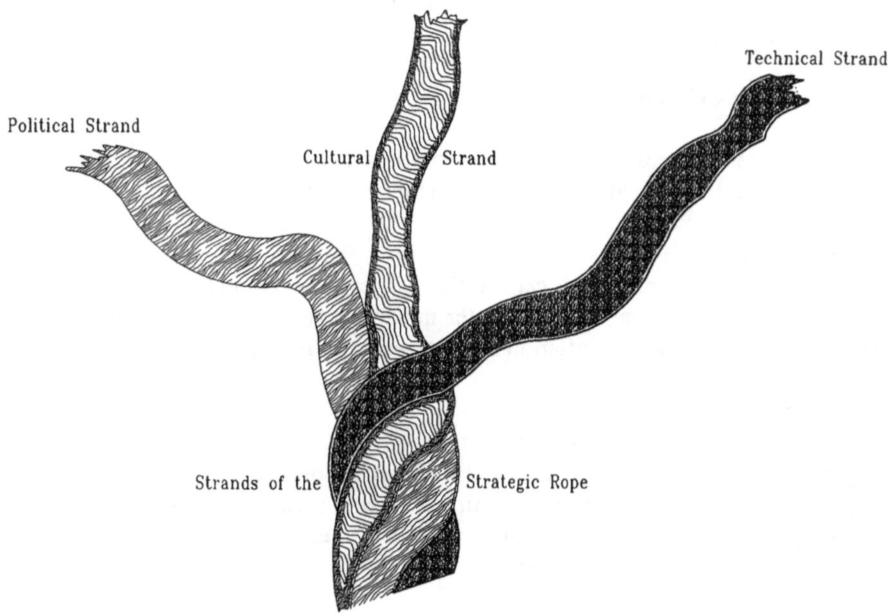

Fig. 1. The strategic rope.

The metaphor of a rope is used for two major reasons. First, from a distance, the individual strands are not distinguishable. This is very true in organizations. It is virtually not clear for a casual observer, what is technical, political or cultural. Second, ropes can become unraveled, and when they do, they become weakened. Similarly, organizations can also become unraveled because the technical, political, and cultural strands can work at cross-purposes. As an illustration, a construction firm that is based in a particular region, and later decides to expand its business into other regions, must change its organizational structure and design so as to be more responsive to the new regional demands. It must therefore make fundamental changes in its political, cultural and technological areas so as to reinforce this regional expansion.

Strategic change management that is proposed in this paper is the task of keeping the rope from being unraveled in the face of these technical, political, and cultural problems. Organizations are perpetually in flux, undergoing shifts and changes and none of these three problems is ever resolved. Strategic change refers to non routine, non incremental, and discontinuous change which alters the overall orientation of the organization. The process of strategic change is started when problems, crises or opportunities are recognized by means of a threshold phenomenon. Wieland and Ullrich, (1976) have determined five areas which may be compared to determine if a certain piece or cluster of information exceeds threshold. These are, historical, planning, extra organizational, expectations and scientific. The determination that a threshold has been exceeded is essentially made by comparing the current data with past.

4 When Strategic change is necessary

Strategic change becomes necessary when problems, crises or opportunities emerge in the environment, diversification, technology and personnel.

4.1 Environment

Increasing competitive pressures can be the trigger for strategic change especially in high-technology industry such as computer industry. However, in the construction industry changes in legislation, inflation and interest rates may easily trigger change.

4.2 Diversification

The diversification of an organization into new areas of business generally requires a strategic change. This may include a shift from its present structure to a divisional structure. For example, a construction firm which is planning to expand its operation from one geographic location to another, may find it necessary to have a small office so as to maintain its presence in that market. It may chose to develop a specific know-how for that particular market.

4.3 Technology

Changes in technology will give rise to new data and behavior requirements on organizations. As an illustration, with increase in use of computers, it is now common to have project managers work with laptop computers on site and personal computers are now available in many site offices.

4.4 Personnel

Another important trigger for strategic change is change in personnel. That is, the types of personnel entering the organization may change in terms of education, expectations, or status, such as previously excluded minorities. Also, the people already in the organization may change as a result of retirements.

5 A Case Study with a midwestern firm:

The company used in this illustration is a midwestern firm which was founded in the 1930s. Approximately seven years after its formation, the founder died. At the time of his death, there were no offsprings to take over the management of the firm. Consequently, the firm was put on trust for the family, under an outside leadership. The firm operates mostly in the midwest and maintain substantial management staff in the major cities where it has operations. It does sometimes do projects in the South Eastern states.

In the sixty plus years that the firm has been in business, it developed and built a strong reputation of quality, excellence and leadership in the industry. They have won numerous awards for excellence in construction. Some of their awards include, Build America Award (this is the highest honor given nationally to construction firms that perform to the highest standards of excellence in construction), Build Kentucky Award for innovation and Build Ohio Award for new construction. The firm's volume in sales is about one hundred and eighty million. It is in the top 40 in volume for Construction Management and in the top 120 in volume for General Contracting. All rankings are by Engineering News Record (ENR).

The firm defines its business as Construction Management and General Building Construction. This includes Hospitals, Institutional Buildings, Office buildings, Shopping Centers, Industrial plants and Air and Transit terminals. A major asset for the firm is the loyalty of its personnel. The firm has a policy of developing its management staff from "within." The total employment including first level supervisory and tradesman regularly reaches over three hundred.

6 Analysis of change

The last surviving son of the founder passed away almost ten years ago. During this period approximately 50 plus years, the firm was under outside leadership. The surviving family members were ready for a change and a group of senior executives in the firm decided to explore the possibility for change in ownership and leadership. After months of meetings and deliberation the employees decided to buy the firm from the family under 100% Employee Stock Ownership Plan.

Employees Stock Ownership Plan (ESOP) is a unique form of passing along some measure of ownership to employees. Hence, it can be used as a financing vehicle for transferring ownership of a privately held company to employees. This provides employees with both retirement benefits and the added incentives inherent in holding an ownership interest in the business. The implication is that it may improve employee motivation, through increased financial incentives and a new set of attitudes and team work. Research indicates that ESOP also have beneficial effects on other measures of success. According to a study, sales and employment grew at a significantly greater rate at ESOP companies that offered participation opportunities to their employees than at ESOP companies with no participation plans. In essence, the purpose of ESOP is to form a partnership among share holders, management and employees by creating a trust vehicle through which employees can purchase meaningful equity stake in the firm.

Although, the firm did not use the proposed TPC model as a guiding principle for its change management, the process it went through, however, can best be described using the proposed TPC model. What follows is a brief illustration.

6.1 Technical

This firm, because of their proposed employee stock ownership (ESOP), had to seriously review and articulate its mission such that it will be consistent with their goals and objectives. To accomplish this, the firm had to reaffirm its mission of leading the building construction process for sustained growth and profitability. The firm hopes to accomplish its mission through professional performance for its clients, ethical business practices and loyalty to its important asset-personnel. Another important action taken by the firm is its organizational design. Rather than maintain a top-down management approach, it created twenty six senior managers who are actively involved in strategic leadership. These "partners" as they are referred to share in all information on the company and meet regularly to set goals and define the overall strategies to accomplish those goals. Consistent with its organizational redesign, it developed "point of attack decision making" where its employees are empowered to make decisions once they observe any problems or concerns.

6.2 Political

Prior to its change of ownership, the firm was no different than any other construction firm because only a few of the top executives and the family members were responsible for allocating power and resources of the firm. This small group also decided who will keep the benefits of the organization. The approach the top management used to solve the political issues was to use Employee Stock Ownership Plan (ESOP). After several months of education, discussions, clarifications and meetings, the firm accepted a 100% employee ownership without long term preference based on the status at the purchase date, broad detailed communication with all the share holders, and its strategic planning and decision making based upon a professional partnership model.

6.3 Cultural

Culture consists of the values, objectives and beliefs and interpretations shared by organizational members. The firm has managed the cultural problem by developing ten elements that it defines as its value system. These ten points are posted on all sites and in all premises where this firm has an operation or project. This serves as a constant reminder to its employees and clients of their shared beliefs. These values are built around people, loyalty, commitment, ethics, professionalism, forward thinking, growth from within, team work point of attack decision making and quality. This value system has helped the firm immensely in its business pursuit and earned the respect of other firms because of the commitment and excellence exhibited by their personnel in their various projects. They have had to implement employee lead initiatives at every level of their operations for the sole purpose of growth, leadership and excellence.

In reviewing the firm strategy, it can be observed that the change is dynamic and not static. The organization is constantly fine tuning each of the three systems, technical, political and cultural. The process may be summarized in Figure-2, strategic management matrix.

MANAGERIAL AREAS	MISSION & STRATEGY	ORGANIZATIONAL STRUCTURE	HUMAN RESOURCE MANAGEMENT
TECHNICAL SYSTEM	Assessing and defining what business we are in and determining the business strategy	Defining reporting relationships, and groupings of people and departments to meet business needs	Methods of staffing, development and assessment of people
POLITICAL SYSTEM	Who gets to influence the mission and strategy of your organization	Distribution of power in your organization, both across the organization, up and down the organization	Managing the politics of succession, reward system and appraisal
CULTURAL SYSTEM	Developing a culture, or set of values to support business strategy	Developing managerial strategy and culture to fit your organization	Selection, development and rewards to support your organizations values and culture

Fig. 2 Strategic Management matrix.

The matrix represents the managerial areas on the vertical axis, and the management tools on the horizontal axis. Thus for cultural system, you will have to read across to identify how much change is needed to solve the cultural problem strategically.

7 Conclusion

This paper has presented an illustration of change management based on the TPC model. This is an ongoing study in the Construction Industry. It is hoped that this framework will be used in developing some of the innovations involved in the management of change in the industry.

8 References

1. Argyris, C. and Schon, D., 1978, Organizational learning: A theory of Action perspective, Addison-Wesley, Reading, MA.
2. Beer, Michael, Eisenstat A. Russel, and Spector Bert, (1990) "Why change programs don't produce change", Harvard Business Review, Nov-Dec.
3. Huber P. George & Glick H. William (1993), Organizational Change and Redesign, Oxford University Press, New York, NY.
4. Martin, Donald, D. (1994), "How small business can gain advantages from ESOPS", National Public Accountant, Volume 39, No. 2, Feb.
5. Schein, E.H., 1970, Organizational Psychology, 2 nd Ed., Prentice Hall, Englewood Cliffs, N.J.
6. Smith, William (1990) "Employee Stock Ownership Plans: Motivation and Morale Issues", Compensation & Benefits Review, Volume 22, No.5, Sept/Oct.
7. Tichy M. Noel (1982) Managing Strategic Change Technical, Political and Cultural Dynamics, John Willey and Sons, Newy York, NY.
8. Vaill B. Peter, (1989) Managing as a performing arts, Jossey-Bass Publishers, San-Francisco, CA.
9. Wieland, G.F. and Ullrich, R.A. (1976) Organizations: Behavior, design and change. Richard D. Irwin, Homewood, IL.
10. Wilmont, W. Robb, (1987), "Change in Management and the Management of Change", Long Range Planning, Vol. 20. No.6, pp. 23-26.

EXPLOSIVE MANAGEMENT: DESTRUCTION OR PROPULSION?

P. Barrett and M.G. Sexton
Research Centre for the Built and Human Environment,
University of Salford, Salford, UK

Abstract
At a recent international conference on quality management, the similarities with environmental management and health and safety management were noted. The negative impact on managerial effectiveness of these multiple pressures was articulated as follows "If you are in a minefield what do you do? - Stand Still!" or risk destruction.

This paper views these different pressures as managing against different criteria, but using generally similar mechanisms such that managers are given a clear view and comprehensive technical support to *propel* them and their companies forward.
Keywords: Quality, health and safety, environment, supple systems.

Résumé
Lors d'une conférence récente au sujet du 'quality management' (géstion de qualité), on a les similarités entre ce sujet et al géstion de l'environnement et de la santé et sécurité au travail. L'impact négatif sur l'efficacité de géstion de ces tensions était articulé comme suit: "Si vous vous trouvez dans un champs de mines, que faites vous? - Ne bougez pas!" ou risquez la destruction.

Cet article considère ces tensions différentes comme la géstion selon critères différentes, mais en utlilisant des mécanismes généralement similaires avec le rèsultat que les directeurs reçoivent une impression claire et un soutien technique compréhensif pour les faire avancer avec leurs entreprises.
Mots clés: Qualité, santé et sécurité, environnement, 'supple systems' (systèmes souples).

The Organization and Management of Construction: Shaping theory and practice (Volume One).
Edited by D.A. Langford and A. Retik. Published in 1996 by E & FN Spon. ISBN 0 419 22230 8.

1. Introduction

Construction firms are increasingly being bombarded by a stream of business requirements which have the potential to fundamentally affect their current and future competitiveness. These requirements constitute increasingly pervasive and rapid drivers of change which, unless due managerial vigilance is taken, may render firms' past modes of organisational design and behaviour inappropriate for their prevailing competitive contexts. If the firm does create a discrete reaction to each pressure it risks fragmentation as the pressures push in different directions resulting in conflicting systems. This relentless onslaught of obsolescence or fragmented action prescribes that firms need to create and manage mechanisms which effectively and efficiently scan and filter the business environment for requirements which impact upon the firm and to interpret, translate and internalise their implications in order to direct and generate congruent organisational responses. This paper seeks to address how construction firms can work towards this idealised state of organisational sensitivity, flexibility and renewal through the use of integrated, developmental management systems.

2. Developing pressures

2.1. Context

Organisations can be fruitfully conceptualised as open systems which must effectively interact with their external environment over time if they are to avoid myopia-driven decay in the face of an increasing dynamic and turbulent business environment. The external environment can be viewed as a swirling mass of ever changing and often intangible forces; be they cultural, political, economic, competitive and so forth; which periodically crystallise into transitory dominant configurations which shape organisational contexts and the nature of competition. At present, the issues of quality, health and safety and the environment have emerged as particularly important areas for construction firms to address if they are to survive and prosper, and will be discussed briefly below in the U.K. context.

2.2. Quality

Quality has taken a high profile, especially since 1987 when BS 5750 was first widely publicised. Since then the pressure on firms has grown enormously based on a perception that there is strong client demand. Questionnaire-based research has indicated that around 15% of construction professional firms are third-party certified to BS 5750 so far [1]. A Delphi study led to an estimate that in around five years the percentage will rise to 38% and at around 50% in ten years [2].

There is a preponderance of larger firms involved. The disproportionate cost, and the need for formality has caused problems for small firms. These firms, however, predominate in the industry with around 98% of such firms having less than 20 staff.

2.3. Health and safety

Health and safety has theoretically been on the agenda for years and certainly since the 1974 Health and Safety at Work Act. This gained a further boost in 1991 with the publication of management guidance by the HSE [3] which then set a standard for the management response required as an answer to the 1974 Act. The stakes have risen more recently in the eyes of most firms with the passing of the Managing of Health and Safety at Work Regulations, 1992 and the Construction (Design and Management) Regulations 1994. The UK legislative in this area is now principally putting into effect European legislative requirements.

The major pressure on firms is created through criminal liability which impacts on individuals as well. This exposure has been clearly extended to clients in the CDM Regulations and this is certain to lead to extra pressure.

2.4. Environment

Construction firms are increasingly faced with environment-based requirements which are influencing many facets of their activities, ranging from proactive environmentally-conscious design and construction, through to green marketing and asset portfolio management. These requirements are generated by a diverse range of stakeholders who have an interest in firms' overall environmental performance; be it the ever increasing amount of tougher environmental legislation from the European Community or the growing public alarm about environmental depletion and degradation.

3. Variety of mechanisms

3.1. Context

There are a variety of mechanisms at work in these different areas within the U.K..

3.2. Quality

In the quality assurance area the emphasis, in the UK at least, has been on third party certification against BS 5750 (now ISO 9000) which is orientated towards achieving *consistency* in the process. Because of the emphasis on formal systems and tangible evidence there has been a tendency for paper systems, and in particular the Quality Manual, to take centre stage. Research suggests that, subject to exceptions, professional firms perceive strong client demand for QA-certification, not unreasonably, given that many clients ask whether firms are certified at pre-tender stage [4]. However, it would appear that despite asking this question most clients ignore the answer in favour of first hand knowledge of the firm or person, or the recommendation of a trusted person who has had first hand experience. It would appear that firms and certifying bodies place quite considerable effort into creating paper systems which at best ensure consistency in the process, but are in fact largely ignored by clients who are more interested in evidence of actual performance.

3.3. Health and safety

In contrast, in the Health and Safety area, despite, or perhaps because it is driven by heavy legal sanctions the formal emphasis of the requirements is very much on ensuring management systems exist that really make a difference in practice.

> *"It has to be said that a concern for Safety which is sincerely held and repeatedly expressed but, nevertheless, is not carried through into action, is as much protection from danger as no concern at all"* [5]

The systems in question, and the way the requirements are policed, are both very much *developmental.* Risk assessments are the fulcrum of the requirements, so that people get to *think* health and safety. The HSE have also supported research to create accessible *measures* so that a site can be rated for how hazardous or safe it is relatively easily and a proactive improvement approach achieved [6]. There is still a tendency for companies to create paper systems, but they know this is the means not the end and, in fact, firms of under five people do not have to have documented systems at all!

A clear problem with Health and Safety enforcement is the attendant cost and the HSE has to target its effort. This appears to involve dealing with flagrant breaches and then concentrating on high profile cases. Thus, in practice it is impossible to fully police the various regulations. The CDM Regulations attempt to address this by injecting client pressure and getting the participants to check each other.

3.4. Environment

A variety of mechanisms have been developed to enable firms to improve their environmental performance. The British Standards Institution's standard on environmental management systems [7], the European Eco-management and Audit Scheme [8] and the European Community-driven ecolabelling of building materials and building products (e.g. see [9]) are arguably the most important.

The uptake of these mechanisms by firms is being hindered by their lack of universally agreed managerial systems or objective measures of what constitutes good environmental performance. Furthermore, these (and other similar) prescriptions, although a positive move in the right direction, tend to prescribe desired change *outcomes*, and do not take due account of the required change *processes* to achieve them [10]. In unison, these constraints tend to render these mechanisms as 'facades', laid on predominantly unchanged and fragmented systems, generating reactive and 'end-of-pipe' organisational responses to environmental requirements.

3.5. Summary

In summary, it can be seen that quality requirements tend to be client-driven and are being addressed through the promotion of service or product consistency, rather than trying to cultivate continuous improvements. In contrast, health and safety requirements are predominantly legislation-driven and are historically accommodated within firms through mechanistic systems, rather than culture-based systems which can integrate other, non legislation-based, requirements. In both of the above small firms are not effectively addressed. Finally, environmental requirements are stakeholder-driven, a dimension which requires firms to create multiple perspective goals; a skill which many firms are failing to get to grips with.

With this seemingly insurmountable mish-mash of requirements and organisational responses, it is temptingly easy to suggest that firms are doomed to be saddled with expensive and adversely fragmented management systems.

4. Implications for firms

Generally, the fewer the business requirements being placed on firms, the easier it is to stabilise activities in support of firms' core product and service offerings and to maintain continuity of core business focus across organisational functions and levels. However, it can be seen that the prevailing waves of requirements being injected into organisational domains are characterised by diverse, and often debilitating, complexity. Adopting the law of requisite variety, firms can only adequately manage this level of externally-induced complexity by creating a corresponding level of internal complexity [11].

In reality though, firms tend to try and simplify their internal environment and protect their core activities from such environmental complexity through the setting up of systems which, in effect, seal the firm off [12]. This 'buffering' manifests itself in a range of responses: imperviousness, selective imperviousness, adaptation and action-adaptation systems [13]. First, firms which adopt *impervious* systems attempt to seal the organisation completely off from the effects of business requirements. Second, firms employing *selective impervious* systems tend to concentrate only on those specific requirements which have an immediate impact on firms' core businesses. Third, firms creating *adaptation* systems aim to change themselves through adapting to the contingencies in the requirements emanating from the business environment. Finally, firms which embody *action-adaptation* systems strive to not only adapt to the requirements, but shape the requirements in their own image to change them from burdens to opportunities.

The prevailing response of construction firms to business requirements tends to be towards the impervious end of the above continuum. This proposition is supported by a recent RICS Workshop where practitioners expressed the view that the best they could to do is to *react* to the present deluge of requirements being placed upon them. No one really seemed to feel they were in control or that they could do more than the absolute minimum [14]. Moreover, when a similar issue was discussed at an international conference two years ago a delegate posed the question "What do you do if you find yourself in a minefield?" The answer was "stand very still!" or, again, do the absolute *minimum* [15].

Firms which adopt this minimalist strategy, in the short term, may very well be able to concentrate all their resources on providing a strong technical orientation. However, over time, this preponderance of front-stage activity will disengage the firm from business requirements, leading to strategic drift, stagnation, myopia and eventual crisis and demise. Moreover, those firms which try to become more action-adaptation orientated tend to so through channelling requirements through existing and new, often fragmented, systems. Firms undertaking this approach of responding to each pressure by engaging fully in the disparate mechanisms are going to be confronted with huge overhead costs. The balance between the effort to do work and the effort to demonstrate management control will shift dramatically towards the latter, potentially creating goal displacement [16], as firms elevate the goal of compliance with requirements above their primary goal of providing client-orientated service and product offerings.

This leads to the central question in this paper. How can a firm provide a high level of service by efficiently and effectively addressing the issues of quality, health and safety and environmental matters - rather than despite them?!

5. Formal responses and their limitations

On the quality front there have been developments which start to address some of the limitations of the QA-certification approach. Building on decades of development through the Japanese Deming Award, to the US Malcolm Baldridge Award, to the European EFQM Award [17] and now the British Quality Award, a model has emerged which takes a very broad view together with a self-auditing scheme and an annual competition. This model is drawn from the quality domain, but really is about a particular view of what constitutes good management. In the area of standards a proposal has emerged in Canada for a general management standard which is underpinned by the assumption that it is desirable for firms to have "... a simple set of management principles which can be used to support the management of any function, for any objectives, in any size of organization, and at any level of an organization" [18].

The positive aspect about these approaches is the wider view they take. The danger, however, is that they claim too much. Rather than presenting a way of managing, they propose *the* way and if adopted across the board they will inevitably lead to inappropriate emphases in many situations, especially if they become audit documents. Quite simply they present a simplistic view of management, not a bad view, but a limited perspective when you think of the mass of management debate on so many topics. Here the danger is that you would get what you measure - and very little else ! It is our view that it is more fruitful to create contingency-based approaches that reject the all-purpose principles and constructs embodied within the above models, and adopt a more behavioural, adaptive and flexible view of organisational activities.

The Vision 2000 strategy for quality standards [19] is interesting in this context. It would be possible to argue for many standards tuned to various sectors, however, the Vision 2000 document argues strongly against such proliferation and calls for core generic quality management and assurance standards, supplemented by interpretative documents in more specific industries, such as services. This approach seems to keep standards focused on the control functions within organisations in a flexible way, but without straying into the general management area.

The idea of general management initiatives and core generic standards with interpretative guidance could, in principle, be extended to health and safety and environmental issues. Indeed in the health and safety area the idea of core requirements and then guidance material becoming, at the same time, more detailed and less prescriptive is the generally accepted approach. The problems in the quality area identified above indicate that although the ideas could translate to other organisational activities, the actual mechanisms being pursued so far do not seem to have sufficient capability or capacity to appropriately address the broad range of issues emanating from the business environment.

6. An integrated, developmental way forward

6.1. Context

It can be seen that the complexity confronting firms provides a huge challenge and that the current approaches, or more recent developments, all carry problems.

The following section endeavours to effectively and efficiently 'nest' the management of business requirements within an approach which focuses on the *improvement cycle* within firms. Thus, it can complement steady-state quality management / assurance, or be a viable option in the absence of such activity. It addresses an area where systems can contribute - the hinterland between individual action and corporate strategy - but does not pretend to address the whole of management activity.

Relating the discussion to earlier parts of this paper, it will be seen that the emphasis on key measures of tangible achievement fits well with the health and safety approach, at least as it appears in theory. In terms of quality there is increasing pressure to achieve improvements as well as consistency. In the environmental field an approach which allows incremental progress to be made could break the stalemate between facade management and meaningful action which encompasses stakeholders' views.

It is suggested that the key to achieving improved organisation responses to a range of business requirements is to introduce balanced cybernetic-based feedback mechanisms so that construction firms actually learn from occasions where it could have done better or from opportunities / ideas from other sources. In simple terms, the construction firm collects feedback externally, as well as internally, then analyses the information and integrates it into future service provision in order to improve performance levels. It should be stressed that the process is iterative and thus encourages continuous improvement.

Based on this sort of thinking and aiming for systems which are client-responsive and facilitate a continuing cycle of improvements, an approach has been developed, through debate and observation, which endeavours to meet these criteria. The approach has been styled "supple systems" and is discussed next.

6.2. Supple systems

The supple system approach originated from the quality management domain [20], but can be fruitfully enlarged and refocused towards the continuous improvement of how organisations address multiple requirements generated from the business environment. The key features of supple systems are given in Table 1. There is not space here to describe the approach in detail, however, in summary, the approach advocates that a strong, but flexible *audit* system is developed which ensures that improvements in the quality of the service are being achieved. The audit system identifies sources of feedback, assesses if action is required, and at what level, prioritises between alternatives, allocates responsibility, checks later that action was taken, tries to objectively assess the impact of the actions and finally feeds these findings back to those involved. It should be noted that the above facets of supple systems should be embedded within a strategic framework that directs firms' inputs and processes towards improved output performance (see [21]).

It is the authors' view, argued in more detail elsewhere [22] that this is a fruitful way forward, with the "supple systems" approach having the potential to assist firms:

- faced with a constant stream of requirements, to sort out those which are relevant, from those which are not;
- identify those requirements which can be absorbed by existing operating systems, and which requirements need to be addressed from a more strategic perspective;

- to not only reconcile often disparate and conflicting requirements, but synergistically integrate them; and,
- to create generic systems which have the capacity to manage multiple demands against different criteria.

Table 1 - Key features of supple systems

Feature	Comment
Client / Stakeholder orientated	Above all the systems are tested against stakeholder, and especially client requirements by actively seeking feedback through both hard and soft data.
Minimalist / holistic	"As much as you must, as little as you may", that is, not having systems for their own sake, but rather targeting high risk / gain areas. Better to have made some progress on all important fronts than to have a patchy provision.
Loose-jointed	The systems operate at an audit level: clarifying objectives, checking performance and integrating efforts. At an operational level different styles and approaches can be accommodated, especially when they have proved themselves over time.
Evolutionary	Allow incremental and continuing progress to be made from whatever base.
Symbiotic with social systems	Build on the norms and culture of the organisation, for instance allowing self-control or group pressure to operate where appropriate.

7. Research Agenda

There are many areas demanding further study, but a key cluster of related issues has become clear:

- Firms should conceptualise that the addressing of a requirement-driven problem is "...in fact a whole series of 'nested' problems, each alternative solution leading to a new set of problems at the next level" [23]. The holistic sensing and interpretation of these requirements is therefore crucial if firms are to formulate and implement performance-orientated actions (for example see [24]).
- Towards this aim, it is suggested that systems should synergistically interact with, and integrate, other external and internal subsystems. Adapting and synthesising the notion of boundary-spanning units and lateral relations ([25] and [26] respectively), it is proposed that supple systems should be energised and supported by cross-functional teams which scan and assess the external environment for current and anticipated requirements.
- Furthermore, these requirements should be decomposed into issue content and issue process dimensions (see [27]). In this way, injecting the language of integration/differentiation [28], the function-specific issue content (for example, measures for health and safety) can be identified and directed towards the relevant function in order to create optimal differentiation.
- Likewise, the generic issue process elements of incoming requirements (for example, the actual measurement process which will be used to operationalise the health and safety measures) can be abstracted to form a core of enabling skills which can be drawn upon to more effectively manage future requirements, thereby optimising the integration of otherwise disparate requirements.

8. Conclusions

It has been seen that firms are confronted with a bewildering array of demands and that taking each as a discrete problem with an isolated solution is not really a feasible approach. This is especially so when it is put in a wider context of a continuous stream of new demands of which we have addressed prominent examples.

An approach is needed which develops the firm's capacity to adapt to changing demands, or better still to anticipate and mould those demands.

Drawing upon the metaphor presented by Lewis, firms will not be able to capture the organisational meaning of business requirements if they have no net to catch them in [29]. Supple systems are proposed as an integrated, developmental net which can provide positive benefit to construction firms, however small, to do something, start somewhere, in what will become a continuous development in the capability to identify and satisfy business requirements. As time goes on a robust, but flexible framework of systems and processes will be created. At the same time staff should develop an understanding of their stakeholders and the requirements from them and of the key inter-relationships within the organisation, leading to a successful service-orientated culture pervading the firm and ensuring the effective and efficient delivery of product or service offerings.

Ultimately, the "supple systems" approach has the potential to create a synergistic and symbiotic nexus between business requirements and organisational activities through the nurturing of systemic thinking that views requirements as being fluidly embedded within dynamic contexts which require correspondingly holistic, flexible and ever progressive solutions. To take this forward will require the further research identified.

In practical terms professional firms are under pressure to address key issues that are congruent with their professional status. As leading practitioners in construction they must aspire to produce high quality work which is achieved with due regard to the health and safety of those involved and which also exhibits a responsible attitude towards the environment. A supple systems approach is proposed as a way in which firms can grow into a role where they react, adapt, but also generate positive change. This is to be achieved through integrated, developmental management systems that propel them forward rather than tear them apart !

9. References

[1] Barrett, P. & Holling, J., (1994a), Survey of QA-certified Professional Firms in UK Construction Delphi Study of Longer Term Developments with UK Quality Management Experts, *SERC Project Report No.1*, University of Salford: UK.

[2] Barrett, P. & Holling, J., (1994b), Delphi Study of Longer Term Developments with UK Quality Management Experts, *SERC Project Report No.5*, University of Salford: UK.

[3] HSE, (1991), *Successful Health and Safety Management*, HMSO, London.

[4] Barrett, P. & Grover, R., (1995), *A Quality Strategy for Chartered Surveyors*, to be published Autumn 1995.

[5] Hidden, A., (1989), *Clapton Junction Railway Accident Report*, Report of Anthony Hidden QC, Tribunal Chairman, 27 September.

[6] Duff, A.R.; Robertson, I.T.; Cooper, M.D. & Phillips, R.A., (1993), Improving Safety on Construction Sites by Changing Personnel Behaviour, *HSE Contract Research Report No. 51*.
[7] British Standards Institution,, BS7750: Specification for Environmental Management Systems, *British Standards Institute*, UK.
[8] Council of European Communities, (1993), *Council Regulation (EEC) for a Community Eco-Management and Audit Scheme*, Brussels.
[9] Atkinson, C.J. & Butlin, R.N., (1993), Ecolabelling of Building Materials and Products, *BRE Information Paper 11/93*, Building Research Establishment: Watford, UK.
[10] Sexton, M.G. (1995), The Prajna Approach to Sustainable Construction;, *in Proceedings of the International Workshop on 'Environmental Impact Evaluation of Buildings and Cities for Sustainability*; Florence, 13-15th September.
[11] Ashby, R., (1963), *Introduction to Cybernetics*, John Wiley, New York.
[12] Thompson, J., (1967), *Organizations in Action*, McGraw-Hill, New York. Pp.18-21.
[13] Leavitt, H.J.; Dill, W.R. & Eyring, H.B., (1973), *The Organizational World*, Harcourt, Brace, Javanovich, New York. Pp. 306-310.
[14] RICS, (1995), *QA Research Project Workshop*, RICS: London, 6th June.
[15] Vision Eureka Conference, (1994), *Quality Management in Building and Construction*, Lillehammer, Norway, 13-16 June.
[16] Etzioni, A., (1964), *Modern Organizations*, Prentice-Hall: Englewood Cliffs, N.J.
[17] EFQM, (1992), *The European Quality Award*, EFQM: Brussels.
[18] Canadian Standards Association / Deloitte & Touche, (1993), *General Principles of Management Systems*, Canadian Standards Association / Deloitte & Touche: Canada. P.3.
[19] ISO, (1992), *Vision 2000: A Strategy for International Standards' Implementation in the Quality Arena During the 1990's*, ISO, Geneva
[20] Barrett, P. (1994), Supple Systems for Quality Management, *RICS Research Paper Series*, RICS, London
[21] Barrett, P.; Sexton, M.G. & Stanley, C., (1995), Cultivating Continuous Improvements in Facilities Management, Paper presented at the *COBRA '95 RICS Construction Research Conference*, 8-9 September.
[22] Barrett, P. & Sexton, M.G., (1995), Integrated, Developmental, Management Systems, Paper presented at the *COBRA '95 RICS Construction Research Conference*, 8-9 September.
[23] Cyert, R.M.; Simon, H.A. & Trow, D.B., (1956), Observation of a Business Decision, *Journal of Business*, 29: 247.
[24] Thomas, J.B.; Clark, S.M. & Gioia, D.A., (1993), Strategic Sensemaking and Organizational Performance: Linkages Among Scanning, Interpretation, Action, and Outcomes, *Academy of Management Journal*, 36: 2: 239-270.
[25] Thompson, J., (1967), *Organizations in Action*, McGraw-Hill, New York. Pp. 66-67.
[26] Galbraith, 1973), *Designing Complex Organizations*, Addison-Wesley, Reading, Mass. Pp. 46-64.
[27] Dutton, J.E.; Walton, E.J. & Abrahamson, E., (1989), Important Dimensions of Strategic Issues: Separating the Wheat from the Chaff, *Journal of Management Studies*, 26: 4: 381.
[28] Lawrence, P.R. & Lorsch, J.W., (1967), *Organization and Environment*, Harvard Business School, Boston.
[29] Lewis, C.I., (1956), *Mind and the World Order*, Dover, New York.

LONG-TERM SUCCESS OR FAILURE OF CONTRACTORS' STRATEGIES OF DIVERSIFICATION

P.M. Hillebrandt

Department of Construction Management and Engineering,
University of Reading, Reading, UK

Summary

This paper considers time series data and evidence from two studies of large contractors which examined the market strategies adopted by large contractors in the boom of the second half of the 1980s and those of the recession in the first half of the 1990s.

The general findings are that;

- contractors did not by their diversification policies avoid construction cycles;
- contractors benefitted from housing activity but not from property development;
- contractors could have better anticipated the change in the housing and property markets;
- material producers and contracting businesses were less volatile than housing, property and other businesses;
- external market factors were very important in success or failure but internal processes also played a part;
- in retreating to core businesses in the recession contractors often had no choice but in many cases were right to do so;
- contractors should not totally exclude diversification in the future.

The Organization and Management of Construction: Shaping theory and practice (Volume One).
Edited by D.A. Langford and A. Retik. Published in 1996 by E & FN Spon. ISBN 0 419 22230 8.

Sommaire

Basé sur les donnés de "temps sénel" et deux études de grandes entreprises du bâtiment cet exposé examine les stratégies de marché adopteés par les grandes entreprises du bâtiment pendant la prospérité de la deuxième moitié des années quatre-vingts et la récession de la première moitié des années quatre-vingt-dix.

Les conclusions générales sont que:

- les entreprises du bâtiment n'ont pas évité par leur politique de diversification les cycles de construction;
- les entreprises ont bénéficié de l'activité du logement, et non pas du dévelopement des biens immobiliers;
- les entreprises auraient pu mieux anticiper les changements des marchés du logement et de l'immobilier;
- le commerce des matériaux et de la construction étaient moins changeants que le logement, l'immobilier et les autres commerces;
- les facteurs de marché extérieurs étaient très importants pour les succès et les échecs mais les procédés internes ont également joué un rôle;
- les entreprises du bâtiment n'avaient pas d'autre choix que de se replier sur le commerce debase pendant la récession mais dans beaucoup de cas, elles ont eu raison de le faire;
- A l'avenir les entreprises ne devraient pas exclure totalement la diversification.

Keywords: Strategy, diversification, contracting housing, property, materials, fluctuations, cycles.

1 Introduction

In the period from the early 1980s to the early 1990s large construction companies changed their market strategies considerably.

A study undertaken in 1986 by Hillebrandt and Cannon [1] of twenty large UK contractors. It is estimated that they accounted for 60% of the contracting work of the top 35 UK contractors UK contractors. The study found that market strategies were dominated by growth and diversification and construction companies invested heavily in property, housing, building materials, mining, plant, marketing and a whole range of miscellaneous activities. The reasons are well documented in the study.

A later study undertaken in 1994 by Hillebrandt, Cannon and Lansley [2] of eighteen similar companies, supported by published data on seventy of the top eighty companies found a dramatic retreat to core businesses brought on by the recession.

Were the contractors making wrong judgements of their markets in the 1980s or were other factors responsible for their change of strategy? This paper attempts to assess how far the strategies, particularly those of diversification, when formulated made sense in relation to market conditions and to see how far they succeeded or failed because of external market conditions or because of factors internal to construction companies.

The 1986 study identified five main reasons for diversification: 1) to avoid dependence on the cyclical nature of construction and particular clients and markets 2) to increase profitable growth 3) to increase activity or enhance security by controlling supplies 4) to make good use of positive cash flow and 5) to increase fixed assets.

Large contractors diversified into a whole range of activities but the most important were construction related: housing development, property development and material production, especially aggregates and sand and gravel. When the housing and property markets collapsed and the general recession developed, they retreated to their core business of contracting. To a large extent this retreat was not voluntary. In order to grow and to diversify companies had borrowed large amounts of money so that their gearing was very high. This had to be reduced and one way of doing this was to sell assets. Construction companies tended to sell those of their peripheral businesses which had a good market value. However in spite of the reduction in diversity of activities being somewhat imposed, contractors generally are now reacting against any idea of future diversification.

2 Avoidance of fluctuations

There is sense in diversifying out of construction to avoid the fluctuations of the construction cycle but housing, property and material production have shown themselves to be as cyclical as construction itself and this was to be expected. Contractors have in the past diversified into some totally different businesses such as deep mining for minerals, boat building and, in one case many years ago, meat processing. Generally these diversifications have not been a success in the long run.

What evidence did the contractors have before they diversified and, if they had a choice would de-diversification be sensible in terms of the objective of reducing the vulnerability of the businesses to fluctuations? Table 1 shows indices of output of contracting, proxied by new construction output minus new private housing; private housing, which is nearly all speculative; private commercial building which is some indicator of the ups and downs of new property development; crushed rock aggregate production and sand and gravel production for the main materials production.

Studying this table it can be seen what view the contractors could have taken of the fluctuations in contracting compared with housing or property development. Looking back in 1986 - which was the date of most of the first study - contracting was seen to have fallen considerably from the high levels of the early 1970s and recovered only slightly by 1986. The fall from 1971 to 1981 was 37% and still 29% by 1986. Housing in 1986 was down by 31% on 1973 but it had shown more extreme ups and downs than contracting. By contrast property development was relatively stable with a long steady almost uninterrupted growth from 1976 to 1986 and by 1986 was above the 1973 level. On this basis, if contractors wanted to avoid fluctuations they should have concentrated on commercial development and contracting and left housing alone. Indeed to some extent they did do this. Figure 1 shows that in current prices from 1987 to 1990 contracting turnover of 70 of the top 80 companies nearly doubled, housing increased by just over a third to a half and property development increased by approaching a half.

Table 1. Indices of output for Contracting, Private Housing, Commercial Building and Aggregates, Sand and Gravel 1971-1994. 1989 = 100

Year	New Construction Minus Private Housing	Private Housing	Commercial Building	Crushed Rock Aggregates	Sand and Gravel
1971	104	126	48	na	85
1972	99	133	47	59	89
1973	96	144	48	75	98
1974	88	104	44	70	86
1975	86	93	40	65	89
1976	85	100	35	58	84
1977	83	94	35	56	76
1978	85	107	40	58	78
1979	80	99	37	61	78
1980	74	78	37	61	75
1981	66	71	40	54	68
1982	67	81	45	61	70
1983	67	97	46	66	77
1984	71	94	48	66	76
1985	73	89	52	68	77
1986	74	99	58	73	80
1987	80	110	69	84	85
1988	87	123	79	95	99
1989	100	100	100	100	100
1990	109	78	111	96	89
1991	107	68	97	88	75
1992	104	68	79	85	68
1993	100	75	65	na	66
1991	101	80	68	na	68

Notes
1973 was the peak of the previous boom in all work by contractors including new private housing and new commercial though new work as a whole peaked earlier. 1989 was chosen as the base year as it was the beginning of the downturn.
na not available

Sources:
Figures for construction, private housing and commercial building are based on DOE's constant price series of value of output all corrected to 1990 price levels. Figures for aggregates and sand and gravel are based on production or sales. All original data came from various issues of DoE Housing and Construction Statistics.

Profits from housing, however (see Figure 2) increased more than profits from property development and were in absolute terms very high.

Suppose contractors in 1986 were looking ahead, as they should have been, to determine their strategy, how much of the changes could they have foreseen? They were on the up-and-up in all types of work. All three types of work were buoyant and from 1986 to the peak of output contracting increased by 47%, private housing by 24% and commercial work by 91%. In 1986 it must have seemed that they couldn't go wrong. The profit figures also showed that they were right to be in all three businesses. But could they have foreseen the fall? To some extent they should have done. The NEDO forecasts first talked of the danger of an oversupply of office accommodation in December 1986 and made a negative output forecast for 1989 in December 1987 years before the fall in orders or output and in time to avoid some new developments. However the gestation period of the large property schemes was long and so also was the construction period. This warning would have come too late for the very large schemes.

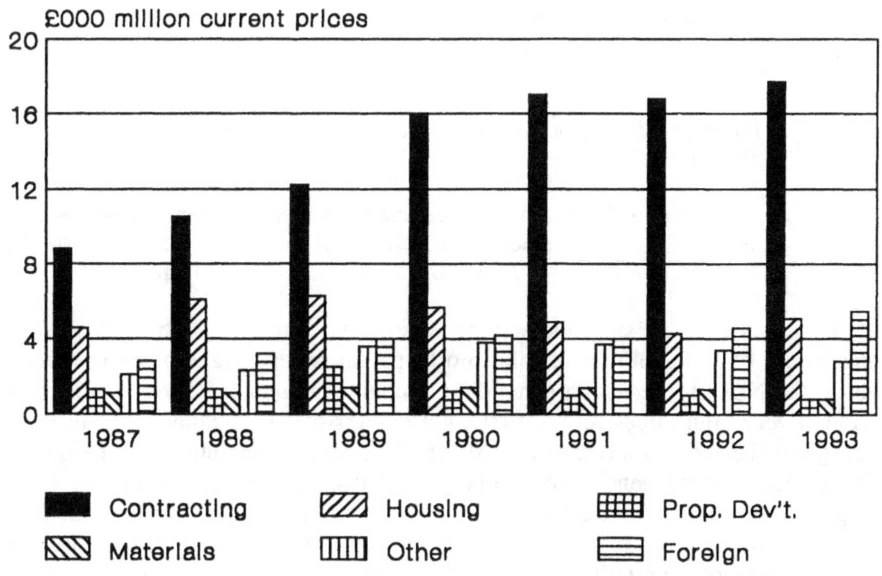

Fig. 1. Turnover by activity of 70 of the top 80 UK construction companies 1987-93

Source: Construction Forecasting and Research Ltd based on company accounts for and quoted in Hillebrandt, P. M., Cannon, J., and Lansley, P. R. (1995) *The Construction Company In and Out of Recession*, Macmillan; London.

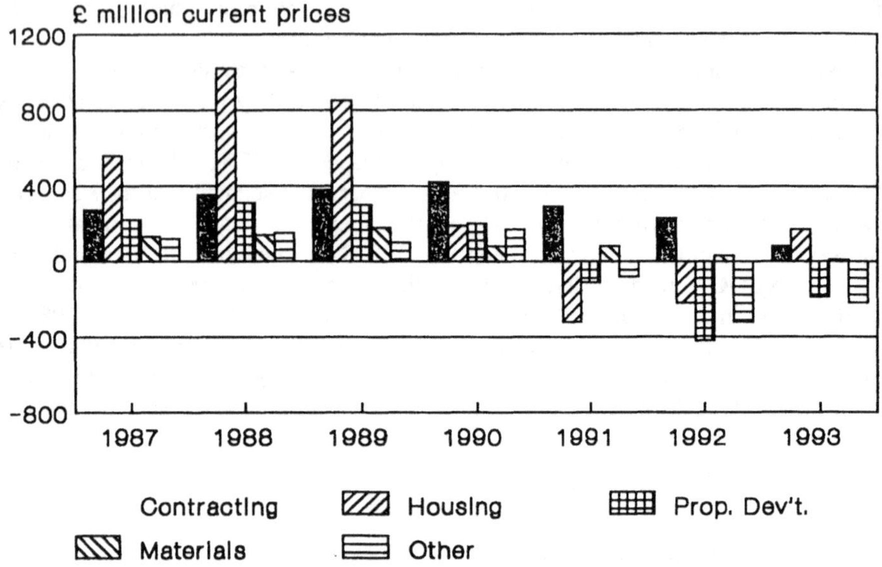

Fig. 2. Pre-tax profits by activity of 70 of the top 80 UK construction companies 1987-93

Source: Construction Forecasting and Research Ltd based on company accounts for and quoted in Hillebrandt, P. M., Cannon, J., and Lansley, P. R. (1995) *The Construction Company In and Out of Recession*, Macmillan; London.

In the case of housing, the writing was on the wall in March 1988 with the announcement of the abolition of joint mortgage tax relief to take place in August. However, the NEDO forecasting panel did not comment and contractors continued to buy land at very high prices for the first eight months of 1988. Some companies did not recognise the recession even in 1990. In the case of contracting the reasons for the fall in output were general recession exacerbated by the specific factors of the housing and property market collapses. Public sector expenditure had a moderating influence on the downward trend with spending on health, education, infrastructure and even housing being relatively steady. The demand for aggregate, sand and gravel follows that for construction. Road building is an especially large consumer of aggregates. Unfortunately very recent data for crushed rock aggregates are not available but fluctuations in demand for them have been greater than for construction though for sand and gravel they have been about the same.

3 Sector by sector consideration of diversification

3.1 Contracting

Contracting companies do not of course diversify into contracting but contracting must be the benchmark by which other sectors are judged. It has already been seen in Table 1 that contracting suffered from fluctuations, although they were not as violent as in housing or commercial building. However, Figure 1 shows a very different picture for large contractors in that, at constant prices, turnover in contracting rose steadily to 1991 and remained at about that level in 1992 and 1993. As prices fell during the recession in real terms turnover probably increased. Profits did not do so well as seen in Figure 2 and were, in any case, not high compared with housing or property profits on turnover, but they were at least still positive in 1993. However, the cash situation became tighter during this period, partly because of the element of cash flow in profit which was falling, but also because the more difficult trading conditions brought squeezes wherever the client or suppliers could achieve it. This affected the business as a whole and the capital intensive businesses in particular.

3.2 Housing

Diversification into housing certainly, in the short run, achieved the objective of increasing profitable growth. Figures 1 and 2 show that in 1988 the profit from housing was about level with the profit from contracting, though the turnover was only about 60% of that in contracting. Thereafter profit fell and became negative in 1991 and 1992 while turnover fell back only to 1987 levels. But over the period 1987 to 1993 as a whole large contractors benefitted from a large net housing profit.

A major reason for diversification into housing was to make good use of positive cash flows. Housing does this well by investing the surplus cash in land. However, the contractors considered the matter wrongly. They thought about the cash but not that it was only a flow going through their hands which did not belong to them. Moreover, flows dry up, and when contracting became more difficult cash flow diminished. It is possible that it will be reduced permanently as different forms of financing for projects and contractual arrangements become more prevalent.

The reason for diversification to avoid cyclical fluctuations has not worked and could not have been expected to work in housing. If, in 1986, the contractors looked at the past fluctuations in housing and said that it was worth putting up with fluctuations in order not to miss out on the vast profits which could be made, they were acting logically. In fact it is doubtful whether they did look at past fluctuations. They were concerned with day to day operations and these were going well. They were not acting logically in failing to realise in March 1988 that the boom would end. Internal factors then took over in the companies. The main boards could not curb the housing arms of their companies while they were making such large profits even if they realised that it could not last. When the market collapsed it was too late. They had bought land at high prices and had difficulty in selling houses. Some realised relatively early, but others went around with blinkers on till 1990.

Were they right to reduce and in some cases to get out of housing in 1990 to 1995? Many of those who sold early had to do so. But even in 1995 companies were selling their housing businesses. Probably they are right. It has been seen that housing fluctuations are very great over a long period and the outlook for housing in

1995 looks bleak, with less security of employment for house buyers, a key long term factor in confidence. Many observers consider that house prices will never boom again as in the past. I am less certain. People have very short memories and at least in some segments of the market a substantial rise in prices over and above the rate of inflation cannot be ruled out. Contractors should probably withdraw from housing now, but a move back into it might be appropriate in some years or decades ahead. They wouldn't want to, and shouldn't, miss out on any future profits bonanza.

There is however another factor to consider. In the last few years specialist housebuilders have performed rather better than contractors in their housing activities. It may be that in order to follow and react to the market situation single minded attention must be devoted to housing. It has also been suggested that a first class housing development manager will not wish to work for contractors as opposed to housing specialists. Contractors' housing managers are often building specialists who happen to work on housing.

3.3 Property

Again growth and profits were major incentives to develop in property. However, while neither the contribution to turnover nor the absolute profit were as great as in housing, the rate of profit related to turnover was about the same in the good years and proportionately much greater than in contracting. Unlike the situation in housing over the years 1987 to 1993 the losses from property of the large contractors were nearly as great as the profits so neither long term growth nor profits were achieved, while contracting profits continued to be positive throughout. A second reason to go into property was to make good use of positive cash flow and, in so far as the property was being retained, to increase fixed assets. This it did until the fall when property which had not been let became unsaleable. The evidence from past years is that property, as shown by commercial building, was a stable area of activity relatively free from cyclical factors. That idea was demolished as the boom collapsed.

So the contractors would have been better off not to have gone into property. However, it is difficult in 1986 to see how they were to know that although by 1987 or 1988 when it was clear that an over supply of offices was building up, they should have pulled out. Some contractors are still trying to rectify the results of their property exploits. But to say that they should keep out of property development in the future is too sweeping. The period to about 1987 was one of good solid development. Only later did it get out of hand.

3.4 Materials

The move to material production often had very different reasons from the development of other businesses. A major aim was to control supplies and in 1986 this was important, as the increase in activity was seen to be jeopardising supplies. To own and operate your own material supply gave greater security. At the same time it increased fixed assets. Although Table 1 shows that aggregate output fluctuated, output by the large contractors of materials did not fall greatly from 1989 to 1993. Production of materials did not, however, directly make much impact on growth and was only reasonably profitable until 1991. On the other hand, overall losses were not made and in some cases profits for materials are in 1995 better than from any other source.

Contractors made a logical decision to have control over sources of supply while that was a problem but thereafter each operation had to be considered on its merits.

3.5 Other activities

Other activities increased considerably in the period 1987 to 1990 and then fell back again. Their contribution to profit was small and the losses incurred in 1991 to 1993 were greater than the profits made in the previous four years. Why was this? The reasons for going into these other activities may have been partly to have some contra-cyclical activities, but if so it did not work. In some cases the activities were acquired "by mistake", with another acquisition, or were the cherished idea of a respected staff member. In some cases they were deliberate board policy.

Some companies made very substantial investments in other non-contracting activities. Mining is a case in point and one company had around a third of its business in mining. There had been connections with contracting in the beginning and each successive change in the type of business had some links with the previous business. First the company was in the business of opencast coal production where, in the post war years the machinery for removing the overburden was normal large contracting earth moving equipment, although now it is specialised. However, that link and experience in opencast coal brought knowledge of the coal industry and the next step was deep coal mining. The know-how of deep mining facilitated an entry into the mining of other minerals.

It seems it is necessary either to have some know-how which is transferable from one business to another or to be so large in each business that the company can attract top specialist managers for each business. This is what conglomerates such as Trafalgar House do and it is also the policy of Bouygues, the French contracting and television group.

Many of the adventures by contractors with other products were small scale. The reasons for their fall may have been that they were peripheral to the activities of the company as a whole and did not get enough attention or even investment or that they, and the related market, were not properly understood even by their managers. All these are internal factors.

4 Summary of overall conclusions

The conclusions may be summarised as follows:

- Enhanced activity in housing increased turnover and profits in the short run and over the period 1987-93 as a whole.
- Enhanced activity in property increased turnover and profits but, over the period 1987-93 as a whole, the net benefit was negligible in terms of profit.
- Both housing and property development adversely affected companies' balance sheets and forced sales of assets.
- Materials businesses were not especially profitable but have been more stable than other businesses. They fulfilled for the boom period the objective of increasing control over supply.

- Other activities were never very profitable as a group and were loss making over the period 1987-93 as a whole. The reasons are probably internal to the companies.
- None of the activities discussed fulfilled the requirement of off-setting the contracting cycle. Contracting turnover and profits of large contractors were less volatile than those of any of the other businesses..
- Contractors could have used past data to support their diversification to property and aggregates and sand and gravel for contra-cyclical reasons but not to housing.
- Contractors could have foreseen earlier the fall in the office property market and the fall in the housing market.
- Preoccupation with internal factors and a lack of attention to the market as a whole was probably the reason why contractors did not anticipate the decline in property and housing.
- Housing and property development were good uses for positive cash flow only while the cash flow continued at a high level. Thereafter it was disastrous.
- Property development was theoretically a way of increasing fixed assets held by contractors but it went badly wrong, because of the amount of unlettable and hence unsaleable property.

Contractors must accept that not only contracting, but also housing development and property development are subject to wide fluctuations, largely because all of them are investment goods which are required irregularly. Materials are so closely connected to construction that demand for those too is unstable. The skill in diversification policy probably lies in anticipating the market changes and being prepared to increase or decrease activities in various markets in advance of the market changing. Nearly all the problems of the diversification policy arose from external market factors but the effects of many could have been minimised by internal vigilance on how the market was likely to change and by placing less reliance on current profitability as an indicator.

Diversification into other products has not been very successful. Contractors are perhaps not good at other businesses and unless the operation is a large one it is difficult to attract the best managers in the business.

Overall, therefore, contractors should diversify or expand where there are opportunities in a business they know well and be prepared to shrink again as necessary. Future diversifications or expansions in housing development, property development or material production, should not be permanently ruled out, whatever the expedient actions of the present time.

5 References

1. Hillebrandt, P.M. and Cannon, J. (1990) *The Modern Construction Firm*, Macmillan; London.
2. Hillebrandt, P.M., Cannon, J. and Lansley, P. (1995) *The Construction Company In and Out of Recession*, Macmillan; London.

ANALYSING CONSTRUCTION CONTRACTORS' STRATEGIC INTENT FROM MISSION STATEMENTS

F.T. Edum-Fotwe, A.D.F. Price and A. Thorpe
Department of Civil and Building Engineering,
Loughborough University of Technology, Loughborough, UK

Abstract

A construction company's strategy formulation usually culminates in action programmes that are presented in mission statements which form an integral part of the company's annual report to its shareholders. Mission statements of construction companies express not only the intended strategies of these organisations, but also present an assessment of previous performance upon which the long-term vision of the company's executives, in terms of what it aspires to be, who it wants to serve, and how it wants to compete in its chosen markets, is based. Analysis of mission statements highlight how construction contractors respond to changing conditions in order to ensure their survival within their perceived future business environment. Detailed analysis of the mission statements present a view from within the company of the strategic intent arising from their planning, which is aimed at addressing the future of the company. This paper presents an analysis of the mission statements of twelve large construction contracting companies. The analysis draws on nine principal factors that characterise the missions of large corporate bodies to evaluate the mission statements of the selected contracting companies with a view to identifying variations in the nature of strategies that were 'intended' to be pursued. The analysis shows that construction contractors' strategies embody five main factors.
Keywords: Mission statement, construction contractor, content analysis, content factors, strategy.

Sommaire

La formulation de stratégy d'une entreprise s'exprime par des programmes d'actions qui sont présentées comme declaration de mission et qui forme une partie intégrale du rapport annuèl de l'entreprise. La déclaration de mission des entreprises de construction présente non seulement les stratégies projetées de ces organisations mais aussi une évaluation de la performance antérieure sur laquelle sont basés les visions futures des executives, a voir, les aspirations, les services offers et les termes de compétitions dans les marches. Une analyse des plans de mission demontre comment les entreprises repondent aux changements de conditions dans le but d' assurer leur survies dans l'environnement d'affaires. Une analyse detaillée des déclarations de mission presente une vision interne des stratégies dans les entreprises, qui resultent de leurs planification qui a pour but d'adrésser le futur de l'entreprise. Cet article presente une analyse des declarations de mission de douze grandes entreprises. L'analyse deduit de neuf facteurs pricipaux qui caracterisent les missions de grands conseil d'entreprises en vue d'evaluer les déclarations de mission des entreprises choisis, et d'identifier les changements dans la nature des stratégies projetées. L'analyse montre que les stratégies des entreprises sont basées sur cinq facteurs principaux.

The Organization and Management of Construction: Shaping theory and practice (Volume One).
Edited by D.A. Langford and A. Retik. Published in 1996 by E & FN Spon. ISBN 0 419 22230 8.

1 Introduction

The focus of a construction contractor's strategy is usually reflected in its mission statement. The outcome of its past strategy and the future direction for the contractor's business is usually presented as executive summaries contained in the annual report. There was great enthusiasm for developing organisational mission statements in the 1980's. Piercy and Morgan [1] attributed this interest to the fundamental work of Drucker [2] on the importance of business mission to efficient management. Brabet and Klemm [3] equally recognised the importance of business mission for attaining profitability. The statements were considered as valuable tools whereby managers can motive, focus and provide leadership for their staff towards the objectives of the company, as well as presenting a coherent programme of strategy for the consumption of shareholders and third party interests.

This paper examines the mission statements of construction companies to identify their content and variation over time. A carefully prepared mission statement was recognised by Drucker [2] as the first step in strategic management. Analysis of such statements therefore should provide information on what is addressed by construction contractors in their strategic planning, and the changing patterns of how construction businesses have been perceiving and functioning within their business environment in order to ensure their survival. Considerable diversity exists in the composition and application of their mission statements among companies in different business sectors. Analysis of the mission content of manufacturing and service companies has been undertaken by David [4]. There is, as yet, no similar analysis undertaken for companies in the construction industry. Such an analysis should therefore establish the focus for construction companies, in their strategic planning from a perspective internal to the construction business. Mission statements are almost universally presented qualitatively in a descriptive format, and therefore, to effectively analyse together these disparate mission statements from different organisations, a method that ensures the effective translation of qualitative and textual data into quantitative format is employed.

1.1 Definitions for mission statement

A company's mission statement has been defined by David [4] as that enduring statement of purpose which distinguishes one organisation from other similar enterprises. It is considered as a declaration of the company's 'reason for being'. Other forms of phraseology which have been employed for mission statement include creed statement, statement of purpose, statement of philosophy, statement of beliefs, and statement of business principles. All the above phrases connote a long-term perspective, a basic requirement for formulating effective strategies. These various definitions put forward for the mission statement of organisations can therefore be summarised as *a common management tool that addresses an organisation's 'theory of business', and forms the basis of its corporate strategy.* The use of the terminology in this chapter therefore adopts this explanation.

1.2 Components of mission statements

Broadly speaking, all organisations have a mission or 'reason for being', even if corporate planners and senior executives have not consciously put it into writing. Results of recent research suggest that a key aspect of strategic planning - getting senior executives to agree on, and putting into writing a definition of the purpose and scope of the company can actually translate into profits [5]. However, what a mission statement should contain is still open to debate. In his earlier work Drucker [2] outlined the components of missions statement as involving a company addressing itself to three key questions; 'who is our customer?', 'where is our customer?', and 'what is the value of our customer?'. The three key questions of Drucker were viewed by Pearce [6] to be less comprehensive for providing a balanced and effective mission for corporate entities. Pearce [6] subsequently put forward six factors that characterised missions of high performing companies, and suggested that mission statements of organisations should comprise six factors. The factors were presented as:

- product or service, market and technology;
- survival goals outlining growth and profitability;
- company philosophy;
- company self-concept;
- public image; and
- responsibility to stakeholders.

The factors of Pearce [6] were empirically determined from mission statements surveyed from the Fortune 500 firms. David [4] also undertook a comparative analysis of the mission statements of manufacturing and service firms. Both Pearce and David [4, 6] achieved their objectives by applying a content analysis methodology.

According to McGinnis [7], a mission statement, in addition to defining what the organisation is, and aspires to be, should be limited enough to exclude some ventures and broad enough to allow for creative growth, distinguish a given organisation from all others, and serve as a framework for evaluating both current and prospective activities. In line with this, Want [8] considered the primary components of a mission statement to be of five categories. These were outlined as purpose, ultimate business aims, corporate identity, policies of the company, and values. The various categories employed by Want [8] are explained below.

The *purpose* includes the primary justification for the organisation's existence, which translates as the primary lines of business for the company, the products and services provided or markets to be served, and manner in which they are to be served.

Ultimate business aims address the market share, profitability or size, strategies for achieving growth and optimal productivity, and impact on competition.

Corporate identity embodies how the company wants to be recognised in the markets in which it undertakes its business by clients, competitor companies, other interested corporate entities like the financial institutions, the social community, and the its employees so that all these interested parties can develop a sense of identity and commitment with the company.

Policies of the company refer to the philosophy and style of leadership of its senior executives, the relationship between top management and the company's board of directors, owners and shareholders, and the company's overall decision making structure.

Values include consideration for customers, competitors, employees, regulatory agencies, and the general public.

The various submissions on content of mission statements by different analysts overlap. In essence, they contain many of the same issues, with Pearce [6] favouring greater detail and specificity, while Drucker [2] showed a preference for a more general coverage. The most comprehensive list of categories for the content of mission statements was outlined by David [4], who put forward a preliminary list of nine categories in his comparative analysis of the mission statements of firms in two business sectors, which is presented in Figure 1. The nine mission content categories brought together the most comprehensive content list, and embodied all the different factors presented by Drucker, Pearce and Want [2, 6, 8].

1. Clients/Customers
2. Services/Markets/Products
3. Location/Geographical spread
4. Technology
5. Economic survival/Profitability
6. Philosophy/Values/Aspirations
7. Self-concept/Strengths and weaknesses
8. Concern for public image
9. Concern for employees

Figure 1: List of mission content categories *[taken from David 1989]*

2 The content analysis method

Content analysis has been employed for the successful examination of historical artefacts. Easterby-Smith et. al. [9] recount examples where this method was employed in analysing historical accounts. The method basically involves the counting of certain key phrases or words or ideas embodied in written materials, and the frequencies are then analysed. This is achieved by selecting the written statements, developing categories, and measuring the frequencies of appearance for the categories in the various statements. By applying appropriate statistical assessments, conclusions can be drawn on the various content factors. The method is commonly employed to analyse qualitative and unstructured data, and leads to an appreciation of the concepts involved in managerial activities which are set down as written documents [10]. This can be established by identifying non-random variations in their textual data. The information presented in corporate literature is considered as corresponding to an objective reality that reflects the strategic intentions of construction contractors.

This method was employed in this research to analyse the mission statements of construction contractors because of its suitability for the purpose. Construction contractors' mission statements are presented as qualitative information, which are all too often unstructured. To be able to capture the scope and coverage of its content, the method provides a means of isolating the important issues addressed in the statements for a number of contractors, in order to arrive at a unanimity on construction contractor mission statements. This research analysed annual mission statements of twelve construction contractors over a five-year period using the content analysis method. The annual examination identified the main mission factors that featured in their executive summaries. Each of the identified mission contents was then analysed over a five year period to identify any variability. The results of the analysis have been presented in section 4.

3 Nature of contractors' mission statements

The mission statements of construction contractors are usually embodied in their annual reports. These are presented as part of the executive statement of the Chairman, and the in some cases of the Chief Executive Officer. The executive statements usually review the past performance and then define a mission for the future. The mission statements of contractors vary in content, and address different factors which can be identified with the content categories presented by David [4]. Excerpts from the mission statement of one of the major organisations in the construction industry, which is referred to in this paper as Company A, is provided to exemplify how construction companies include the categories identified by David [4] in their mission statements. This is based on the executive statement which forms part of their 1992 annual report.

Clients and customer
No reference

Services, Products and Markets
The Company A Group includes the United Kingdom's largest construction materials supplier, the largest private house builder and the largest building and civil engineering contractor. A balanced range of other construction materials and services complete the group's portfolio, which all lies in the construction industry.

Location
...to take advantage of the good market conditions in the UK.
...our confidence in the long term prospects of the North American markets.

Technology
No reference.

Concern for Survival
One certainty is that the organisation and commitment which have made our Housing Division such an effective and profitable competitor will not fail in the new conditions.

Philosophy
The group operates on a rigorously decentralised basis in seven autonomous divisions.

Self-concept
...mount a large-scale, keen and effective response to the opportunities that exist.
Concern for Public image
Company A is the biggest UK controlled building materials and construction group.

Concern for employees
Individual success is at the heart of our corporate philosophy and our corporate success.

The excerpts from Company A's 1992 mission statement presents a comprehensive approach in their mission. This broad scope allows for devising and considering a wide range of feasible alternative objectives and strategies that does not limit managerial innovation. The Company A Group comprises several business units in different sectors of the industry. Such a broad approach to their mission statement also enables the corporate centre to reconcile differences among the various units of the company that have to confront diverse stakeholders in their various business sectors.

To illustrate the presentation and scope of construction contractors' mission, two such statements for leading companies in the industry, referred to in this paper as Company B and Company C, are hereby reproduced.

Company B
The built environment is essential to our daily life, providing shelter and the means to create economic activity. Company B supplies a wide range of products and services to meet these needs- from the building of homes to the construction of major infrastructure projects. The group endeavours to undertake these activities responsibly and in the best interests of its shareholders'.

Company C
'Company C plc is a major force in the construction, housing, mechanical engineering and technology related markets in the UK, with operations in a growing number of overseas markets and an increasing involvement in investment-led projects. The Group's successful and flexible blend of services, skills and products is tailored to meet the needs of clients, and is complemented by its continuing caring concern for the community, the environment and its own employees and pensioners. The Group's strategy remains to build on its strength. It continues to focus on the core businesses of Construction and Homes, developing its ancillary skills and activities to support them. In parallel there is a continued drive to increase the productivity, efficiency and cost-consciousness of all parts of the Group, and to maintain the proven high standards of quality, safety, and reliability to the benefit of customers, shareholders and employees'.

4 Analysis of the mission statements

The nine elements that featured in the work of David [4] were adopted as a basis for analysing the content of the contractors mission statements. Each mission statement was judged as including a particular factor if reference was made to that factor either directly or implied by a phrase contained in the executive statement of the annual report. A value of one (1) was assigned to that particular factor for the company in the year considered. Where no reference is made to a content factor in the mission statement, either directly or by implication, a value of zero (0) was assigned that particular factor. The analysis was performed utilising the mission statements of twelve major construction contractors. The basis for selecting the twelve companies was primarily the ready availability of their annual reports. The statements from the selected companies covered the period 1987 to 1991, and 1993. The selected period offered different conditions of business environment ranging from recovery, through an economic boom, to a recession. This way it ensured that the contents resulting from the analysis, will reflect the changes in the focus of strategic responses for contracting companies in the industry in different phases of the economic cycle.

4.1 Content of contractors' mission statements

The analysis began with the composition of matrices that summarised the annual content categories in the statements of the twelve companies. For each factor, a mean annual value was determined. The mean values ranged from: *0.00*, in which case no company within the sample included that factor in their mission statement; to *1.00*, for which all the companies included that particular factor in their mission statement. A factor is rated as forming part of the content of contractors' mission statements if it exhibits a consistent mean value closer to the value of one (1). Consistency here was established by the value of the t-statistic with zero as the control mean value. Table 1 outlines summary statistics of the overall content factors derived from the matrices. A low value for the statistic indicated a consistent non-inclusion, with the converse holding true. Figure 2 shows the plot of the corresponding histogram of content factors to ascertain the factors that reflected consistent inclusion or non-inclusion in the mission statements.

4.2 Analysing annual variation in content of mission statements

The variation in content of mission statements was assessed by pairings of the distributions for the annual means, and evaluated by paired t-test. The pairings were performed by matching a particular year's content factors with all the rest in succession. This way it was ensured that gradual annual changes which result in a major long-term variation is not overlooked. Table 2 presents the result of the paired t-test for the data from 1987 to 1991.

Subsequently, the five-year mean for the period from 1987 to 1991 was paired with the mean value factors for 1993. Table 3 presents the results of the paired t-test to evaluate variation in mission content between the two distributions.

Table 1: Summary statistics of overall content factors

Content factor	5-yr. mean 1987-91	S.d.	t-statistic	Comment
Clients/Customers	0.066	0.071	2.091	0.00†
Services/Markets	0.932	0.093	22.382	1.00*
Location	0.667	0.131	11.345	1.00*
Technology	0.032	0.044	1.633	0.00†
Economic survival	0.900	0.109	18.410	1.00*
Philosophy	0.867	0.175	11.092	1.00*
Self-concept	0.868	0.126	15.407	1.00*
Public image	0.200	0.094	4.740	0.00†
Employees	0.300	0.075	8.885	0.00†

1.00- strong evidence of consistent inclusion* *0.00† - strong evidence of non-inclusion*

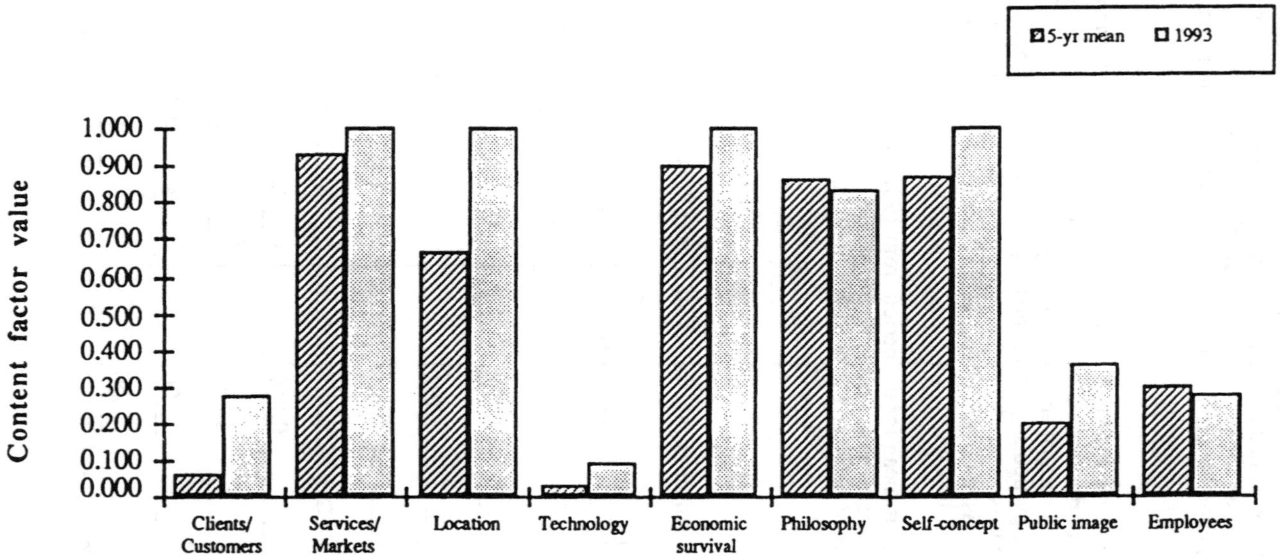

Figure 2: Degree of inclusion/non-inclusion of content factors in contractors' mission

Table 2: Summary statistics for mission variation 1987-1991

Annual pairings		Mean for annual values		t-statistic for
Year 1	*Year 2*	*Mean Yr1*	*Mean Yr2*	paired means
1987	1988	0.537	0.583	1.049
1987	1989	0.537	0.537	0.000*
1987	1990	0.537	0.592	1.264
1987	1991	0.537	0.434	2.330+
1988	1989	0.583	0.537	1.114
1988	1990	0.583	0.592	0.286
1988	1991	0.583	0.434	3.410*
1989	1990	0.537	0.592	1.400
1989	1991	0.537	0.434	2.053
1990	1991	0.592	0.434	3.878*

** -p<0.01, + -p<0.05*

Table 3: Summary statistics for mission variation between 5-year mean (87-91) and 1993

	5-year mean (1987-1991)	*Mean value factors- 1993*	*t-statistic for paired means*
Average of mean value factors	0.537	0.647	2.868
Standard deviation	0.382	0.388	

5 Discussion

This section addresses the outcome of the content analysis for the sample contractors. Although it will be inappropriate to generalise with the sample for an industry-wide perspective of the focus of contractors' strategic planning, the analysis still provides an insight into the issues that receive the long-term attention of the companies in the sample, from which other companies can learn.

5.1 Strategic content of contractors' mission statements

From the results of the analysis it can be observed that, for the sample contracting companies, five factors dominate their mission statements and hence the focus of their strategic survival. These are:

- services and markets;
- location and geographical spread;
- economic survival and profitability;
- philosophy and values, or aspirations; and
- self-concept, or strengths and weaknesses.

The dominant factor of *location and geographical spread* in the strategy of construction contractors is only logical, as the products of their business operations are not transportable. By defining market sectors for their business operations, contractors attempt to achieve a balanced corporate portfolio. All the companies in the sample could be classified as industry leaders, and were therefore aware of their position within the industry, and clearly articulated business *philosophy* to exploit their position of strength.

The general absence of four major factors in the strategic focus of construction contractors present a case for an improvement in the scope of their strategic concerns. The factor *concern for employees* featured in some of the mission statements of the contracting companies in the sample, nevertheless, the level of its mean value and t-statistic indicated a general non-inclusion in the missions of the whole sample. No meaningful inference could therefore be drawn on this factor. It can be inferred that for the sample of companies, the focus of their strategic planning does not cover the three factors *clients and customers*, *technology*, and *concern for public image*. The non-inclusion of the factor of *technology* in the mission of contractors is perhaps consistent with the general view of an industry which is still labour intensive. However, the non-inclusion of the factor *clients and customers* in the mission of the sample companies does not reflect the project oriented nature of the industry, where long-term relationships with clients can significantly influence levels of turnover and minimise business risks. It would have been expected that contractors' strategies would focus on the employers who generate the majority of the industry's workload. However evidence from the analysis point to the absence of such relationships. Exploiting the development of such relationships as a strategic option, should provide opportunities for construction contractors to minimise their business risks. There has been an increasing awareness in recent times, of the influence exerted by the built environment on nature by the general public. This has prompted some construction contractors to include environmental concerns in their strategic focus. However, the low mean value for the factor *concern for public image*, could be interpreted as indicating that a significant proportion of construction contracting organisations are lacking in this area.

5.2 Variation in the focus of contractors' strategies

The paired annual mission contents revealed that between 1987 and 1991, the focus of contractors' strategic planning did not undergo any appreciable change. This is evidenced by the values returned from the t-test of the paired data. The low values of the resulting t-statistic was consistent for all the paired data making up the sample. In The business environment in which contractors operate is characterised by cycles of change, influenced by several economic factors, such as funding shifts, deregulation, and foreign competition through globalisation of the industry's markets. Construction contractors' mission statements, however, indicate a stability in the focus of their strategic orientation. This perhaps should be the case, as strategy, should denote an enduring character. However, the consistent non-inclusion of the factors of *clients*, *technology*, *public image* and *employees*, over the five year period offers some insight into areas for which could be addressed to provide greater scope for competitive advantage. Analysis of the statements for 1993 did not reveal any change in the number of factors that played a dominant role in the strategic focus of construction contractors. However, there was a modest change in the mean factor values for the five dominant factors, suggesting a wider appreciation of these factors for their strategic survival by construction contractors.

6 Summary

The chapter has employed a content analysis method to evaluate the strategic focus of construction contractors' mission. The analysis revealed that five factors dominated the mission of construction contractors. These were *services and markets*, *location and geographical spread*, *economic survival and profitability*, *philosophy and values (aspirations)*, and *self-concept (strengths and weaknesses)*. The evidence from the analysis shows that contractors give little or no consideration to the other factors of *clients/customers*, *technology*, *concern for public image*, and *concern for employees*, normally associated with mission statement.

One of two inferences could be drawn from the outcome of the analysis. First, that construction contractors give consideration to the factors of *clients (customers)*, *technology*, *concern for public image*, and *concern for employees*, in their search for corporate objectives, however, no strategic advantage is seen in these factors and so they do not feature in their strategic options. Second, and the most obvious, is that these less dominant mission factors do not receive any attention from construction contractors in their search for strategic options.

7 References

1. Piercy, N.F. and Morgan, N.A., 1994. Mission analysis: an operational approach. *Journal of General Management*, **19**(3), pp. 1-19.
2. Drucker, P.F., 1973. *Management: Tasks, Responsibilities, and Practices*. New York: Harper & Row.
3. Brabet, J. and Klemm, M., 1994. Sharing the vision: company mission statements in Britain and France. *Long Range Planning*, **27**(1), pp. 84-94.
4. David, F.R. 1989. How companies define their mission. *Long Range Planning*, **22**(1), pp. 90-97.
5. Rarick, C.A., and Vitton, J. 1995. Mission statements make cents. *Journal of Business Strategy*, **16**(1), pp. 11-12.
6. Pearce, J.A. II, 1982. The company mission as a strategic tool. *Sloan Management Review*, Spring, pp. 15-24.
7. McGinnis, V.J. 1981. The mission statement: a key step in strategic planning. *Business*, November/December, pp. 39-43.
8. Want, J.H., 1986. Corporate mission. *Management Review*, August, pp. 46-50.
9. Easterby-Smith, M., Thorpe, R. and Lowe, A., 1991. *Mangement Research: An Introduction*, London: Sage Publications.
10. Lacity, M.C. and Janson, M.A., 1994. Understanding qualitative data: a framework of text analysis methods. *Journal of Management Information Systems*, **11**(2), pp. 137-155.

UK NATIONAL TECHNOLOGY FORESIGHT: THE CONSTRUCTION INDUSTRY PERSPECTIVE

B.A. Young
Bartlett School, University College London, London, UK

Abstract
The first UK National Technology Foresight Programme was conducted in 1994/95. The construction industry was one of fifteen sectors taking part in the exercise. The objectives of National Technology Foresight are presented along with the methodology and criteria for selecting the key priorities. The Construction Industry Panel identified four engines of change and five key opportunities. The paper discusses each of these priorities in detail. A summary of the major tasks and organisations needed to implement them is included in the paper. The conclusion focuses on the extent by which the Technology Foresight Process meets its objectives and, in so doing, the lessons to be learned for the future.
Keywords: Technology Foresight, Construction Industry.

L'Abrégé
Le premier programme nationale de technologie de prévision de la Grande Bretagne a eu lieu en 1994-1995. L'industrie de la construction a été l'un des quinze secteurs qui ont participé à cet exercise. Les objectifs du programme sont presenté en même temps que la methodologie et les critères choisis pour indiquer les priorités clefs. Le jury de l'industrie de la construction a identifié quatre voies de changement et cinq opportunités principales. Le document étudie chacune de ces priorités en detail. Un resumé des tâches majeures et de l'organisation requises pour la mise en place de ce programme est inclus dans ce document. La conclusion se concentre sur la mesure dans laquelle la technologie de prévision atteint ses objectifs et, par consequent, les leçons qu'on peut en tirer pour le futur.

1 Introduction

The central thesis of "Realising our Potential", the government White Paper published in May 1993 was that the UK science and engineering base has the capacity for being more aware of and responsible to the needs of industry and other research users. Equally, firms and other organisations should be aware of and receptive to the research being done by the science and engineering community. The Government conceded that it would take steps to encourage a greater level of communication and raise the level of mutual understanding.

The Organization and Management of Construction: Shaping theory and practice (Volume One). Edited by D.A. Langford and A. Retik. Published in 1996 by E & FN Spon. ISBN 0 419 22230 8.

Many decisions were being taken about the relative merits of strategic research, without the benefit of any systematic and comprehensive assessment of the match between potential research outputs and take-up by firms. Research programmes which may be successful in the quality of the research may not necessarily offer economic reward to the potential recipients, the end users. Such a waste in resources must therefore be minimised by producing a better match between publicly funded strategic research and the needs of industry and other end users.

Having considered the experience of other advanced nations and assessed best practice in the UK, the Government decided that the establishment of a UK Technology Foresight Programme (TFP) was essential. ....."This country (UK) could and should benefit from the application of Technology Foresight, not only as a means of gaining early notice of emerging key technologies but also as a process which will forge a new working partnership" (p17). The idea that partnerships and networks which would emerge from the TFP was perceived as a vital mechanism for encouraging a greater common understanding of the trends and uncertainties surrounding further technological developments.

It is the purpose of this paper to describe, in some detail, the process by which the Technology Foresight Programme took place and how the construction industry participated in the process. The results from the construction sector survey will also be reported. The final part of this paper addresses the question as to whether the process was a success and the way forward.

2 Principal Objectives of Technology Foresight Programme

Following in the footsteps of Japan, Germany, the Netherlands, USA and Australia's experience of a national TFP, the Office of Science and Technology (OST) (1994), who is responsible for achieving the successful completion of the Programme, deliberated the objectives of the Programme which are as follows:-

That the process can be designed to

- break down barriers and create contacts across organisational boundaries between small and large firms and across industrial sectors;
- develop a consensus of future technological scenarios, their likelihood and importance;
- raise and stimulate awareness among the science and technology community of the long-term potential vis-à-vis technology and market opportunities.

A number of secondary objectives have been identified as:

- forging a consensus on national goals;
- purchaser/supplier chain link involvement
- assisting government departments in clarifying their role in wealth creation and enhancing quality of life.

The outputs or results can be used to:

- identify particular generic technologies which society recognises as important goals over the next decade;
- identify fields and targets which will be important in the long-term(30 years)
- set priorities between broad fields of science and technology
- identify fields of technological fusion which cut across disciplinary boundaries.

The objectives considered by each of the national approaches are described in Table 1, together with their strengths in actually realising those objectives.

How do the OST's objectives compare with the overseas commitments? Firstly, the Government was of the opinion that the exercise should provide a systematic means of identifying and assessing generic technologies and their long-term economic and social benefits. Secondly, priority setting both within and between fields of science and engineering must also feature strongly. Thirdly, other government departments should be informed of the priorities set for applied science and technology. Lastly, industrial exploitation of science and technology will be strengthened by contacts and networks between industry and the science and engineering research community. When comparisons are made with Table 1, it can be seen that the OST objectives are broadly similar to those of the overseas experience.

3 Technology foresight process

The task of overseeing the Foresight Programme was devolved to a Steering Group chaired by the Government's Chief Scientific Advisor. The organisation structure for the Programme is shown in Figure 1 (IPRA 1994).

Consultants were appointed to work closely with OST. Fifteen expert panels were established to represent the sectors taking part in the programme. The Construction Panel was not included in the original selection of sector groups. Only after the Construction Sponsorship Directorate, argued successfully for its inclusion did approval take place in February 1994. Each of the fifteen sector groups are listed in table 2.

The composition of the expert panel was devised from an initial survey conducted by PREST consultants requesting respondents to identify their level of expertise in a given area. In addition, each individual was asked to nominate others who met the criteria in question. Using this mapping tool, a database of experts was constructed and the process of co-nomination was adopted for generating the pool of experts.

Table 1. Principal objectives of overseas technology foresight
Source: OST January 1994

Country	UK			Germany			Neths	USA	Japan	Aust/NZ
Project or organisation	ACOST ET Cttee	DTI	PRISM (OSO)	Tech for 21st Century	Ctte Basic Science	Delphi	Ministry Economic Affairs	Critical Technologies	STA 30-year Forecast	Basic Research Priorities
PROCESS										
Raises awareness widely of different vision of future	✘	✓	✓	✓	✘	✔	✓	✘	✔	✘
Encourages networks and contacts	✘	✔	✓	✓	✘	✘	✔	✘	✘	✘
Develops broad consensus on ways forward	✘	✓	✔	✓	✓	✔	✔	✓	✔	✓
OUTPUT										
Detailed priorities within fields	✔	✔	✔	✓	✓	✔	✔	✓	✔	✓
Broad priorities between fields	✘	✘	✘	✓	✔	✓	✘	✘	✓	✔
Priorities for Technology Policy (generic technologies)	✓	✓	✓	✔	✘	✘	✓	✔	✘	✘
Identify fields with long-term (up to 30 years) potential	✘	✘	✓	✘	✓	✔	✘	✘	✔	✓
Identify fields of technological fusion	✘	✘	✘	✔	✔	✔	✓	✓	✔	✔

Key: ✔: Primary effect ✓: Secondary effect ✘: Little impact

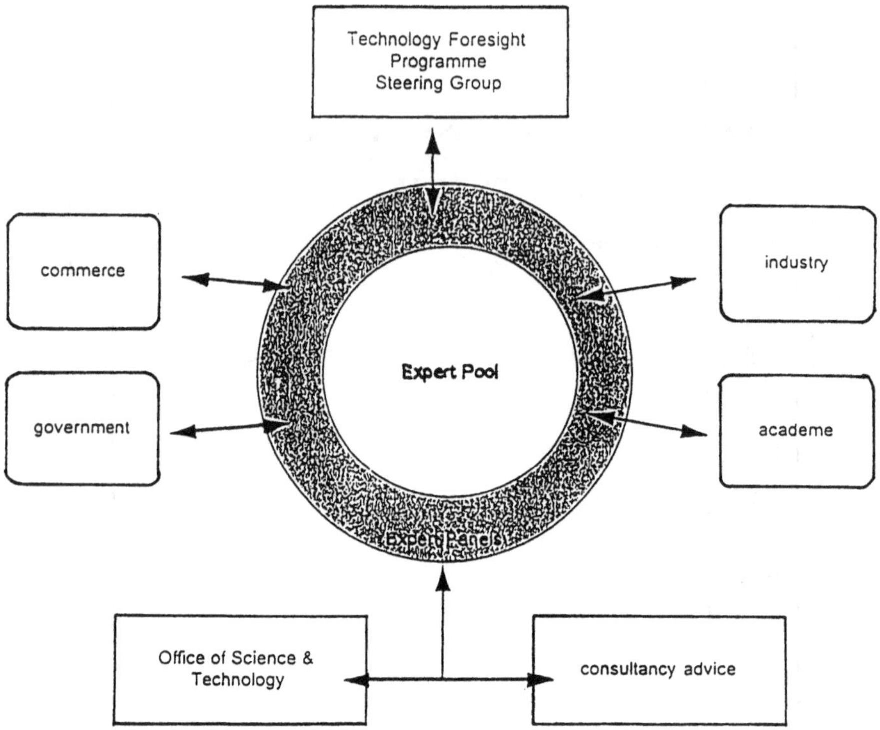

Fig. 1. Technology foresight programme: organisation chart
Source: IPRA Ltd March 1994

Table 2. List of sector groups participating in technology foresight programme

Sector Groups	
Agriculture	Construction
Food and Drink	Materials
Energy	Chemicals
Defense	Leisure
Manufacturing	Finance
Communications	Retail and Distribution
IT/Electronics	Transport

Over 3,000 forms were sent out to all sectors. A 40% response rate was recorded. Just less than 40% of the respondents reported that they were engaged in academic research, whereas the remaining categories are dominated by industrialists.
(Figure 2)

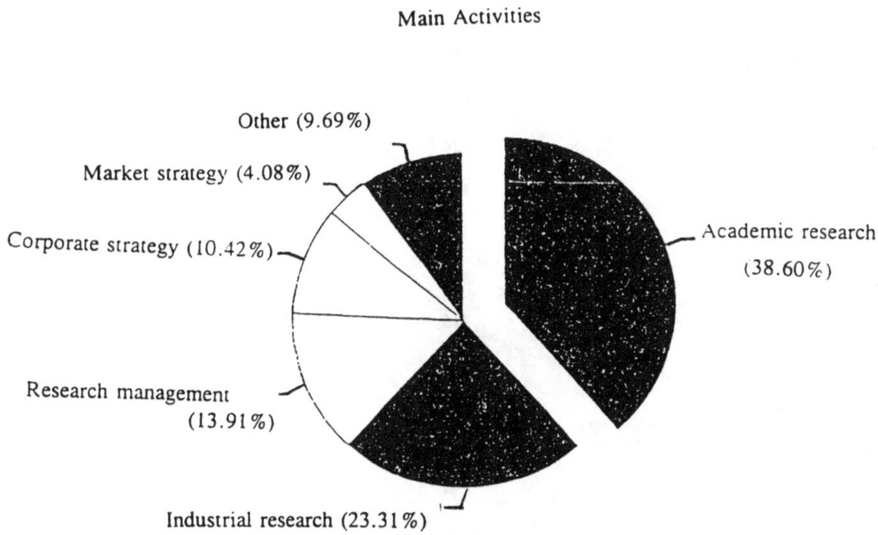

Fig. 2. Main activity of respondents
Source: Georghiou et al University of Manchester 1994

When the numbers are categorised according to 'Expert' by Field, only 131 individuals responded from the construction industry claiming to be knowledgeable or expert in a particular field (see Table 3).

Georghiou et al (1994) advanced a hypothesis to explain these phenomena. They claim that well organised research communities with a clear stake in the future distribution of research resources are more likely to respond. It is widely understood that because the construction industry is highly fragmented, forming a cohesive research community is difficult to achieve (Young 1995). However, another explanation may be that the generic nature of some sectors is reflected by the broadening of the skill base. This would imply that the construction industry skill expertise is highly specific.

Table 3. Numbers 'knowledgeable' or 'expert' by field
Source: Georghiou et al, University of Manchester 1994

Field	No. knowledgeable or expert
IT and/or electronics	515
Materials	441
Environment and safety	381
Manufacturing and production	370
Business processes	341
Chemicals	324
Energy	308
Life sciences	306
Communications	253
Aerospace	215
Defense	199
Food and drink	193
Natural resources	180
Health	175
Transport	141
Construction	131
Retail and distribution	63
Lifestyle and culture	53
Financial services	51

From the 131 construction experts the OST selected 18 members for the Construction Panel and an assessor who represented the construction industry on the Steering Group. Once in place, the Panel nominated further members from the expert pool to extend the breadth of coverage.

The Construction Panel met for the first time in May 1994 and over the following ten months, nine Panel meetings took place, an extensive literature review was conducted and the methodology included eleven workshops and a two-stage Delphi survey. In all, over two hundred people attended the workshops. Five hundred and fifty experts were consulted during the two-stage Delphi survey and one hundred and fifty one people replied. A twenty eight per cent response rate was recorded. A detailed breakdown of all the results are given in the Construction Panel Report (1995). The key findings and priorities are presented below.

The Construction Panel , through its literature review, workshops and the Delphi process were able to identify and prioritise the **key opportunities** and **engines of**

change against the criteria set by the Steering Group. These criteria include;

- economic and social benefits arising from market opportunities - wealth creation and quality of life;
- the ability of the UK industry to capitalise on specific opportunities - UK capabilities;
- the ability of the UK science base to contribute to specific science and technology opportunities - UK capabilities.

By analysing the data, using this selection process, nine main priorities comprising four engines of change and five key opportunities were arrived at. Tables 4 & 5 list the headings. These nine key priorities are then classified according to main tasks and the organisations needed to implement them (see Table 6).

Table 4. Engines of change

1.	Promoting learning and learning networks through improved, and more appropriate education and training
2.	Setting up mechanisms to ensure that all players in the construction process are kept well informed and their activities fully coordinated by means of advanced information and communication technology (ICT).
3.	Introducing a national fiscal policy that encourages and fosters long term investment and economic growth.
4.	Creating a culture of innovation through the joint efforts of government, education, industry and institutions.

Table 5. Key opportunities

1.	Using standard components to achieve customised solutions.
2.	Applying advanced business processes to construction (working together efficiently).
3.	Adopting a 'constructing for life' approach.
4.	Benefiting society and the environment through construction developments.
5.	Creating nationally competitive infrastructures.

Table 6. Major tasks and implementing organisations
Source: OST, Key Points 1995

Task	Lead organisations
Education/Learning networks	
Establishment of a forum charged with the development of education policies and curricula designed to produce world class constructors	Education departments
Promotion of learning networks	EPSRC, DoE
Reaping the benefits of the information revolution	
construction industries	EPSRC, ESRC, DoE, DTI
Establishment of a virtual reality centre for construction and development of a managed programme	EPSRC
Fiscal changes	
Research into realtionships between changes in fiscal policy and effects on construction investment	EPSRC, ESRC, DoE, DTI
Innovation culture	
Promotion of innovation, for example by appointing directors and professors of innovation in construction	Industry, EPSRC, Universities
Fiscal policy/legislation in construction	central Government
Customised solutions from standard components	
A managed programme aimed at developing a world leading competence in design, application and supply of customised solutions	Industry, DoE, EPSRC
Components of lightweight materials	
Business processes in construction	
Multidisciplinary research into development and application of improved processes in construction	
Constructing for Life	
Research and establish relationships between performance, expected life and costs of materials and components	EPSRC, ESRC
Environmental and social consequences of development	
New managed programme on holistic approach to social and environmental benefit analysis	EPSRC, ESRC, NERC, DoE
Competitive infrastructure	
Study group to identify UK strengths in the engineering and research base	EPSRC
Competitive infrastructure working group on research required for improvements in UK and export markets	Construction Industry Board, DoE, EPSRC

4 Conclusion

Did the UK National Technology Foresight Programme meet its objectives? As far as the actual process is concerned, the Construction Panel achieved all its objectives. The other fourteen panels are still deliberating the outcome. The breakdown of barriers and the networking across organisational boundaries were actively pursued. A consensus of technological scenarios were developed and priority setting supported four Engines of Change and five Key Opportunities. In order to increase awareness and for momentum to gather pace amongst participants of the UK construction industry, the Construction Panel findings cannot be ignored. It is evidently clear that the construction industry relies on other industrial sectors to provide technological inputs into products and processes. Where technological fusion cuts across industrial boundaries the construction industry must grasp the opportunity for joint ventures, sharing the risk and gains with its industrial and academic partners.

Notwithstanding the government's important role during the implementation process, it must ensure that the next UK National Technology Foresight Programme considers the lessons which have been learned from the 1994/95 exercise. For example, from start to finish, the process was by far too demanding in such a short time frame. If it had not been for the commitment and enthusiasm shared by the panel members, the positive response from the expert pool and the enormous effort the OST exercised, the Programme would have faltered in the early stages. It, therefore, seems reasonable to suggest that given this concerted effort, the very first National Technology Foresight Programme conducted in the UK, from aconstruction industry perspective albeit a late entrant, was a resounding success.

Acknowledgements: Construction Panel, OST Staff

References

1. CMM 2250. (1993) *Realizing our Potential*, HMSO, London.

2. OST (January 1994) *UK Technology Foresight Report.*

3. IPRA Ltd. (March 1994) *A briefing paper to inform discussions at two industry workshops, 13 & 14 April 1994.*

4. Georghiou L. and Loveridge D. and Nedeva M. Co-nomination in Foresight (1994) *A report on the use of co-nomination to identify participants for the UK Technology Foresight Programme,* University of Manchester.

5. Young B.A. (1995), *Technology Foresight No.5: Research and Development - Interfirm Collaboration,* HMSO, London.

6. OST (1995), *Progress Through Partnership, Construction No 2*, HMSO, London,

7. OST(1995), *Key Points. Construction No 2*, HMSO, London,

STRUCTURAL ADJUSTMENT FOR A COMPETITIVE ENVIRONMENT: SURVIVAL STRATEGIES FOR MEDIUM SIZED CONSTRUCTION FIRMS IN ZAMBIA

Peter M. Mukalula

School of Environmental Studies, Copperbelt University, Kitwe, Zambia

ABSTRACT

The institution of the Structural Adjustment Programme in Zambia has had serious repercussions on the construction industry. There has been a dramatic drop in the volume of work in recent times. Competition for the few jobs available has been fierce by both local and foreign companies.

Survival has therefore been an issue to contend with. Medium sized construction firms (MSCF) have felt the impact most during this process of adjustment. The study examines organisation structures before and after the advent of economic liberalisation. It further unravels tangible solutions to problems of management in having to face emerged complexities in terms of formalisation, centralisation and decentralisation, power factors etc, and the business strategy that firms adopted.

KEY WORDS

Organisation structure, strategy, Structural Adjustment Programme

SOMMAIRE

Le calendrier de travaux, ainsi que tout autre outil créé au cours de la phase de planification et de programmation, constitue une carte routière pour la gestion des travaux dans sa phase de mise en oeuvre. Ces outils forment un système de contrôle visant à aider les chefs de projet à anticiper les problèmes suffisamment tôt et à les traiter au mieux. L'avancement-même du projet se traduit en termes physiques et de coûts, qui permettent de mettre à jour le calendrier des travaux et du budget, et de définir le pourcentage de réalisation effectivement effectué et d'en déduire les affects sur la partie des travaux restant à réaliser. Ce procédé permet d'identifier les problèmes qui pourraient entraver la bonne marche des travaux et en conséquence de penser des solutions pour les résoudre ou les amoindrir. Une fois les solutions choisies, les outils de programmation des travaux communiquent et contrôlent les changements apportés. La planification, le contrôle et le suivi des travaux dans le contexte socio-économique particulier de la bande de Gaza posent le problème d'une gestion spéciale pour atteindre les objectifs définis pour le projet. Le présent papier tente, en un premier temps, d'apporter une vue d'ensemble des procédés de contrôle et de suivi et, en un deuxième temps, d'analyser les systèmes de contrôle et de suivi des projets d'infrastructure dans la bande de Gaza.

The Organization and Management of Construction: Shaping theory and practice (Volume One). Edited by D.A. Langford and A. Retik. Published in 1996 by E & FN Spon. ISBN 0 419 22230 8.

INTRODUCTION

Since the last decade, the Zambian economy was characterised by high inflation, currency devaluation, and low levels of investment (Times Of Zambia, 1994). Zambia was among ten other countries in sub-Saharan Africa whose change in macro-economic policies during the adjustment period 1987/91 was described as 'deteriorated' with index scores between 0.0% and -0.2%. The construction industry also declined during this period registering a growth rate of 5.2% in 1986, -4.1% in 1987 and -13% in 1988 (Development Plan, 1989).

The Structural Adjustment Programme was instituted to try to improve the performance of the national economy. In trying to reduce inflation and therefore improve productivity, the government has:-

(i) reduced spending on capital projects.
(ii) reduced grants to local authorities.
(iii) triggered cost saving measures taking place in Zambia's copper conglomerate, ZCCM.

The above measures have substantially reduced the level of construction activity and consequently increased the level of competition. This has forced MSCF's to employ survival strategies in a bid to survive. Owing to their small medium capital bases, these firms have responded, in the main, by scaling down their operations that have led to substantial alterations in their organisational structures.

The challenges brought about by the so called 'enabling environment' on the operational responses of MSCFs, are the subject of this paper.

SAP EFFECTS ON THE CONSTRUCTION INDUSTRY

The inevitable result has been one of aggressive competition and the following aspects can be singled out as the observed effects of the existing situation:-

(i) **Liquidation and foreign competition**
Free market policies have brought both a reduction and an emergence of new building firms in Zambia. Some building firms have gone in to liquidation due to the high competition. The lack of construction activity has only left the tough going. The emergence of foreign companies who seem to be given favourable conditions (such as tax free benefits) over the local ones, has not seemed to help the situation much but has 'cowered' local potential (Turner, 1995; Mashamba, 1996).
(ii) **Changed procurement strategies**
Free market policies have led some building firms to change their strategies by adopting better approaches to job procurement. With reduced government funding to both government departments and parastatal bodies, these institutions have had to adopt alternative methods of procurement to the traditional RIBA Plan of Work system such as target cost and prime cost, etc., in order to rationalise their resources in the changed environment (Mukalula, 1996).

(ii) **Unemployment**
Free market policies have accelerated the rate of the employment retrenchment process as they are very few jobs to occupy all contractors. Employment levels are at their

lowest for the construction industry. Statistics available for the construction formal sector employment for the years 1990 to 1994, show a significant reduction of those employed.

1990	1991	1992	1993	1994
33.4	33.1	27.8	22.1	18.5

Source: Employment Trends, Construction Industry, Central Statistical Office, p17,1994

Unofficial figures for the year 1995 stand well below 15.0. In the period between March 1992 and December 1993, both the parastatal and private sectors lost 7,600 employees coming third after the manufacturing and agricultural sectors who lost 9,900 and 7,800, respectively (CSO, 1994). This is because firms have been constantly trimming their staff by scrapping off certain job positions and through the elimination of departments or divisions. Such changes have led to alterations in spans of control, levels of formalisation, sizes and arrangements of departments, authority relationships etc.

(iii) **Expansion projects**
Exploring new areas by opening branches and offering new construction products so as to increase their market locally, nationally and internationally, although Turner (1995) says that this has taken place at a very small scale. Diversification in terms of product mix, has increased the complexity of the organisation structure as expansion entails spatial or geographic differentiation which have in turn changed formalisation and power distribution in the firms. Opportunities to expand have been made possible in that money that once was utilised as security for foreign currency bidding purposes is now having to be reinvested in improving the operations of the company. Construction companies who are finishing their projects early are making huge savings in overheads unlike before. As most major building materials have always been imported, stringent foreign currency purchasing measures introduced during the mid '80's contributed to their delayed delivery and therefore late completion of contracts. Laid down procedures required huge amounts of money to be deposited in support of currency bids that only tied up the contractor's capital. Hefty advance payments were a common feature in contract documentation which were consequently 'frontloaded' when priced so as to offset such payments. Careful management was therefore ensured which made most firms adopt highly centralised type of organisation structures.

(v) **Information Technology in workplaces**
Mechanise their operations so as to promote higher productivity and remain viable. Information technology is fast being adopted by most firms to cater for various tasks such as payroll preparations, interim valuations and other administrative documentation whose formatting is easily standardised by available word processing packages. Workplace structures, following after revolutionised, innovative organisation structures, have been adopted.

Turbulent times

The 'turbulent' (Burns in Pugh and Hickson, 1989) construction environment has affected the operations of existing firms in that their business spheres and processes have changed and brought about the need to alter their organisational structures. This has been done in a bid to have their operations become *"more resourceful and productive"* as contained in the following comment in one of our country's tabloids which stated that:

> *"...These are certainly tough times for Zambia as employment shrinks and belts are tightened even further. We are all finding it tough going. But to abandon market forces is no solution. Whether we like them or not, they are an inescapable fact of life. Realism requires that we acknowledge this by becoming more resourceful and productive. By making ourselves competitive we can use market forces to our advantage instead of mourning and groaning"*(Times Of Zambia, 1994).

ORGANISATION STRUCTURES BEFORE AND AFTER LIBERALISATION

Almost 80% of the work coming on to the construction was provided by state owned enterprises in the first fifteen years of Zambia's independence (Development Plans, 1964 - 1980). Government jobs were distributed to building firms following a rota that was kept at the Buildings Branch of the Ministry of Works and Supply. At the time, the number of large and medium sized contractors was almost limited and jobs were guaranteed as foreign firms were not allowed to operate on the same basis unless they were attached to a donor funded project (Golson, 1995). There was therefore less competition and firms tended to develop and maintain formal, rigid organisation structures.

With the changed business environment, the firms studied were eager to become knowledgeable of opportunities available to them. The effectiveness of how these firms have responded has been seen in the way they have altered their structures for specific tasks or functions that they would want to achieve in order to survive. Beishon et. al., (1976) argues that,

> *"If [a] system is to survive, maintenance substructures must be elaborate to hold the walls of the social maze in place. Even these would not suffice to ensure organisational survival, however. The organisation exists in a changing and demanding environment and it must adapt constantly to the changing environmental needs. Adaptive structures develop in organisations to generate appropriate responses to external conditions."*

This is what Burns' (Pugh and Hickson, 1989) 'organismic' type of organisation advocates for an unstable environment ; the flexibility of having to understand it is absolutely essential for there to be effective organisation of the internal structures to suit such.

Study Criteria

The change in organisation structures before and after the institution of the SAP is of interest. In order to study the changes occurring before and after the institution of the

IMF-monitored economic measures, a number of distinct organisational features were singled out that consisted of the following:-

(a) Organisational size
(b) Organisational complexity
(c) Formalisation of Work procedures; and
(d) Centralisation and decentralisation of authority.

These features are a part of extensive studies done by researchers such as those of the Aston group and found in works by Weber and Henry Fayol (Pugh and Hickson, 1989). The survey was conducted in the two major towns of Kitwe and Lusaka where , by virtue of the location of these cities, construction works are going on all the time. A total of 15 major MSCF's were studied and the responses received revealed features (b) to (d), above, as the most common factors that were crucial considerations to what firms understood to be an 'organisation's structure'. As the size of an organisation depends on the number of people, there was great disparity in the number of people employed in the firms studied that the basis of comparison was substantially eroded for the purpose needed.

(i) Organisational Complexity

As a structural condition, organisational complexity is exhibited through a number of factors i.e. what separate parts there are of an organisation. This is shown by the division of labour existing, the number of hierarchical levels, and the spatial dispersion of the organisation.

The most commonly used form of structuring in MSCF before the economy was liberalised was the traditional functional approach consisting of departments that made them and were based strictly on the achievement of the firms' objectives. Departments encompassing staff support and techno-structure were very rare and though, specialisation of labour existed, this was loosely defined. Most firms preferred employing "jacks of all trades" which they saw as positively contributing to their urgent need for restructuring in this post economic era as it obviated the recruitment of specialised staff. The most likely departments one could find MSCFs before economic liberalisation were:-

(i) Constructing Division
(ii) Purchasing and Stores
(iii) Estimating and Quantity Surveying; and
(iv) Accounts.

Within the existing divisions of the MSCF's, alterations done have reduced or increased the degree of internal segmentation of the firms studied. The following are the results of the survey on how the changed economic environment affected the firm's organisational complexity for survival:-

Altering departmental arrangements/composition	34%
Laying off workers thus reducing the sizes of departments or units.	46%

Creating more departments	20%

Source: Field Survey

The results indicate that much change in organisation complexity has occurred through the reduction of staff with departmentation alterations following on. Such changes have affected organisation structures in the following ways:-

(i) The shedding-off of staff and changed size and compositions of departments has brought about lean vertical differentiation in the organisation hierarchies; reducing the spans of control. Reasons for laying off workers were:-

(a) to reduce costs
(b) to avoid paying idle workers as there was little to do; and
(c) to maintain a small but motivated labour force

(ii) The creation and elimination of departments either increased or reduced the horizontal differentiation. It was revealed that although some firms introduced some departments, the vertical differentiation did not tally with that of the horizontal. This means that one head of department would have been heading two to three departments which further widens the span of control thus weakening organisational efficiency.

(ii) Formalisation

By definition, formalisation includes statements or procedures, rules, roles and operation of procedures dealing with decision seeking, the communication of those decisions and instructions, together with the conveying of that information. In other words, increased organisational formalisation is a means of controlling the behaviour of the members of the organisation by limiting individual discretion.

The survey revealed the following:-

Rules used to apply strictly and even with the changed situation	14%
Rules used to be applied strictly and still do	43%
Rules used to apply fairly and still do	43%

Source: Field Survey

These results indicate that only 14% of the respondents changed organisational rules. It is therefore recognised that while structures have changed, a commensurate change in rules to govern the effectiveness of these structures, has not occurred in the majority of the organisations. The following observations consist of the reasons why the situation is as it is:

(i) Standardisation of the work process - Work standardisation processes depended on the departments that felt the need to re-organise themselves. Evidence of this was seen in standard formulations of data sheets and submitted reports.

(ii) Standardisation of Skills - Although firms stipulate the type of qualified manpower they need from job seekers, they have not strictly adhered to the rules set. The majority of MSCF's favoured personnel who were multi-disciplinarians as they could perform the work of the civil engineer, quantity surveyor or estimator.

(iii) Codification of rules - Rules did not seem to be a major way of ensuring conformity to organisational requirements because top executives had direct supervision over the operatives in most cases.

It appears that although there does exist some forms of formalisation in the organisations studied, it is most of the time by-passed and in its place, a 'dictatorial' approach to managerial leadership adopted.

(iii) Centralisation and Decentralisation of authority
Although being a formalisation component, its considerable attention during the period of 'adjustment' is of significant interest. The survey results were as follows:-

Highly centralised authority system	42%
Highly decentralised authority system	35%
Both at play	23%

Source: Field Survey

The survey revealed that decision making authority was vested in the highest man which made delegation of authority, a rare phenomenon and where it existed, it involved non-financial matters.

However, the study also showed that the current trend seems to have favoured the adoption of highly centralised systems since the 'hostile' business environment has forced top most executives to involve themselves in making all important decisions concerning the running of their firms. This has been a negative response to the prevailing situation as high levels of centralisation prove to be detrimental to a firm's survival through unjustifiable delays. Decentralisation, therefore, would prove to be the only viable alternative in a competitive environment. Delegation of authority is a **must** to speed up the decision making process. As construction works normally occur away from head offices, delegation of authority can definitely speed up operations.

STRATEGIC IMPLICATIONS OF STRUCTURAL CHANGE

Chandan (1991) cites Guelick's (1972) four types of strategies:

(i) **The Stability strategy** - one that a satisfied organisation in its business sphere would embark on and *"...may not be motivated and adventurous to try new strategies to change the status quo."*(Ibid)

(ii) **The Growth strategy** - as one that seeks to expand as well as diversify the organisation's activities. Owing to the *volatility* of the business environment, growth is reflective of effective management of the organisation.

(iii) **The Retrenchment strategy** - simply stated, this is the reduction in product, services, and personnel. It is mainly aimed at stemming wastefulness in the face of adverse competition by re-organising the company.

(iv) **The Combination strategy** - is a merger of various strategies mainly used by large complex organisations.

The study has shown that strategies (i) to (iii) have been utilised by MSCF's for surviving the competitive environment. All these strategies affect the structure of organisations. Fellows et. al.,(1983) states that;

> *"strategy and strategic management do not occur in a vacuum: they occur within a specific business which has a unique organisation structure. The organisation structure is the facilitating framework through which strategy is implemented, and will therefore influence strategic decisions and be affected by those decisions."*

As the study has shown, the basic problem for any organisation is how to strategically cope with uncertainties. The answer is in how these MSCFs have co-ordinated activities within their organisation structures while adjusting to their outside business environments.

As strategy is clearly geared towards the firm's external environment, other factors have evolved by virtue of the competition existing in the industry. There is now, as a result, greater focus on the personal needs of members in the organisation as well as their having to adopt work cultures existing in other companies as means of gaining a competitive edge. Such power factors (in that they enforce their presence from the outside) have become crucial factors of consideration during any structural changes made to MSCFs especially as they are associated to individuals working in those firms.

CONCLUSION AND RECOMMENDATION

In conclusion, construction work consists largely of an unstable volume and product mix and at times involves the geographical distribution of that work, which must be managed through a congenial administrative apparatus. Adopting rigid administrative structures proves to be very uneconomical to maintain. Changes in the economic environment must therefore be followed by changes in the overall strategic approach of firms, more especially the assessment of the impact of the changes on the people in the organisations. Widening the involvement of individuals and groups during the strategic choice processes, will help ensure easier implementation of change.

In further recommending the way forward, it is clear that the idea of 'strategy' in the MSCF's studied, seems to be at variance with the actual 'structural changes' that have taken place. There is therefore greater need for these firms to understand their changed environment and that in turn, to effectively shape their internal organisation structures in order to remain viable in a competitive environment.

REFERENCES

Beishon, J and Peters, G (1976) Systems Behaviour (London: Harper and Row)

Central Statistical Office (1994) Employment Trends, Government Printers, Lusaka.

Chandan, J S (1991) Management Theory and Practice, Vikas Publishing House Ltd

Development Plans (1964-80) Government of the Republic of Zambia, Government printers, Lusaka.

Fellows, Richard, Langford David, Newcombe Robert and Urry Sydney (1983) Construction Management (Part 1), New York: Longman Inc

Golson, Paul (1995) Construction As An Industry In Zambia, The Construction Industry In A Changing EnvironmentConference Proceedings, University Of Zambia, School Of Engineering, Lusaka.

Guelick, W F (1972) Business Policy: Strategy formulation and Action, McGraw Hill

Mashamba, S (1996) "Construction - An Industry Under Siege" in Profit Magazine, No. 4/8 January

Mukalula, P M (1996) The Effects Of The Structural Adjustment Programme On Maintenance Procurement Contracts, CIB W92 Procurement Symposium proceedings, Durban, South Africa.

Pugh, D S and Hickson, D (1989) Writers On Organisations, Penguin Books

Times of Zambia (1994) "The Tough Economic Environment", June 2.

Turner, P (1995) Opportunities in the sub region, The Construction Industry In A Changing Environment Conference Proceedings, University Of Zambia, School Of Engineering, Lusaka.

ECO-MANAGEMENT AND CONSTRUCTION

George Dunlop
Bell College of Technology, Hamilton, UK

Abstract

Eco-management is widely accepted as essential for the well bieng of the earth. Development through construction leads all human expansion and as such contraction will require to incorporate the concepts of eco-management.

The paper proposes a model for this incorporation, examines legislative, conomic and social drivers of change. Presents the conclusions of an environemntal survey of house builders in Scotland and reaches a number of general conclusions for the future.

Sommaire

On accepte partout que l'écomanagement est essentiel au bien-être général du monde. Le développement, par le construction, mère toute l'expansion humaine et par conséquent le bâtiment devra incorporer les éco-management.

Le papier propose un modèle par cette incorporation et examine le èlèments législatifs, économiques et socious par le changement. On présente aussi les résultats d'une enquête par rapport, a l'environnement parmi les constructeurs des maisons en Ecosse et arrive à plusiers conclusions par l'avenir.

Eco-management: the concept

The concept of eco-management has been derived from the resolution by the Council of European Communities Regulation (EEC) No 1836/93[1]. It is a wider concept than is implied by the majority of current thinking on environmental issues. The concept subsumes most of this thinking with the addition of promotion of continuous improvement, audit and publication of performance.

The Organization and Management of Construction: Shaping theory and practice (Volume One).
Edited by D.A. Langford and A. Retik. Published in 1996 by E & FN Spon. ISBN 0 419 22230 8.

Under the Regulation all industrial activities were covered.

This is clearly stated in Article 1 point 2 of the Regulation.

"2 The objective of the scheme (eco-management and audit scheme) shall be to promote continuous improvements in the environmental performance of industrial activities by:

(a) the establishment of and implementation of environmental policies, programmes and management systems by companies in relation to their sites;

(b) the systematic, objective and periodic evaluation of such elements;

(c) the provision of information of environmental performance to the public."

The introduction of such a wide ranging set of objectives to the construction industry would seem to be fraught with problems. These being of sufficient magnitude to delay further development. The United Kingdom government are one of the few member states who have complied with Article 6 by establishing a "system for the accreditation of independent environmental verifiers and for the supervision of their activities". The UK competent bodies was established in April 1995, within the required 21 month period in Article 6.2. The remit of this body indicates a range of industrial areas specifically covered by EMAS[2] which does not contain construction, possibly in recognition of the problems associated with construction.

The EMAS proposals do however cover "woodworking industries" which appears to include timber-frame house producers who are part of the construction industry.

The paper will endeavour to define terms within which the construction industry can and should adopt the practice of eco-management.

A brief review of published literature on environmental issues will indicate a dearth of material directly relevant to construction. Exceptions are "Environmental Management in Construction" by Alan Griffith[3] and "Guidelines on Environmental Issues" by the Engineering Council[4] both of which were published in 1994 which is post Regulation 1836/93 but pre date the implementation date of April 1995. With this lack of academic analysis it might seem reasonable to assume that the practice would lag behind even further. This is not, as the paper will show, the case in fact. The construction industry has for a number of years been actively involved in environmental performance improvement. This has been in large part to comply with legislative requirements and reaction to consumer pressure. Examples which have been reasonably well recorded are the environmental impact assessment for major civil projects such as the Eurotunnel route from London to the Tunnel, various motorway projects including the M74 southern extension in Glasgow. At a different level house builders in the UK have been continuously improving the energy efficiency of their product partly in response to changes in regulation, partly in response to client demands but almost all these changes have resulted in significant environmental benefit.

In the absence of a suitable scheme for eco-management measuring, monitoring, recording and publication these positive steps have not been recognised.

The above concept of eco-management in construction will be driven by three underpinning factors - legislation, economic and social which will result in consequential environmental achievements. To think that significant changes in practice will be driven by ecological or environmental considerations would seem unrealistic in a commercial free enterprise world and it is also at odds with previous experience. It is the authors contention that eco-management practice in construction will become a major aspect of the industry by a similar mechanism as public health engineering evolved in the last hundred years. The analogy can be argued using the Maslow hierarchy of Physiological needs as the structural model. Figure '1' shows graphically the relationship on the right of the development of Public Health engineering to Maslow's hierarchy. On the left of the diagram is a representation for the future global incorporation of eco-management in construction projecting the recent steps towards a future global built environment.

The Maslow type analogy has some support in recent published works which underline the growing legislative and economic material. The systems for development of environmental management exist in the developed world through structures such as BS 7750[5] and its new ISO partner. Other aspects of industry are recognising the need to become ecological/environmentally friendly. The economics of sustainability have been developed by authors such as Peter Bartelmus[6], David Pearce and Kerry Turner[7] and Michael Common[8]. Such texts argue the economic case for environmental improvement which if successful will no doubt percolate through to primary industries such as construction. Richard Welford[9] has his text Environmental Strategy and Sustainable Development subtitled The Corporate Challenge for the 21st century which reflects the need for industry to adopt business ethics which recognise environmental requirements on a global manner.

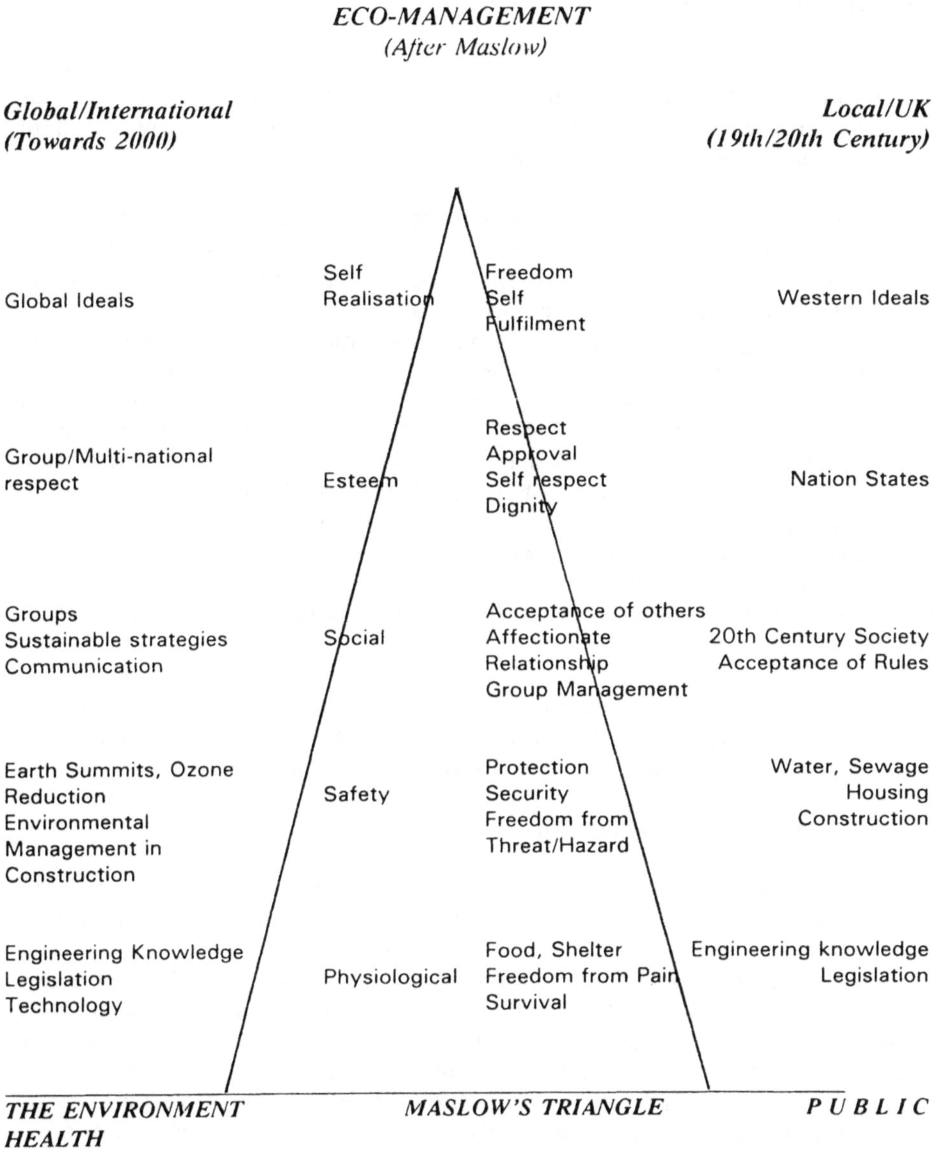

Figure 1

Public Health (19/20th Century) and The Environment (Towards 21st Century) by Dunlop After Maslow
Maslow: AH (1968) Towards a Psychology of Being
Van Nostrand Reinhold, New York

Legislative Economic and Social Framework

In the first section it has been stated that the legislative process will drive eco-management in construction. It has also been claimed that as environmental impacts are global such legislation will be developed at International European and National level.

Unlike the development of public health engineering where practice in many cases preceded national legislation the environmental issues practice has been in many cases preceded by international statements. The Bruntland Report of 1987[10] popularised the concept of "sustainable development" and highlighted the global conflict between conservation of the environment and continued global development on the "western" pattern. It was apparent that a balance was necessary between the quality of human development and the consequential degradation of the natural environment. This concern prompted the Earth Summit in Rio 1992[11]. The United Nations Conference of Environment and Development (UNCED or Earth Summit) which met in Rio de Janiero in June 1992 was the largest diplomatic event in history with 172 governments represented and 118 of them by heads of state. The summit propagated Agenda 21 a huge, wide ranging Programme of Action on Environment and Development to be monitored by a Commission for Sustainable Development.

Whilst many of the delegates at the conference left disappointed with the outcome it provided a major foundation stone for the various regional and national legislative bodies to build credible environmental policies. It also required the signatories to set in process procedures to achieve a number of environmental improvement targets.

In the European context the European Union have since its inception in 1972 incorporated legislative measures designed to improve environmental policy and practice. There are three types of legislation which are binding, Regulations, Directives and Decisions. Regulations lay down the same law throughout the EU and apply directly to all Member States, they do not have to be translated into national laws but apply in the same way as national laws. Directives set out aims to be achieved, but unlike Regulations, these require the Member States to translate these requirements into their national laws within a stated period of time. Decisions, like Regulations, are directly applicable in law but can be addressed specifically to named individuals, companies or Member States.

In the area of eco-management and environmental issues pertinent to construction the following are examples of EU legislation.

Regulation 880/92, which introduced a voluntary eco-label award scheme for consumer products across the Member States.

Regulation 1836/93, which introduced a voluntary eco-management and audit scheme for industrial sites across Member States.

Directive 85/337 on the assessment of the effects of certain public and private projects on the environment, which sets out the (largely procedural) requirements for developments such as large scale industrial and infrastructure projects.

With this raft of legislative material many areas of the construction industry and its supplier industries will require to implement eco-management principles.

At a national level the UK, as a Member State, has to comply with all the EU Regulations, Directives and Decisions. It does however raise its own legislation and a key section of environmental legislation is the Environmental Protection Act 1990 (EPA90)[12]. This tackles a wide range of environmental problems which impact on the construction industry.

The economic drivers will continue to influence environmental policy. Personal, company, national and even international economic decisions will always have an impact. In construction the client bodies whether they be individuals, businesses or national agencies have historically always opted for the cheapest offer for the same specification. This philosophy will always tend to mitigate against the environmental option which has historically been more expensive.

Social attitudes influence the behaviour of industries towards more environmental conscious procedures. Examples are the use of CFC's and halons. Public attitudes towards the ozone deplete properties of these gases have resulted in construction moving towards considerable reduction and use of alternatives. It is factually correct that in 1991 EU Regulation 594/91 addressed a phasing out of the production of such gases however the perception within the industry is that the change was a consequence of social pressure in other related applications. Social pressure does however create problems for the construction industry which can be exemplified in the following ways. One problem is the differentiation between the social impact of the vocal, active minority of protest groups and the silent majority of beneficiaries in cases such as motorway projects. Direct action groups can have a major impact on a project however in the event if the project proceeds the vast silent majority reap the personal benefits of improved transportation facility in silence.

Similarly house builders are under social pressure to produce the 'green' house which is energy and environmentally friendly. However given the choice between extra insulation to produce energy efficient house or the money spent on "white goods" the latter option is the more marketable choice. This personal economics of a short-term nature mitigates against environmental attitudes.

These three drivers legislation, social and economic frequently act in a co-ordinated way to influence environmental design. The increased energy efficiency achieved in housing due to improved insulation levels provides a good example of which has the greater effect. In the Guidelines on Environmental Issues on page 6 four key principles of environmental protection are identified as.

- the Prevention Principle
- the Precautionary Principle
- the Polluter-pays Principle
- the Principle of Integration

All four of these principles are achieved by improved insulation.

By improving energy efficiency the owner "prevents" the release of wasted energy, it also meets the "precautionary principle" by decreasing the release of CO_2, the "polluter-pays principle" is achieved by having the owner pay only for the minimum energy use, in many ways it provides an "integrated" solution as it provides social and economic benefits with no complex trade-off effects in other areas. However as previously stated the highly insulated house is not necessarily the preferred option amongst buyers. It is however a reality through government legislation to improve the insulation levels through changes in Building Regulations. In this way the producers will improve insulation, pass the economic initial cost to buyers who will in the long term reap both the economic and environmental benefits.

It is by this complex inter-relationship of drivers that the construction industry will become more eco-management conscious.

Local Research

As part of the research undertaken for production of this paper and in partial completion of a European Project a number of housebuilding companies were surveyed on their environmental attitudes. This survey was by requesting the companies to complete a fairly substantial questionnaire.

The questionnaire was designed in four sections. Section one required a yes/no response to fourteen questions on elements of environmental/energy efficiency improvement in housing provision. Section two contained two questions on the use of environmental assessment techniques. Section three consisted of four questions assessing the use of specific procedures within the organisations which may be seen as indicative of companies actively pursuing improved performance in terms of quality and environmental practice. The final section was five questions relating to specific general environmental issues.

Section one of the survey tested if Scottish Housing was implementing environmentally beneficial strategies. The results indicated levels of well in excess of 75% support for the majority of aspects. The second section on implementation of standard procedures showed a lower support with generally only half the companies giving support. Similarly section three which dealt with national standards was only supported by fifty per cent with the following specific issues being raised.

The use of Environmental Impact Assessment (EIA) methods on site was supported by six out of the twelve who were in the main large volume builders. This seems a reasonable response as it is highly unlikely that builders with less than 300 units per annum are liable to develop estates where EIA is appropriate.

The use of BS 5750 Quality Management Systems was again only supported by four companies with none of the large volume builders supporting this type of initiative. The use of BS 5750 may be seen by many builders as not being appropriate to their existing control systems, also as being an unnecessary extra endorsement when they already subscribe to the National House Builders Registration Council scheme which is a recognised quality standard with marketing potential. The BS 5750 standard may be more appropriate for companies targeting corporate clients as opposed to the normal individual client which is the prime customer for speculative builders.

BS 7750 Environmental Management which is the environmental clone of BS 5750 suffered even less support with only one company indicating that they would consider adoption of BS 7750 and the other eleven showing no support for the standard. This type of high negative response could be rooted in a number of factors such as: little perceived economic benefit; previous experience of application of BS 5750; perception that it is just another bureaucratic system; little or no knowledge of possible advantage or some other reason. Irrespective of the actual reasons for the high negative response this particular standard appears to be of little perceived value.

The final question on the considerations of having environmental issues as part of a purchasing policy was only supported by five out of the twelve. These five were a mixture of three large volume producers and two relatively small producers both of whom have quite distinctive local markets. The rationale behind these decisions is obviously much more complex than with the simpler issues in the previous three questions and probably warrants further investigation.

The final section on green issues and training confirmed the tenor of the full response in that companies recognised that "green issues" were and would in the future be very important. They also recognised the need to englighten staff and provide staff training in such issues. This would seem to agree with the earlier discussions in this paper.

Conclusion

Incorporation of Concept and Legislation

In section one the concept of eco-management was developed, a rationale for its incorporation was put forward by way of an analogy with public health engineering. In section two the existence of fairly extensive legislative, social and economic drivers were developed. The impact of such forces was tested in the Scottish Environmental Research survey of house builders which by implication indicated a degree of substance in the working of the analogy. In this final section the future potential for eco-management in construction will be further developed.

The Maslow type analogy appears to hold in terms of public health engineering. In almost every developed or developing country the "Western Ideal" for construction in terms of public health is accepted practice. Few construction companies or developers of built environment would proceed without including items such as adequate sewage and water supply systems. It is considered inappropriate to design developments without such considerations the good practice first developed late in the nineteenth century in UK and other western societies is now accepted globally. In terms of eco-management there has been an extensive development of essential foundation legislation which has stared out from a global perspective as opposed to the local and national aspect in public health.

International legislation, national directives and the associated published literature, as previously identified, would appear to confirm that the construction industry will adopt business ethics which recognise the environmental requirements in a global way.

Undoubtedly the construction industry is moving rapidly in the direction of becoming more eco-friendly. The evidence is apparent in the contents of the paper. As electronic communication and computer applications grow the potential for rapid transfer of knowledge will increase the speed of development of eco-management.

Designers will use IT to make ecological design decisions in the near future with the same enthusiasm as they use the technology in architecture or structural design.

Despite the optimism of this text real environmental/ecological benefit will only develop when individuals apply such principles to all their decisions.

Bibliography

1 Official Journal of the European Communities, 10.7.93 No L168/1 - 18 Council Regulations (EEC) No 1836/93 of 29 June 1993

2 EMAS Regulations - Eco Management and Audit Scheme Department of Environment, London 1995

3 Environmental Management in Construction - Alan Griffiths MacMillan 1994

4 Guidelines on Environmental Issues - The Engineering Council - September 1994

5 BS 5750 Quality Management Systems - BSI London - 1990

6 Environmental Growth and Development - Peter Bartelmus - Routledge, London - 1994

7 Economics of Natural Resources and the Environment - David Pearce & Kerry Turner - Harvester Wheatsheaf - Hemel Hempstead - 1990

8 Environmental and Resource Economics - Michael Common - Longman, London - 1988

9 Environmental Strategy and Sustainable Development - Richard Welford - Routledge - 1995

10 Bruntland Report - United Nations - 1987

11 Rio Summit - United Nations Conference of Environment and Development - Rio de Janiero - 1992

12 Environmental Protection Act 1990 (EPA 90)

2 The firms and strategic planning

CONSTRUCTION FIRMS' STRATEGIC PLANNING

P. Venegas and L.F. Alarcon
Department of Construction Engineering and Management,
Catholic University of Chile, Chile

Abstract
A methodology to analyze construction firms' long term strategies is described. This methodology provides a systematic approach to study and analyze external and internal scenarios for a construction firm doing strategic planning. A conceptual model that is a simplified model of the variables and interactions present in the analysis of strategic decisions in the construction industry is first built. A modeling approach, designed to predict the impact of strategic decisions, supports the analysis process by integrating expert knowledge and assessments of the strategic planning team into a mathematical model. This model uses concepts of Cross-Impact Analysis and probabilistic inference to capture risk and uncertainties present in strategic scenarios. Strategic options for long term company development such as organization, resource allocation, R&D investment, and others can be evaluated using criteria selected by company management. The model allows management to test different combinations of company long term strategies and predict expected sales, market share, or other measures of company performance. The application of the methodology in a Chilean construction company is used to illustrate the methodology. This company has implemented an aggressive strategic planning program using this methodology. The planning process used in this methodology demonstrated that it has a value itself by allowing manager to structure, enrich and validate their perceptions about the company and its external environment, and facilitating detection of company's competitive advantages.
Keywords: strategic planning, construction firms, decision analysis.

Sommaire
Cet article décrit une méthodologie pour l'analyse des stratégies à long terme dans des entreprises du bâtiment. Cette méthodologie offre une approche structurée et systématique pour l'étude et l'analyse du cadre interne et externe d'une entreprise du bâtiment qui déroule sa planification stratégique. Cette analyse s'appuit sur un modèle conceptuel qui représente un modèle simplifié des variables et des interactions présentes dans l'analyse des décisions stratégiques, par l'integration des connassainces des experts et des estimations de l'équipe de planification stratégique dans un modèle mathématique. Ce dernier modèle utilise des notions de l'"Analyse d'Impact Croisé" et de l'inférence des probabilités pour saisir le risque et l'incertitude présents dans les cadres d'intérêt. Des choix stratégiques pour le développement à long terme d'une

The Organization and Management of Construction: Shaping theory and practice (Volume One).
Edited by D.A. Langford and A. Retik. Published in 1996 by E & FN Spon. ISBN 0 419 22230 8.

entreprise du bâtiment, tel que l'investissement en recherche et développement, la pénétration de nouveaux marchés et autres, peuvent être évalués en utilisant des critères selectionnés par l'administration de la compagnie. L'exercice de ce modèle permet à l'administration supérieure d'essayer de différentes combinaisons de stratégies de long terme pour la compagnie et de prédire les ventes, la part de marché ou autres objectifs plus complexes, à la suite de la mise en pratique de quelconque de ces initiatives. Pour illustrer les avantages de cette approche, on montre l'application réelle de cette méthodologie dans une compagnie chilienne du bâtiment, laquelle en a fait usage pour mettre en place un plan stratégique agressif. L'expérience indique que le processus de planification utilisé dans cette méthodologie a de lui-même une valeur importante, étant donné qu'il permet aux gestionnaires engagés dans ce processus de structurer, d'enrichir et de valider leur perception de l'entreprise et de son entourage, ainsi que de faciliter la détection des avantages concurrentiels.
Mots clé: planification stratégique, compagnies du bâtiment, décision stratégique.

1 Introduction

Most, if not all, engineering/management decisions are ventures because prevailing environments of uncertainty in which these are formulated and implemented [8]. In particular, selection and evaluation of strategic alternatives imply to make critical decisions for the profitability, growth and life of the construction firms. Nevertheless, applications of multiple objectives, multiattribute, or multiple criteria decision-making to select and to evaluate problems have been very limited even in simplified forms, and extensions to dynamic and uncertain situations have still to be tackled [6]. Many authors have pointed out [8] that models to support the strategic decision-making process need to recognize and integrate the uncertainty as part of the actual world and the human knowledge.

As a response to these challenges, the authors, working together with COMIN S.A., a Chilean construction firm, developed a decision-theory based methodology which represents a powerful tool to help decision-makers in the strategic alternatives evaluation process. The information obtained from the use of this methodology becomes the firm performance prediction, as a consequence of a strategy or a devised portfolio. The strategies seek to affect corporate goals such as: (1) return on equity; (2) sales; (3) gross margin; or another complex objectives that require structures which don't exist [6] [7]. The methodology is inspired on a modeling approach called General Performance Model (GPM) which has been successfully applied to some complex decisions and situations [2] [3] [4]. A description of this approach is presented in the next section.

2 General Performance Model

The General Performance Model (GPM) was originally developed in a Construction Industry Institute (CII) research project. Initially, Alarcón and Ashley [1] used this methodology to forecast and analyze the effects that project options, such as construction incentives, project organization and team building alternatives, have on project performance. GPM applies concepts of cross-impact analysis and probabilistic inference to capture the uncertainties and interactions among variables of the causal model, allowing the explicit incorporation of these elements in the decision analysis process.

The methodology consists of two principal structures: first, a conceptual qualitative model that requires the identification of relevant variables and its causal structure; and second, a mathematical model that uses concepts of cross-impact analysis and probabilistic inference to capture uncertainties and interactions among model variables. The use of these models allows predicting potential impacts of the decisions.

3 Model structure

A real-world macromodel was developed to exhibit the variables and interactions that influence the strategic decisions in a construction firm. The model was developed by a COMIN's strategic team which was integrated by five executives and owners of the firm. Figure 1 shows the conceptual model developed with COMIN to identify how a strategy impacts the firm results. It has five levels: ***external factors, strategic options, drivers, processes*** and ***corporate goals***. In the ***strategic options***, each layer represents a single option set such as capital investments, human resource management systems, policy and procedure development and implementation, and market penetration programs. There are many alternatives and combinations of them for each option set. A specific human resource management system can be combined with a particular market penetration program and choices among the other options areas to form a long term strategic plan.

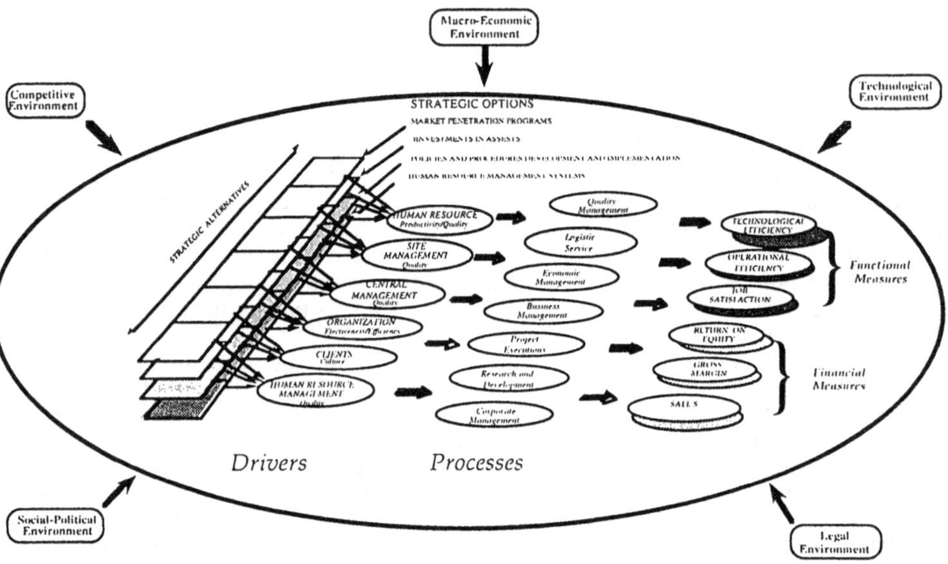

Fig. 1. Conceptual strategic model

Following the ***strategic options*** there is a set of variables that are directly affected by these options; these are called ***drivers***. Each combination of strategic alternatives within an option set is assessed as to its probable impacts on ***drivers***. Those variables, in turn, propagate the effects through interactions among themselves and with the ***processes***. The model then, is defined as a set of variables whose effects propagate from left to right, where each variable is modeled internally as a set of five mutually exclusive and collectively exhaustive events. The events represent the existing range of performance for each variable.

The ***external factors*** are variables whose effects propagate from the external to the internal variables. Some examples are: macro-economic environment, technological environment and competitive environment. Their impact affect the initial probability distributions of the internal model variables, therefore, in the execution algorithm of the mathematical model they occur in first place.

The model descriptions and definitions that follow are very simple in nature because they had an illustrative purpose. The definitions only represent the common understanding of the strategic team members about the meaning of the variables and elements of the model and they don't pretend to be accurate or general definitions.

3.1 Identification of strategic options

Strategic options represent the decisions under evaluation in the model. An option can be characterized by the presence or absence of certain characteristics. The strategic team work must specify the main characteristics which will define the option. In the example case, strategic team members, after a long process, selected a group of four options for the next planning cycle. One of these options is described below as an example.

3.1.1 Market Penetration Programs

During the strategic planning process, one at most ambitious strategies was the program of market penetration. In special, COMIN's strategic team members believed that Chilean's north mining projects could establish a market with stable perspectives. This opportunity needed a strategic plan for the firm to take a good position in this market. The characteristics of this option were grouped in four main types of decisions: (1) temporal or permanent resource location in the area; (2) sophisticated or basic level of infrastructure to allocate in the market; (3) independent or dependent management from the central office; and (4) independent or dependent human resource management from the central office.

3.2 Drivers

Drivers are variables directly impacted by the implementation of the strategic alternatives which spread their effects to the corporate goals. The following were considered:

- *Central Management (Quality)*: This driver refers to the management team that participates in the main office of COMIN.
- *Human Resource (Quality and Productivity)*: production personnel including craftsmen working in the projects and employees who accomplish management functions.
- *Site Management (Quality)*: the management teams on sites, who are in charge of coordinating and organizing projects' executions.
- *Clients (Culture)*: the characteristics of the customer pool of the company.
- *Organization (Effectiveness and Efficiency)*: the characteristics that describe the way in that the company is organized for production.
- *Human Resource Management (Quality)*: This driver refers to all functions in charge of attracting, training, selecting, controlling, promoting, developing, motivating and rewarding the human resource.

3.3 Processes

The processes are variables which, impacted by the drivers, affect directly the corporate goals of the firm. The following processes were selected and considers its own factors:

- *Logistic Service*: in charge of supporting the construction process by managing key resources.
- *Project Execution*: refers to the all activities involved in the construction process.
- *Corporate Management:* in charge of planning, directing and verifying the corporate goals.
- *Business Management*: in charge of supporting the company profitability.
- *Economic Management*: in charge of improving the portfolio of economic resources.
- *Research and Development*: in charge of developing, comparing, adopting and implementing technologies and knowledge generated inside and outside of Chile.
- *Quality Management*: in charge of planning, controlling, coordinating, directing and organizing activities that guarantee the quality projects.

3.4 Corporate Goals

The corporate goals are useful to reflect the impact of strategies on the firm performance. The strategic team members selected a set of goals that have an equilibrium between financial and functional performance. This set of goals is an essential part of the strategy, because they provide company executives a comprehensive framework that translates COMIN's strategic objectives into coherent set of performance measures. The strategic team members defined six outcome measures which represent the corporate goals they wanted to use to evaluate strategic decisions. These outcome measures and their corresponding assessments are shown in section 4.4 Table 1.

3.5 External Factors

The external factors are environmental variables which directly impact the performance of the model internal variables. It is necessary to incorporate this influences, because any action from COMIN is conditioned by the environmental characteristics. The following are the external factors included in the model:

- *Macro-Economic Environment.* This element considers all economic variables that have an impact on the next planning cycle.
- *Competitive Environment*: This refers to the characteristics of the competition pool.
- *Social-Political Environment*: considers all aspects derived from political and social system.
- *Legal Environment*: refers to factors that could stimulate or inhibit the COMIN's behavior in front of the market.
- *Technological Environment*: reflect the level of applied knowledge systematization in the COMIN's market for controlling, arranging and altering the physic and social environment.

4 Assessment process

4.1 External factors

To run the model it is necessary to assess the external conditions for the period under evaluation. These assessments are used to update the initial probabilities of the model variables through the cross-impact matrix.

4.2 Impact of options on drivers

The scenarios or decision alternatives whithin an option can be defined by the presence or absence of certain characteristics. For instance, one characteristic of a market penetration program is the level of infrastructure, it can be basic or sophisticated. Other characteristics can have two or more possible states, by combining option characteristics in differents states it is possible to build scenarios or alternatives for that option. The effects of each scenario on the model drivers can be assessed using a probabilistic definition. A high impact on a driver could be defined as one in the upper 20% of the impact scale. In this example, such effect was assessed by the strategic team members.

4.3 Cross-Impact Information

The mathematical model uses concepts of Cross-Impact Analysis [5] and Probabilistic Inference. Cross-Impact Analysis (CIA) is a technique specifically designed to study how the interactions of events, present in a mathematical model, affect the probabilities of those events. It is used to analyze numerous chains of impact that can occur, to determine the overall effect impact that can occur and the total effect of these chains on the probability that each event will occur. A detailed discussion of the mathematical model is developed in [1].

Figure 2 shows a partial representation scheme to collect information in the form of a simplified cross-impact matrix. The strategic team members answered the following question: "If changes were to occur in the column states, how would this affect row states?". The reply indicate the strength and direction of the "impact" according to the scale shown in the right of Figure 2. This simplified questioning process described above is used to collect the basic information for the mathematical model. A method to convert the assessment to the numerical level, first proposed by Alarcón [1], is used in this methodology. Details on the procedure and testing of this assumptions can be found in [1].

	COMPETITIVE ENVIRONMENT	MACRO-ECONOMIC ENVIRONMENT	TECHNOLOGICAL ENVIRONMENT	SOCIAL-POLITICAL ENVIRONMENT	LEGAL ENVIRONMENT
HUMAN RESOURCE	MOD-	SLI-	MOD+	SIG+	NO
SITE MANAGEMENT	SLI-	NO	SLI+	SIG+	SIG+
CENTRAL MANAGEMENT	SLI-	SIG+	SLI+	SLI+	SLI+
ORGANIZATION	NO	NO	NO	NO	NO
CLIENTS	MOD-	SIG+	SLI+	SIG+	SLI+
HUMAN RESOURCE MANAGEMENT	SLI-	SLI-	MOD-	SIG+	NO

SIG+	Significantly in the same direction
MOD+	Moderately in the same direction
SLI+	Slightly in the same direction
NO	No effect
SLI-	Slightly in opposite direction
MOD-	Moderately in opposite direction
SIG-	Significantly in opposite direction

Fig. 2. Cross impact matrix

4.4 Corporate measures assessment

The strategic team members defined the outcome measures they wanted to use to evaluate their strategic decisions (Table 1). Once the measures were defined, they provided assessments about the variability of the corporate goals in the next three years, using an approach similar to that used in PERT. These assessments were approximated to a Beta distribution in order to represent the variability of the outcomes in the mathematical model calculations. Some assessments were made in U. F. a monetary equivalent to the Chilean currency (1 U.F. = US$30)

Table 1. Balance Scorecard and Assessment for the Next Cycle

CORPORATE GOAL	ASSESSMENT	1996	1997	1998
1. Sales	Lowest	167.000 U.F.	167.000 U.F.	200.000 U.F.
	Likely	220.000 U.F.	270.000 U.F.	300.000 U.F.
	Highest	290.000 U.F.	350.000 U.F.	400.000 U.F.
2. Return on Equity	Lowest	0%	0%	0%
	Likely	16%	22%	22%
Aftertax Profits / Equity * 100	Highest	44%	46%	42%
3. Gross Margin	Lowest	0%	0%	0%
	Likely	7,5%	7,5%	7,5%
Aftertax Profits / Sale in Year	Highest	15%	15%	15%
4. Job Satisfaction	Lowest	43%	43%	43%
	Likely	80%	80%	80%
	Highest	90%	90%	90%
5. Technological Efficiency	Lowest	25 U.F. / 100*M.H.	25 U.F. / 100*M.H.	25 U.F. / 100*M.H.
	Likely	36 U.F. / 100*M.H.	36 U.F. / 100*M.H.	36 U.F. / 100*M.H.
Sale of Net Service / Total M-H	Highest	40 U.F. / 100*M-H	44 U.F. / 100*M-H	50 U.F. / 100*M-H
6. Operational Efficiency	Highest	13 ITEMS/1000 M-H	13 ITEMS/1000 M-H	13 ITEMS/1000 M-H
	Likely	10 ITEMS/1000 M-H	9 ITEMS/1000 M-H	8 ITEMS/1000 M-H
Number of Defects / Direct M-H	Lowest	8 ITEMS/1000 M-H	6.5 ITEMS/1000 M-H	4 ITEMS/1000 M-H

5 ANALYSIS CAPABILITIES

5.1 Impact of strategic option alternatives on corporate goals

The methodology was used to evaluate every existing strategic option alternative. GPM provides the analysis capabilities to help decision-makers evaluate the relative effects of the strategies implicit in the option alternatives. For each strategic option alternative, the methodology provides an estimate of the expected value for each corporate goal (Figure 6) and information on its probability distribution (Figure 8). The purpose of this section is to illustrate the analysis results rather than obtain conclusions about the options; therefore, only a limited discussion is carried out. Nevertheless, the discussion tries to show how the model results can be used to identify the best strategic actions and the most important characteristics of those options for the firm.

5.1.1 Example: Market penetration

Figure 3 shows a rank order of market penetration alternatives in descending order of benefits for 1998 (in this year the program should be implemented). The alternatives were ranked according to their expected Sales and compared with the Return on Equity. Both results are a deviation measures from planned performance for 1998. Alternative (12) was selected as the base program, it has temporal resource location, basic level of infrastructure, dependent central management and dependent human resource management. The base case program represents the actual characteristics of the strategy in the market, and have neither positive nor negative impact on Drivers.

In terms of Return on Equity, the top 3 programs have common characteristics, they have: basic level of infrastructure and dependent central management. Dependent or independent human resource management does not make any difference at this level of performance. For the Sales, the characteristics are similar. The Sales and Return on Equity worst performance is predicted for alternative (14), which has the same characteristics of the present strategy in this market.

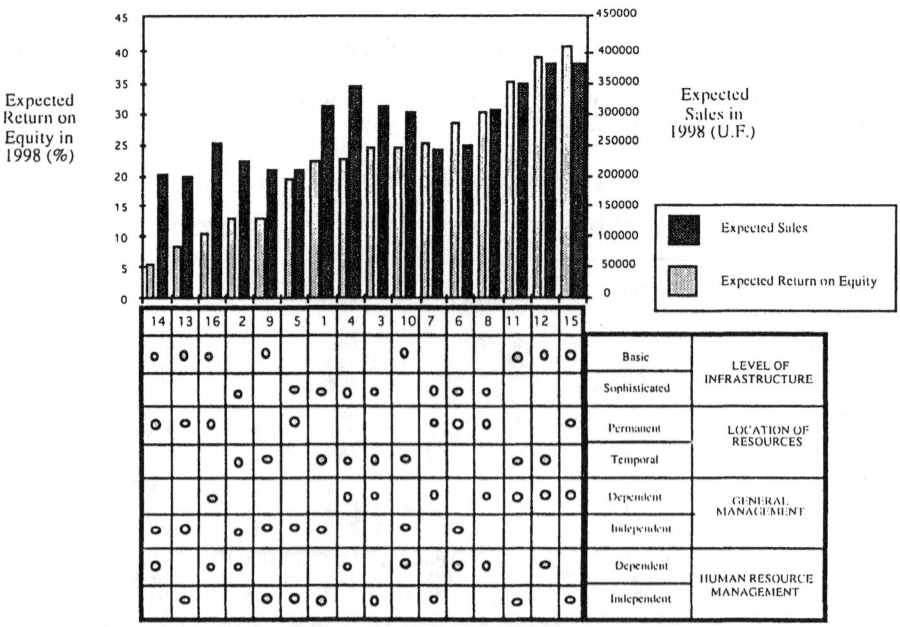

Figure 3. Benefits of market penetration programs

Figure 4 shows the probability distribution of some scenarios shown in Figure 6. In this way, it is possible to see that scenario (10) represents the worst sales performance; however, it has the smallest variation of results. This alternative has a basic level of infrastructure, temporal location of resources, independent general management and dependent human resource administration. The scenario (1) represents better sales expectations, but it has a big variation in the probability distribution. The implementation of this alternative would generate high conditions of uncertainty and risk on the final results. Maybe, scenario (1) could be a better alternative if the decision-makers are adverse to risk.

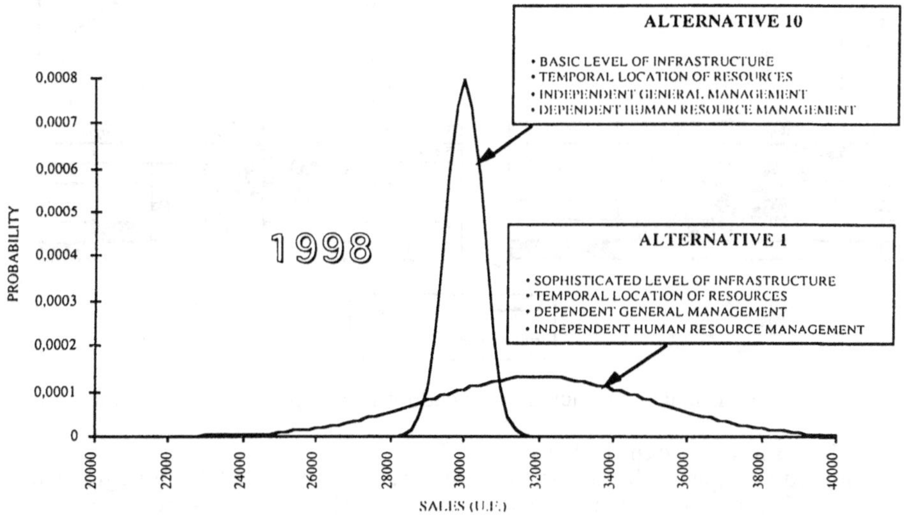

Fig. 4. Sales probability distributions in 1998 for market penetration programs

A careful review of the above results can help the strategic team members to gain a better understanding of the implications of program characteristics and the gains or losses that may result from making changes in the penetration strategy. At the same time, this information can help to design a more efficient monitoring and control procedures within the company.

5.3 Dynamic impact of strategic alternatives on corporate goals

Figure 5 shows another type of analysis made in the strategic decision-making process. This figure shows the performance of corporate goals during a period of time. This performance is affected by the implementation process of a strategy. In this case, Figure 5 shows the sensitivity of Technological Efficiency and Operational Efficiency to policy and procedure development and implementation. This analysis, is obtained from the impact assessments on the drivers for each implementation program step of the alternatives in the selection process, made by strategic team members. In this way, it provides a control structure, which includes a set of expected indicators for each step in the implementation process. With this tool, it is possible to analyze the expected change in the organization, the reaction capacity, and adaptation velocity when prompt response is required.

Figure 5 shows that alternative (2) causes, for the first steps of implementation, an important reduction in the operational efficiency, because the programs requires many individual efforts. Nevertheless, for the program start-up (Step 2.6), the expected results are very positives. On the other side, alternative (1) shows that the implementation impacts are not dramatics, but, at the same time, the expected results are not ambitious.

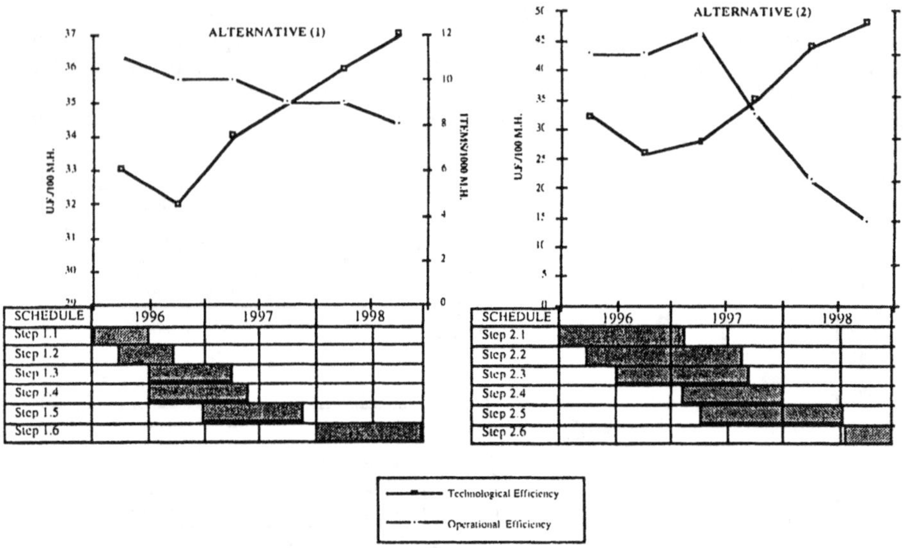

Figure 5. Dynamic Impact of Policy and Procedure Development and Implementation

5.4 Other analysis capabilities

This a limited example of the analysis capabilities of this approach to strategic planning. The results of numerous simulations can be used to perform several other analyses in the strategic planning process. One of the most importants is the GPM capability to analyze combined effects of different strategic options acting simultaneously, as shown graphically in Figure 1, this analysis capability is called the analysis of a strategic portfolio. The model can be also used to analyze the effect of external factors on the final result of the strategic decisions. Other examples of analyses are: comparative analysis of the effects of different combinations of strategic alternatives; dynamic sensitivity analysis of corporate goals to the performance of decision factors; explanatory capabilities through the model causal structure; etc.

6 SUMMARY AND CONCLUSIONS

This paper has presented a methodology for the evaluation of strategies based on construction managers' knowledge and experience, decision analysis modeling techniques, and Cross-Impact Analysis Formalism. A conceptual model was developed using the knowledge and expertise of COMIN's strategic team members. A mathematical model was developed by adapting concepts of a previous CII research. The analysis capabilities of the mathematical model were used to analyze the combined effect of strategic options on company performance. The model can incorporate several corporate goals proposed by the strategic team to measure the effect of strategic decisions.

This methodology can evaluate the impact of business, functional or corporate strategies on firm performance to support the strategic planning process on an individual company basis. Some of the features of the methodology are the following: (1) it has predictive capabilities to help establish performance corporate goals; (2) it incorporates risks and uncertainties, usually present in the environment of strategic decisions, in the analysis; (3) it allows a comprehensive performance evaluation; (4) it allows the relative comparison of the effects of alternative strategic actions on individual or combined performance measures; (5) it allows the evaluation of the effects on a corporate goal of several options acting simultaneously; (6) it provides sensitivity analysis capabilities to identify the most important variables for company performance; (7) it can help strategic management to obtain a better understanding of factors that affect performance, to identify opportunities in the environment, and then take effective actions; and (8) it provide important measures for the strategic control of management, in terms of expected values in the long process of implementation.

This research experience also confirmed the cognitive value of the methodology as a problem solving process. The methodology provides a systematic and structured process for a strategic team discussion on relevant planning issues in a construction firm. Members of COMIN, in their initial evaluation of the modeling process, have found an educational value in the process itself.

The model could be easily used to study the effect of new strategic options on firm performance. The analysis of inputs and processed information could provide evidence to support general conclusions about the effect of the individual or combined strategic decisions on firm performance. The ultimate result would be a better understanding of how corporate strategies affect firm performance.

7 Acknowledgments

The authors gratefully acknowledge COMIN for providing funding to support this research effort. They acknowledge the graduate researcher Carlos Ruiz for his enthusiastic collaboration in editing the paper draft. Also, they extend their gratitude to the Fundación Juan Pablo II and the Corporación de Investigación de la Construcción for supporting the work of the authors.

8 References

1. Alarcón L., and Ashley, D. (1992) Project performance modeling: a methodology for evaluating project execution strategies, *Source Document 80, A Report to the Construction Industry Institute*, The University of Texas at Austin, EE.UU.
2. Alarcón, L. and O'Ryan, R. (1994) Environmental policy impacts. *Project Management Institute 25th Seminar Symposium*, Vancouver, Canada, Vol. 1. pp. 995-1000.
3. Alarcón. L., Ashley, D. and Venegas, P., (1995) Evaluation of strategic decisions in construction projects (in spanish), *Boletín de Información Tecnológica*, Vol. 2, No. 1, pp. 32-36.
4. Ashley, D. and Teicholz, P. (1994) Prediction of integration impacts on engineering-procurement-construction (EPC) processes and industrial facility quality, *A Proposal to the National Science Foundation.* Department of Civil Engineering, University of California at Berkeley, EEUU.
5. Gordon, T. and Hayward, H., (1968) Initial experiments with the Cross-Impact Method of forescasting. *Futures*. Vol.1, No. 2, pp. 100-116.
6. Male, S., and Stocks, R., (1991) *Competitive Advantage in Construction.* Butterworth-Heinemann Ltd, UK.
7. Seydel, J., and Olson, D. (1990) Bids considering multiple criteria. *Journal of Construction Engineering and Management ASCE.* Vol. 116, No.4, pp. 609-623.
8. Singh, G. and Das. B., (1995) To enginner is to venture: be a FRONTIER-VENTURER. *Second Congress of Computing in Civil Engineering ASCE.* Vol. 2, pp. 1220-1227. Atlanta, 1995.

SURVIVAL STRATEGIES FOR SMALL PROFESSIONAL FIRMS IN A VOLATILE MARKET

John Boon
UNITEC Institute of Technology, Auckland, New Zealand

Abstract
This paper reports a pilot study to investigate how small professional firms survive in a volatile market. It concludes that survival is dependent on management of the key factors of, market orientation, flexibility in size and productive capacity, efficiency and price competitiveness and financial resources. It then considers how the firm can best structure itself to manage these key factors.
Keywords: professional firm, survival, marketing, structure, networking, temporary employees.

Sommaire

Cette communication rende compte d'une etude pilote menée afin d'examiner les moyennes de survie des petites entreprises professionelle sur un marché instable. Elle demontre que la survie depend des facteurs-clefs suivant : orientation du marché, possibilité de varier l'importance de l'entreprise et ca capacité de production, eficacité, prix concurrentiels et resources financiel. Cette communication etudie enfin les meilleurs moyennes, pour l'entreprise, de ce structurer afin de gérer ces facteurs-clefs.

1 Introduction

During the period 1987 to 1995 the construction market in Auckland New Zealand has been through a particularly volatile boom - bust - boom cycle. Some professional firms have survived this period whilst others have not. This paper is a summary of a preliminary study carried out at UNITEC. The objective of the study is to determine how small professional firms survive in a volatile market. Because of the limited nature of the study (four firms were studied) it is acknowledged that the conclusions must be regarded as tentative, however it is felt that a framework has been successfully developed that will enable further examination of the problem.

1.2 Research Methodology

The study was carried out using loosely structured interviews. The initial intention was to examine how firms used temporary employees to adjust their size to match a varying workload. However during the course of the study it became clear that survival was a more complicated process than this and the ideas set out in sections 4 -7 were developed.

1.2 Definitions

Small professional firm - the study is focussed on firms that provide professional services to the property development industry, such as project managers, architects, engineers and

The Organization and Management of Construction: Shaping theory and practice (Volume One). Edited by D.A. Langford and A. Retik. Published in 1996 by E & FN Spon. ISBN 0 419 22230 8.

quantity surveyors. Because of the volatility of the market defining "small" precisely is difficult as firms varied in size considerably during the period of study. The study is focussed on firms generally comprising between 5 and 20 people.
Survival - the measure of survival adopted is that the firms have remained in business throughout the period of the study.

2 The Market

All firms studied are based in Auckland, New Zealand. New Zealand has a population of 3.4m Auckland is the principal commercial city and has a population of 1.2m. The market within which the firms studied operate varies considerably. One firm's business is confined to the Auckland area, two operate nationally whilst the fourth has derived work from within New Zealand and from Australia and Asia during the period of study. In order to illustrate the extent of volatility in the market national figures are presented here.

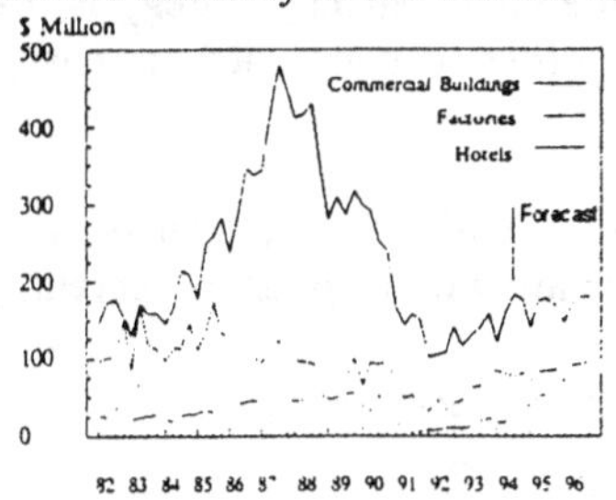

fig 1 Real Non-residential Building Activity 1982 - 96
(source Rider Hunt Forecast)

3 Case Studies

3.1 Firm A

This is a firm of Architects that was established in New Zealand during the mid 80s as a subsidiary of an overseas firm. It specialises in a certain area of commercial development and most of its work has come from that area. Initially it enjoyed considerable success and built its staff up to about twenty seven people but in 1990 it experienced a major fall off in its work load. Since then it has varied in size considerably as is shown in the table 1:

Year	No of employees			
	Total	Technical		Admin
		Permanent	Temporary	
1990	27	20	4	3
1992	4 - 11	4	0 - 7	1
1994	26	14	10	2
1995	15	11	2	2

Table 1 Number and types of employees in firm A during study period

The fluctuations in staff numbers arise from a varying workload. Most of the work undertaken is reasonably large ($2 - $20m construction costs). There is therefore a need for a relatively large team of people while the project is being designed and documented.

The marketing effort is focussed on building long term relationships with clients, particularly those property owners and developers who can use its specialist knowledge. Major projects can take several years of nurturing before a final commission is awarded. Despite its specialist knowledge competitive fee bidding is an important part of the firm's marketing mix.

CAD and general computing is used but has not been developed to the point where a competitive advantage can be gained from it.

The firm has found it has to manage its cash reserves carefully. To a certain extent it uses the profits it makes from individual commissions to carry it over quiet periods. However in recent years fee competition has been such that commissions are at best only marginally profitable. In order for the firm to avoid losses it has found it needs to downsize rapidly when a work shortage is encountered. A side effect of this strategy is that it has moved offices five time during the study period in order to cope with the changing size of the firm.

Its current policy is to operate as a core group with a network of temporary employees. Some further flexibility is provided by interaction with its overseas parent. Within the core group are retained:

- design skills particularly in its area of specialisation
- project management
- base technical knowledge
- marketing activities

The temporary employees generally provide the following services and skills

- technicians engaged in the production of documentation
- CAD operators (some provide their own CAD stations)
- graduate architects who both carry out detailed design and produce documentation

Registered architects have also been employed as temporary employees on occasions. Temporary employees can be with the firm for anything from a few weeks to over one year. Most often they are there for several months. The firm tries to maintain a regular group of temporary employees but finds this difficult as they are often not available when required as they are working with other firms.

3.1 Firm B

This firm is a structural engineering consultancy, formed by its two owners in 1989, although they had both had individual businesses for some time before this. It now comprises the two owners, three full time employees (graduate engineers and technicians) one part time draughtsperson and one part time administrator. Now it does a wide range of commercial and residential work. Its work load tends to be a lot of small jobs interspersed with some large commissions.

It seeks to build long term relationships with its clients and regards regular clients as "friends". Much of the work is as sub consultants to firms of architects. However it has also carried out significant commissions as lead consultant. One recent commission was for architectural and engineering services on a major project in Asia.

It maintains the flexibility to provide this range of services by having a small stable team and a network of sub-consultants. These sub-consultants include architects, soils engineers, surveyors, fire engineers and another structural engineer. All these sub consultants are small firms of 1-3 people; some have offices others work from home. The firm tries to maintain a stable network of sub-consultants that are reliable and know how it works. However there are changes to the network if regular sub-consultants are not available when required.

In order to carry out the Asian project the firm used twelve sub-consultants. They never met as a whole team and each sub-consultant worked from its own base. Coordination required considerable management effort. Communication was partly by meetings but mainly by telephone, fax and courier. The firms sees these last two communication channels as essential for this type of networked working.

The firm only rarely sub contracts its core structural engineering work. It has put a lot of effort into using information technology to improve its efficiency and now has a battery of customised software to assist it carry out its tasks. It has no drawing boards or secretary: "*everything is computerised*". As a result it can achieve high levels of productivity - *"a bridge that would have taken us six months to design now takes only one month"*. It therefore copes with peak loads in its core work by working extended hours (60 -70 hours per week).

Work also comes to the firm through the network when other members of the network obtain commissions they are unable to handle on their own.

Cashflow is an important factor of survival for the firm; it is monitored closely and carefully managed. Sub-consultants are paid as the firm receives payment of accounts from its clients. This can have significant cashflow benefits compared with paying employees on a fortnightly or monthly basis. They have been hurt by clients becoming insolvent in the past. They now use a debt collection agency to collect outstanding debts.

3.3 Firm C

This firm of Architects was started in 1989 when the owner decided to leave the practice he was employed in rather than become a partner. During the first year the firm comprised the owner and one junior draughtperson; it operated from the owner's home. It now comprises two architects, one graduate architect, two draughtspeople and a part time administrator. It operates from rented offices.

It does a variety of work including new houses alterations, renovations to existing houses and various small commercial and community projects. To date they have not experienced a quiet patch of substance. They put this down to the width of the market they operate in and the care and attention they have put into relationship building. Much of their work is now repeat business or comes from referrals.

All architectural work is carried out in-house, temporary employees are not used. Peaks in work load are coped with by working overtime. In addition jobs are staged so that they have break points which enables other jobs to be processed in parallel. They also place emphasis on managing their clients' expectations and making explicit what time scales they can work to at the outset. This means that commissions are sometimes not accepted. They find that the vagaries of obtaining town planning approval are such that jobs can rarely be fast tracked; clients therefore often have reasonably relaxed time frames.

They are now using CAD for some operations and find that it has improved their productivity and flexibility to a certain extent. Computing is also used for administration and accounting purposes.

Management of their cash flow is an important aspect of survival. The business is financed by an overdraft which is limited. Collection of debts on time is therefore important to them. They have recently experienced some difficulties with clients not paying which is a noticeable set back. During the early years of the practice the owner did some part time lecturing work which provided some stability to the cash flow.

3.4 Firm D

This firm of Architects has been in business since the mid sixties. It is now a limited liability company with two owner/ directors and a staff shareholding. It carries out a wide variety of commissions from one off residences to significant commercial and community buildings. It was at its largest in 1987 with a total staff of fourteen including three directors and a high proportion of full professionals. It downsized to a total of five the following year. It now comprises two directors, four registered architects, a computer technician and administrator.

All architectural work is done in-house; as a general policy temporary employees are not used. The firm places a high value on the quality of the work it does and its relationship with its clients. These are essential components of its marketing effort and of the satisfaction the owners get from being in business as architects. It does not feel it can maintain these standards if it uses temporary employees or contracts work out. It has on rare occasions used former employees on a temporary basis to overcome peak workloads. It has also in the past carried out a major commission in association with another firm. In recent years it has found that it is able to expand when required by re-employing former employees.

The firm tries to *"maintain sufficient base capability"* to meet clients' demands. It acknowledges this is an expensive policy and at times it leads to taking work it otherwise might not, in order to keep people employed. However it feels that this enables it to maintain its standards. An increase in fee competition in recent years has made this policy harder to maintain. During periods of high workload overtime is worked. The firm does not pay overtime as a norm but where a significant number of extended hours need to be worked a paid overtime regime is introduced.
CAD and online information systems are maintained, however this is not thought to provide noticeably greater flexibility or a competitive advantage.

Cashflow and profitability are acknowledged as an important component of the firm's survival. It did not move quickly enough to downsize in 1987 and as a result suffered a significant loss. As a result the firm has carried an overdraft ever since. Nevertheless it appears to have a stronger financial base than the other firms interviewed and owns its offices.

4 ANALYSIS

4.1 Key Factors in Survival

Analysis of the interviews with the four firms suggests that there is no one right way to survive in a volatile market. Rather there is an interaction between and a balancing of a number of factors that lead to survival. The experience of the firms suggests that the following are the key factors in survival:

- market orientation
- flexibility in size and productive capacity
- efficiency and price competitiveness
- management of financial resources.

These are each discussed in more detail below. It is argued that these factors are interdependent, movement in one factor affecting the required action within another. For instance if the firm is successful with its marketing in achieving a steady workload, then there is less need for flexibility. Similarly if the firm is sufficiently efficient in carrying out the work and successful with gaining a good volume of work for a time it may generate sufficient financial reserves that it is able to carry its staff over a quiet period without the need for down sizing.

Whilst it is not the focus of discussion in this paper it is important to recognise that the firm's knowledge base forms the core of its ability to provide a marketable service.

fig 2 Key Factors in Survival

4.2 Market Orientation

The firms interviewed all placed emphasis on their relationships with their clients. Promotion (the bringing of the firm's name before the potential client) and securing the commission were naturally seen as being important. However at least as great an emphasis was placed on building a good relationship with the client during the execution of the project. Helping the clients explore their real requirements and then ensuring that those requirements are met were seen as being crucial to the project's success. Delivery of the service within the client's required time frame was also seen as an important part of this market orientation, particularly on commercial projects. This concern for building good relationships led to repeat business and referrals.

The management of customer expectations was also seen as being important, particularly on smaller architectural commissions. If the performance capabilities of the firm (with regard to time frames) is made clear to the client and dates negotiated it is possible to adjust the timing of execution of commissions to smooth out some of the peaks and troughs.

4.3 Flexibility in size and productive capacity

In a volatile market it is by definition difficult to secure a steady flow of work. Of the four firms interviewed firm A had the greatest volatility in its workload; this appeared to arise from the type of work it did. Its commissions comprised fewer, larger parcels of work carried out for commercial clients with high performance expectations. It was therefore less able to juggle its commissions to even out the workflow. Nevertheless all firms to some extent faced the problem of a varying workload. In these circumstances the managers of the firm face a dichotomy. On the one hand they need the productive capacity to be able to provide a good level of service to their clients during times of peak demand. On the other hand carrying staff through demand troughs leads to a rapid erosion of cash reserves and profit. A failure to properly manage either of these problems can threaten the survival of the firm. How this problem is handled seems to require a two stage decision process. One first must decide what level of work the firm gears itself to do, then develop a strategy for dealing with the variations from the geared level. Each of the firms interviewed has a different approach to this problem.

Firm A gears itself for somewhere near the low point in its workload. This has itself meant an almost continual adjustment in the size of the firm. The productive capacity is then further adjusted by using temporary employees.

Firm B has developed the use of computing to enable it to achieve high levels of productive output, from a small base. This can be further expanded by the use of overtime working. However to carry out large full commissions it has the ability to expand its manpower levels by a factor of five using a network of sub-consultants. Its networking arrangements are such that payments to sub-consultants cease as soon as the project is completed. Its size is therefore more or less self adjusting.

Firm C has not to date experienced the same volatility in its work load. It has therefore found that growth in the size of the firm together with the use of overtime has provided it with sufficient flexibility. It does however have a clear contingency plan for downsizing rapidly if its workload dries up and its limited cash reserves become threatened.

Firm D has adopted a strategy of gearing for an output level that is closer to peak demand. This obviously requires less flexibility but greater financial resources. Its need to substantially downsize in 1988 illustrates the difficulty of maintaining this strategy. It is currently pursuing a policy of cautious growth, expanding only when it is confident that the increase in size can be maintained.

4.4 Efficiency and Price Competitiveness

All four firms interviewed operate in a market where fee competition exists. It is generally not the only factor the client takes into account when selecting a consultant, however it is a factor in all cases. The ability to gain work by offering competitive fees bids is therefore an important part of a firms marketing activity. Of equal importance to survival is an ability to carry out the work profitably within the fee income. Efficiency and price competitiveness is therefore an important factor of survival.

All four firms place some emphasis on maintaining a low overhead cost structure. All avoid high accommodation costs and generally minimise expenditure on reception, secretarial and administrative costs. Each firm uses computing and information technology to some extent to achieve efficiency. However only firm B has pursued this matter with vigour to achieve a competitive advantage.

4.5 Financial Resources

Of the firms interviewed only D has significant financial resources. It owns its own offices and the directors own other investment properties. The other firms interviewed all have limited financial resources. Earnings from one job are needed to carry the firm through until the earnings from the next job are received. If the next earnings are not received in time then the firm will experience a shortfall in cash and be unable to pay its employees and meet its bills. Further if the financial resources are eroded the firm may find itself unable to gear up again to meet the next peak in demand. The interviewees were all very conscious of this problem and monitor their cashflow continually. Proactiveness with debt collection and a willingness to downsize quickly in order to maintain financial reserves are seen as key factors of survival by all except firm D.

6 The Structure Of The Firm

This section of the paper considers how the firm can best structure itself to cope with the factors in survival identified above, when it is operating in a volatile market.

Theoretically the managers can structure the firm anywhere on a continuum between a closed firm and an open firm. A closed firm is one that is totally self sufficient and able to meet all the demands of its customers, whilst an open firm comprises a small core surrounded by a network of sub-consultants and temporary workers. The concept of an open firm can be extended to a multi nuclei open firm where the firm has a series of branches each with its own network.

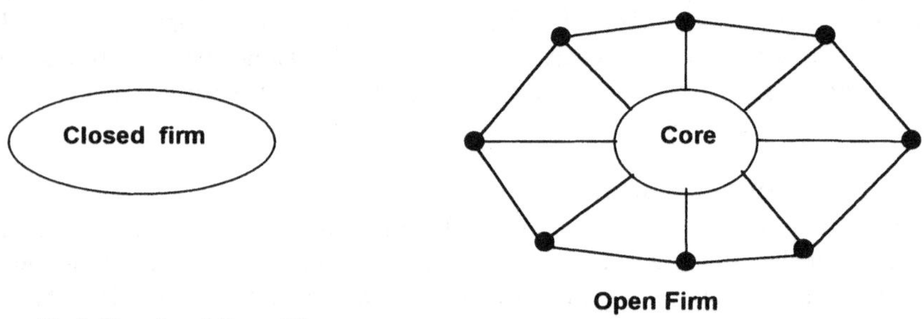

Fig 3 Closed and Open Firms

Firms C and D align with the model of a closed firm whilst firms A and B may be characterised as relatively open firms.

In a truly volatile market the concept of a closed firm seems unsustainable unless the firm can isolate itself from the volatility of the market. Firm C appears to have achieved this by pursuing lots of small commissions and managing the amount of work it takes on. Firm D is also attempting to mange its workload; however its experience of having to substantially downsize in 1988 illustrates the difficulty of this strategy. If it were now to be offered a one off major commission it has three options:

i. remain at its present size and turn the commission down,
ii. enlarge the firm by taking on more "permanent" employees and hope it will secure further commissions to justify its enlarged size,
iii. become an open firm.

If the market is truly volatile option two may be a rather myopic approach, ignoring the inevitability of a later downsizing. It also raises the philosophical question as to whether the firm is being entirely honest to the people to whom it offers "permanent" employment.

If the managers of the firm wish to take advantage of all or most of the business opportunities of a volatile market, structuring the firm as an open firm appears to be the most realistic strategy. This notion has considerable support in current management thinking eg Handy[1] and Peters[2]. Handy describes open firms as shamrock organisations. The first leaf of the shamrock is the core of permanent employees of the firm, the second the firm's network of subcontractors and the third the temporary employees that are brought into the firm when required. If Handy's analogy is applied to firms A and B they both appear as two-leafed organisations, A relying on temporary workers and B on a network of subcontractors (or sub-consultants).

The advantage of temporary employees is that they become for a time part of the firm. They work in the firm's offices and therefore can be easily supervised and required to comply with the firm's quality assurance procedures. As individuals they are less likely to become direct competitors to the firm and if they prove to have skills and knowledge that is particularly useful to the firm they may be taken on as permanent employees. They can also be easily dismissed if they are unsatisfactory. Against this they have the disadvantage that they may lack commitment to the firm, they make acquire specialist knowledge whilst working for the firm and take it to a competitor, and if they have no further work to go to they may spin out the work when the project is nearing completion. In addition they need to be accommodated and equipped, which may mean the firm has to maintain larger offices which are sometimes substantially unoccupied. The cashflow problems of increasing in size when a new project is started fall on the firm.

A network of subcontractors seems advantageous if a wide range of knowledge or competencies is required. Each firm within the network can concentrate on developing its own expertise and the required balance of expertise can be brought to each project. In addition each firm provides its own accommodation and equipment and manages its own financial resources. The overhead costs during slack times are therefore spread around the network, as are the cashflow problems of gearing up. Against this, if the project requires large volumes of a particular expertise (eg architectural design and documentation) the only potential subcontractors may be the firms direct competitors. This may lead to a loss of specialist knowledge or client contact to that competitor. In addition supervision of the project is more difficult with the subcontractor operating from its own premises; the subcontractor may give priority to the work of other clients or have different standards to the core firm.

The decision to network with subcontractors or use temporary employees would appear to require a contingent approach, dependent on the nature of the firm's business and the market opportunities. In addition ongoing management action to maximise the benefits and minimise the disadvantages of the chosen option is necessary. How this might best be done requires further research.

The experience of A and B raises the question of how big should the core be? A in particular has regularly adjusted the size of its core. In theory it may be possible to reduce the firm to a marketing unit, all other work being contracted once a commission is received. Neither A nor B have gone this far, both retaining key competencies within the core. This would appear to be necessary if the firm is to establish good personal relations with the client during the execution of the commission. Interviewees also felt there are problems in marketing if the firm cannot demonstrate it has sufficient in-house expertise and capability to execute the project.

7 Conclusion

Within the limits that are possible with a pilot study the following conclusions are drawn regarding the survival of small professional firms in a volatile market. Firstly in order to survive it is necessary to manage the factors in:

- market orientation
- flexibility in size and productive capacity
- efficiency and price competitiveness
- financial resources.

These factors are interdependent, the effectiveness of managing one factor impacts on the management options available for the others.

Secondly if the firm wishes to take advantage of the opportunities offered by a volatile market, the most realistic strategy is to structure the firm as an "open firm". Within this strategy there are options of using either a network of subcontractors or temporary employees, both having advantages and disadvantages. A contingent approach to the solution of the balance between these options appears necessary. Further research is required on the ongoing management of both options. In addition further research is required on the sizing of the core activities of the firm.

Acknowledgments

The firms who participated in the pilot study are thanked for the time and knowledge so willingly given.
The work of Miles and Snow[3] influenced the advantages and disadvantages of using subcontractors and temporary employees set out in section 6.

[1] Handy, Charles, (1989) *The Age of Unreason*, Harvard Business School Press, Boston.
[2] Peters, Tom (1992) *Liberation Management*, MacMillan, London.
[3] Miles, R. & Snow, C. (1995) *Causes of Failure in Network Organisations*, California Management Review, Vol 34, no 4, University of California, Berkley.

CONSTRUCTION FIRMS' STRATEGIES AND THEIR FINANCIAL PERFORMANCE

M.A.H. Mohammed
Faculty of Surveying and Real Estate,
Malaysia University of Technolgy, Johor Bahru, Malysia
Z.M. Yusof
Faculty of the Built Environment, Malaysia University of Technology, Johor Bahru, Malaysia
V.B. Torrance
Bartlett School, University College London, London, UK

Abstract
This study has two main objectives. The first objective is to identify the relationship between the construction firms' competitive strategies and their financial performance. The second objective is to observe the behaviour of the construction firms' strategies in three different economic periods, i.e., boom (1986-89), recession (1990-93) and recovery (1994-onward). The findings of the first phase indicated that, there were significant relationships between firms' competitive strategies and their financial performance. In the second phase, it was found that the construction firms' strategies were changing drastically during the three different economic periods.
Keywords: Firms' strategies, financial performance, economic periods

Sommaire
La présente étude a deux objectifs majeurs: identifier, d'une part, le rapport existant entre les stratégies de compétition des entreprises du bâtiment et travaux publics et leurs performances financières, observer d'autre part le comportement des stratégies de ces mêmes entreprises au cours de trois périodes économiques distinctes, à savoir le boom des années 1986-1989, la récession des années 1990-1993, et la convalescence d'une période commençant en 1994. Les résultats de la première phase de recherche ont indiqué qu'il existait un rapport d'importance entre les stratégies de compétition des entreprises et leurs performances financières. Pour ce qui est de la seconde phase, il a été montré que les stratégies des entrepreneurs avaient changé à l'extrême au cours des trois périodes économiques respectivement.

1 Introduction

The UK construction firms had experienced unprecedented growth during the strong and boom economic period from 1981 to 1989. It was during this period that most of the major construction firms had expanded their scope of activities and geographical coverage either nationally or internationally. The volume of turnover and the level of profit of these firms rose steadily each year to the peak performance which probably had never been achieved before this period. However, the sweet success ended abruptly by the rapid economic down-turn which began to afflict the construction industry by the end of 1989. Private housing became the first sector which was severely hit by the effect of the economic recession. The recession had continued,

The Organization and Management of Construction: Shaping theory and practice (Volume One).
Edited by D.A. Langford and A. Retik. Published in 1996 by E & FN Spon. ISBN 0 419 22230 8.

deepened and prolonged to 1993 and imparted devastating effects upon the construction industry. The word recession was then changed to "depression" to show the magnitude of the severity of this harsh economic climate. During the depression, many firms incurred heavy losses which eventually sent some of these firms into liquidation.

It was clear that the economic environment had played the most important role in determining the financial success or failure of a construction firm. However, there was room for a construction firm to maneuver in order to be more impervious to the effects of the turbulent economic climate. The ability of a firm to decide on the competitive strategic issues such as the scope of business, the geographical coverage and the resource deployment would increase its competitiveness [1]. However, these strategic decisions were related to the broader aspect of the firms' strategies which include developmental and functional strategy.

2 Objectives

This study aimed to achieve the following objectives:

(a) To investigate the relationships between the construction firms' competitive strategies and their financial performance.

(b) To investigate the appropriate construction firms' strategies in the three different economic periods: boom (1986-1989); recession (1990-1993); and future (1994-onward).

3 Methodology

This study was phased into two phases, i.e. Phase 1 and Phase 2. Both phases have used a questionnaire survey as a mean of collecting the data. Prior to the questionnaires' development, a sample (the UK top 120 construction public limited companies) was contacted to provide their annual reports and accounts for the period of 1986 to 1992. In the first phase, data obtained were analysed by using correlation analysis and analysis of variance (Anova) which were performed by the statistical package SPSS for Microsoft Windows release 5.0 (SPSS® for Windows™ Release 5.0). In the second phase, data collected were analysed by using frequency analysis and comparison of means which were also performed by the similar statistical package.

The sample for this study was drawn from a list of the top 500 construction firms in the UK which was published by Jordan [23]. The turnover of these firms ranged from £10 millions to more than £4000 million per annum.

4 Scope and limitation

As discussed above, this study covered the top 120 construction public limited companies in the UK. It means that the outcomes of the study were biased towards the medium and large construction firms in the UK. Data for the first phase of this study were collected within the period of 1986 to 1992 during which the UK economic condition was changing from the boom to the slump. In the second phase of the study, data were collected to cover three periods of different economic conditions, i.e. 1986-1989 (boom), 1990-1993 (recession) and 1994 - onward (future or slow recovery).

5 Results

The results of the study will be described in two phases following the methodology of the study.

Phase 1: Relationships between firms' competitive strategies and their financial performance

The first phase of this study deals with the relationships between firms' competitive strategies and their financial performances. On the basis of the theoretical development four competitive strategies for a construction firm were identified [1] [2] [3] [4] [5] [6] [7] [8] [9] [10] [11] :

(a) type of activity;
(b) extent of diversification;
(c) extent of internationalisation; and
(d) level of gearing.

From the annual reports of the construction firms and from the previous research, five financial performance measurements were identified:

(a) ROCE;
(b) ROSF;
(c) current ratio;
(d) quick ratio; and
(e) turnover.

The relationships between these two sets of variable were analysed by using correlation analysis and analysis of variance which were performed by the statistical package SPSS for Windows. The summary of the findings is displayed in table 1 and can be concluded as follows:

i. ***Housing and property development*** activities generated a high level of profit during the period of economic boom but incurred heavy losses when the economy turned-down. However, they produced a low level of positive cash flow during both periods. Therefore, a company should undertake massive projects of housing or property development when the economic condition is

Table 1 - A summary of relationship between competitive strategy and financial performance

Competitive Strategy	Financial Performance	
	Boom (1986-89)	**Recession (1990-92)**
Housing development activity	High level of profit but low level of cash flow	Loss
Contracting activity	Low level of profit but high level of cash flow	Low level of profit but high level of cash flow
Property development activity	High level of profit but low level of cash flow	Loss
High degree of Diversification	Large turnover	Large turnover
High degree of Internationalisation	Large turnover	Large turnover
High level of gearing	High level of profit	Loss

strong and stable. When the economic condition turns soft, a company should speedily reduce or withdraw (if possible) from these activities to avoid further losses.

ii. ***Contracting*** activity generated a low level of profitability but gave the highest level of positive cash flow for both periods. During the recession, a company should concentrate on contracting activity to maintain its profitability even though at a low level) and to generate a high positive cash flow which was extremely necessary for the company's survival.

iii. ***Diversification and Internationalisation*** strategies were found to be the main determinants of the size of firms' turnover. The finding indicates that a high level of diversification and a high level of internationalisation were related to the greater turnover. It means that construction furms have to diversify or internationalise their activities in order to increase their turnover.

iv. Firms with a ***high level of gearing*** generated a high level of profit during the economic boom. However, these firms suffered heavy losses and eventually some of them went into liquidation after a long and deep recession. Therefore, construction firms must not burden themselves by heavy borrowing at any time which exceeds 50 % of their equity. The urge to expand rapidly during the economic boom must be curbed to prevent a firm from incurring an excessively heavy debt burden.

Phase 2: Appropriate construction firms' strategies within three different economic periods: economic boom (1986-89); recession (1990-93); and future (1994-onward)

In the second phase of the study, the respondents were requested to give their own past experiences, perceptions, insights and predictions to the various aspects of construction firms' strategies within the three different economic periods. The respondents were asked to give their perceptions for each variable according to the following scale: **1** = not at all important; **2** = not important; **3** = quite important; **4** = very important; and **5** = extremely important.

From the literature search and the annual reports of the construction companies the following aspects of strategies were identified [1] [3] [4] [6] [8] [9] [12] [13] [14] [15] [16] [17] [18] [19] [20] [21] [22]:

- Developmental strategies which included directional, method and generic;
- Diversification;
- Internationalisation;
- Functional strategy which included the use advanced technology, R&D and effective marketing;
- Resource strategy which included management, skilled workers, cash from borrowing, cash from right issues, plant and land banks;
- Financial performance measurements; and

- Profit determinants.

In some ways, the findings of the second phase of this investigation can be used to cross-check the findings of the first phase. A summary of the results is displayed in table 2 and can be concluded as follows:

i. **Directional strategy**
The respondents regarded expansion as an extremely important strategy during the boom period. However, retrenchment was stated as the most important strategy during the recession (regarded as very important). For the future, expansion was again regarded as the most important strategy but was only perceived as very important and not extremely important. This indicated that the industry tended to be more cautious in its expansion programme after its experience in dealing with the severe recession.

ii. **Method strategy**
A clear cut trend can be seen in method strategy where internal expansion was considered as the most important strategy throughout the three periods. It is also interesting to observe that joint venturing was at the second position throughout the periods of study. It means that acquisition had been less important than joint-venturing as a means of expansion. Internal expansion gives more stability whilst joint-venture gives more flexibility for a firm to expand.

iii. **Generic strategy**
Under generic strategy, focus on core business (either contracting, housing, property, or some other activities related to construction) was regarded as the most important strategy during the stable period and for the future. However, during the recession, reducing fixed costs and overheads was considered as the most important. Therefore, it can be said that focussing on core business and reducing fixed costs and overheads will be the basis of a firm's competitive development for the future.

iv. **Diversification strategy**
The findings indicated that the industry considered housing as the most important activity within the three periods. The reason for this trend is quite clear, that is the industry believed that housing would generate the highest level of profitability. However, the findings of the first phase have indicated that housing is not profitable during the recession. Contracting was stated as the second important activity during the recession and for the future.

v. **Internationalisation strategy**
Europe was considered as the main market outside the UK by the respondents. The findings indicate that Internationalisation was not considered as very important by the UK construction industry. Only a handful of the largest companies have significant operations overseas.

Table 2 - A summary of the UK construction firms' strategies during three different economic periods: stable, recession and future (slow recovery)

Strategic variables	Stable period (86-89)	Recession (90-93)	Future (94-onward)
Directional	Expansion	Retrenchment	Expansion
Method	Internal expansion	Internal expansion	Internal expansion
Generic	Focus on core business	Reduce fixed cost & overhead	Focus on core business
Diversification	Housing	Housing	Housing
Internationalisation	Europe	Europe	Europe
Functional	Marketing	Marketing	Marketing
Resources	Cash from borrowing	Management	Skill workers
Financial performance measurement	Pre-tax profit	Cash flow	Cash flow
Profit determinant	Market condition	Market condition	Market condition
Strategic management practice	Measure achievement	Anticipate changes in market condition	Anticipate changes in market condition

vi. Functional strategy

At the functional level, the findings indicated that effective marketing was regarded as the most important strategy in all three economic periods. R&D was not regarded as a very important strategy by the respondents. Without a strong effort on R&D, the UK construction firms might lose their competitive edge over their rivals in the international market.

vii. Resources

During the period of economic boom, cash from borrowing was considered as the most important resource followed very closely by land bank. This indicated that the industry was very active in speculative housing development which was buoyant at this period of time. However, during the recession, management was considered as the most important resource. The strategic change had forced the industry to react appropriately and this action needed good management skills. For the future, skilled workers were perceived as the most important resource. This prediction probably was caused by the believe that there will be a shortage of skilled workers in the future.

viii.Financial performance measurements

It was not surprising if the industry rated pre-tax profit as the most important financial performance measurement during the boom period. However, the deep and long recession had taught them that cash flow was the most vital measurement not only during the recession but also for the future.

ix. Profit determinants

Throughout the period of study, market condition was observed as the most important determinant. There were no other determinants either at corporate level or at project level which were considered as equal to or more important than market condition. Therefore, a construction firm must have a sensitive devise which could enable them to sense any changes in the economic conditions. Then, it must also have a sound decision making process in order to react promptly before it becomes too late.

x. Strategic management practice

Additional information on the level of construction firms' awareness as regards their strategic management practice was also obtained from the questionnaire survey. The responds indicated that there was a steady improvement in their understanding of the importance of strategic management. For the future, the respondents believed that anticipating changes in the market condition, strategic planning and measuring their achievements as the three most important strategic management practices. However, they did not consider that the employment of external strategic management consultant as an important practice.

6 Conclusion

This study showed that many major firms suffered heavy losses because of their inability to anticipate and to assess the magnitude and length of the recession. They took a decision to reduce or to withdraw from certain types of business activities such as housing development, far too late, when they were already in the red. Therefore, a warning mechanism by which they can anticipate and assess the changes in the economic environment at the earlier stage is needed. The major firms also need to devise a procedure to enable them to make a quick response to any severe economic climate.

The results of this study indicated that the UK construction firms regarded internal expansion and focus on core business as their strategy for the future. It was interesting to see that cash flow was regarded as the most important financial performance measurement during the recession and for the future. It was clear that market condition was agreed by the respondents as the most important factor in determining their financial performance.

The above results show two areas of concern for the UK construction industry, these are R&D and internationalisation. These two areas have to be addressed seriously by the industry and also by the UK government. At the individual company level, all construction firms should adopt a policy of continuous and never ending improvement or total quality management. This would not necessarily need a big expenditure.

REFERENCES

1. Ansoff, H.I. (1987) Corporate Strategy. Revised edition. London: Penguins Books.
2. Asenso, H.O. and Fellows, R.F. (1987) Profitability and Size of UK Contractors. Building Technology and Management, pp. 19-20, February/March, 1987.
3. Fiegenbaum, A. (1987) Dynamic Aspects of Strategic Groups and Competitive Strategy: Concepts and Empirical Examination in The Insurance Industry. Unpublished Ph.D Dissertation, University of Illinois at Urbana-Champaign.
4. Hasegawa, F. and Shimizu Groups (1988) Built by Japan: Competitive Strategies of the Japanese Construction Industry. New York: John Wiley and Sons.
5. Grinyer, P.H. et. al. (1978) Strategy, Structure, the environment and Financial Performance in 48 UK Companies. London: The City University Business School.
6. Hillebrandt, P.M. and Cannon, J. (1990) The Modern Construction Firm. London: Macmillan Press Ltd.

7. Kangari, R., Farid, F. and Elgharib, H.M. (1992) Financial Performance Analysis for Construction Industry. Journal of Construction Engineering and Management, Vol. 118, No.2, June, 1992.
8. Male, S. and Stocks, R. (1991) Competitive Advantage in Construction. Oxford: Butterworth-Heinemann Ltd.
9. Porter, M.E. (1979) The Structure Within Industries and Companies' Performance. Review of Economic and Statistics, 61, pp. 214-219.
10. Ramsay-Dawber, P.J. (1992) The Organisational Form and Culture of Successful UK Construction Companies. The Nottingham Trent University.
11. Vernon-Harcourt, T.V. (1982) UK Companies Performance Analysis. Safron Walden: Monks Publications.
12. Ansoff, H.I. and Sullivan, P.A. (1993) Optimising Profitability in Turbulent Environments: A Formula for Strategic Success. Long Range Planning, Vol. 26, No. 5, pp. 11-23.
13. Biggam, R. (1989) Strategic Challenge of the Global Market. Conference for Chief Executives: Competitive Strategies in the Global Construction Industry. The Building Employers Confederation and Construction News, London.
14. Chapman, N.F.S. and Grandjean, C. (1991) The Construction Industry and The European Community. Oxford: BSP Professional Books.
15. Hutcheson, J.M. (1993) Entrepreneurship and Strategy of Building Firms or Why Builders Fail. CIB W-65 Conference. Trinidad, West Indies.
16. Johnson, G. and Scholes, K. (1993) Exploring Corporate Strategy. UK: Prentice Hall.
17. Kay, J. (1993) Foundation of Corporate Success: How Business Strategies Add Value. Oxford: Oxford University Press.
18. Lampl, F. (1989) Illusion or Reality: Can Construction Become a Strategic Industry? Conference for Chief Executives: Competitive Strategies in the Global Construction Industry. The Building Employers Confederation and Construction News, London.
19. Lansley, P. (1989) The Changing Face of Construction: Challenges for the 90s. Building Technology and Management, February/March, 1989.
20. McLea, A. (1991) Report Identifies UK's Failure to Stimulate R&D. New Builder, November, 1991.
21. Newcombe, R., Langford, D. and Fellows, R. (1993) Construction Management 1 - Organisation Systems. Reprinted. London: B.T. Batsford in association with the Chartered Institute of Building.
22. Ohmae, K. (1982) The Mind of the Strategist: The Art of Japanese Business. New York: McGraw-Hill, Inc.
23. "Britain Top 500 Construction Companies." Bristol: Jordan and Sons Limited. 1989.
24. "Investing in Building 2001" 1989. Centre for Strategic Studies in Construction. University of Reading.

CONTRACTORS' STRATEGIES IN HONG KONG'S COMPETITIVE CONSTRUCTION MARKET

A. Walker
University of Hong Kong, Hong Kong

Abstract

This paper examines the competitive edge being developed by a number of Hong Kong's most sophisticated contractors in order to cope with an extremely uncertain business environment. The diversity of Hong Kong's major contractors in terms of corporate structure, organisation structure, market niche, country of origin and hence imported or local strengths is extremely wide. The business environment is affected by the rapid growth in construction work as a result of the Port and Airport Developments Strategy (PADS) and the uncertainty associated with the reversion of Hong Kong to Chinese sovereignty in 1997 which affects the confidence contractors feel about the stability of continuing workloads. Compounding these factors is a tightening contractual situation as the Hong Kong Government place greater risk on contractors. The major contractors have responded to this unstable yet potentially lucrative market in a variety of ways reflecting the companies' perceived strengths. This paper identifies the forces in the contractors' business environment and analyses the contractors' coping strategies and their potential for success within this volatile market.

Keywords: Competitive advantage, strategy, business environment, Hong Kong, contractors.

Sommaire

Cet article examine l'avantage compétitif qu'un certain nombre d'entrepreneurs très distingués de Hong Kong développe actuellement pour faire face à un environnement commercial extrêmement incertain. La diversité des principaux entrepreneurs de Hong Kong en termes de structure de l'entreprise, structure d'organisation, créneau commercial, pays d'origine et donc forces importées ou locales, est extrêmement étendue. L'environnement commercial est touché par la croissance rapide dans le

The Organization and Management of Construction: Shaping theory and practice (Volume One). Edited by D.A. Langford and A. Retik. Published in 1996 by E & FN Spon. ISBN 0 419 22230 8.

domaine de la construction à cause de la stratégie des Développements du Port et de l'Aéroport ainsi que de l'incertitude associée au retour de Hong Kong à la souveraineté chinoise en 1997 qui affecte la confiance que les entrepreneurs ont pour la stabilité des contrats qui continueront. Combiner ces facteurs est une situation contractuelle de resserrement car le gouvernement de Hong Kong met les entrepreneurs en situation plus risquée. Les principaux entrepreneurs ont répondu à ce marché instable, néanmoins potentiellement lucratif, par une variété de façons qui reflètent les forces des sociétés. Cet article identifie les éléments dans l'environnement commercial des entrepreneurs et analyse les stratégies qu'ont les entrepreneurs pour faire face, ainsi que leur possibilité de succès au sein de ce marché instable.
Mots-clef : Avantage compétitif, stratégie, environnement commercial, Hong Kong, entrepreneurs.

1 Introduction

The dearth of research into strategic management in construction was identified by Male and Stocks [1] when they sought to redress the balance in their book on competitive advantage. They pointed to the major importance of managers' ability to develop skills to move away from a production orientation to a strategic orientation where the focus is external and involves the long term relationship between the company and its environment.

The purpose of this paper is to make a small contribution to continuing the process of redressing the balance by presenting the views of managing directors (or their equivalent) of Hong Kong construction companies on their companies' competitive strategies. Whilst much has been written about strategic management in the general management literature, Male and Stocks' book is one of the few to address the issue in construction but even so there is little empirical work aimed at discovering the views of practicing managers.

The orientation of the paper is the competitive strategy of the companies which is defined as the positioning of a company to make the best use of its abilities to segregate it from its competitors [2]. In Hong Kong this encompasses both the competitive advantage in terms of technical and other facilities and advantages in particular markets through experience and/or contacts.

2 Hong Kong's construction market

The years up to 1992 saw a flattening out of total expenditure on building and construction at nearly HK$57 billion per annum at constant 1990 market prices divided in the ratio of about 2:1 for private sector/public sector work as shown in Table 2.1. The type of work is categorized in Table 2.2 for 1992 which shows the distribution of housing (both public and private), private non-residential and civil engineering. Over the last few years this pattern of expenditure and distribution between categories has been maintained but the immediate future is expected to see an increase in expenditure in real terms and an increase in civil engineering's share of the market as the airport and port developments get up full steam. The provisional figures for 1993 indicate a 10% increase in civil engineering's share with commensurate reductions in the other categories.

Housing continues to be a significant part of the construction market. The public sector (the Housing Authority) builds about twice as many flats as the private sector each year but private sector housing has about twice the floor area of the public sector. The Housing Authority also provides Home Ownership (HOS) and Private Sector Participation Scheme (PSPS) flats which allow flats to be purchased. The territories contractors complete one flat every 10 minutes or over 65,000 units per year on average.

Table 2.1 Proportion of public and private expenditure on building and construction

Year	Total expenditure on building & construction	Public sector	% of total	Private sector	% of total
1986	44.244	14.546	32.9%	29.698	67.1%
1987	46.692	15.632	33.5%	31.060	66.5%
1988	48.358	15.075	31.2%	33.283	68.8%
1989	53.088	17.142	32.3%	35.946	67.7%
1990	57.314	18.142	31.7%	39.172	68.3%
1991	57.658	17.190	29.8%	40.468	70.2%
1992	57.650	18.315	31.8%	39.335	68.2%
1993[p]	62.984	25.930	41.2%	37.054	58.8%
Average			33.0%		67.0%

At constant (1990) market prices (HK$ Bn)

Note : p - provisional figures, subject to revisions later on as more data become available

Source : Estimates of Gross Domestic Product (1961 - 1994), Census and Statistics Department, Hong Kong Government.

Table 2.5 Type of construction work in 1992

Buildings		
Housing	38.9%	
Private non-residential	29.2%	
Public non-residential	5.8%	
Sub-total		73.9%
Civil Engineering		26.1%
		100%
Of which decoration, repair & maintenance		12.1%
Total output = 63.056 billion		

Source : Estimation of Gross Domestic Product (1961-1994), 1992 Survey of Building, Construction and Real Estate Sectors, Census and Statistics Department, Hong Kong Government.

The Government's Long Term Housing Strategy (1985-2001) commited the Housing Authority to producing 215,000 flats in the second five year development period from 1991 to 1996 and 135,000 in the third, up to the year 2001. As the result of the recommendations of a Government task force on property prices and housing supply established in 1994, a decision was made to increase the supply of public housing flats by at least 20,000 and private flats by at least 15,000 by 2001.

Overshadowing other public sector projects, which have to go on even so, is the new airport. Whilst parts of it will be privately operated, it is a public sector project. It requires that both the professional and managerial skills of the public sector and similar resources of the construction industry be mobilized on a massive scale. The new airport is part of the Port and Airport Development Strategy (PADS) announced in 1989.

Hong Kong's new airport is one of the world's largest infrastructure programmes. The first phase is expected to cost HK$158 billion at early 1994 prices. The initial phase, due for completion in 1998, will handle 35 million passengers and 2.3 million tonnes of cargo annually. The planned future expansion programme will see the capacity expand to include two parallel runways accommodating 87 million passengers and 8.9 million tonnes of cargo by 2040.

The scope of the new airport project is reflected in the 10 core components of airport-related infrastructure to be completed between 1992 and 1998. Seven of the ten projects - involving the formation of new land and the construction of highways and a new town - are being funded and carried out by the Government as direct capital works projects.

Important and unique features of these airport-related infrastructure projects include:

- 75 per cent of the 1248 ha airport platform will be on land reclaimed from the sea;
- the Lantau fixed crossing will include the longest span suspension bridge (2.2km) in the world carrying both road and rail traffic;
- the establishment of a new town (Tung Chung) adjacent to the airport accommodating 20,000 residents by 1997;
- the completion of Hong Kong's third harbour tunnel;
- 35km airport railway, comprising eight km of tunnels and six km of elevated structures will accommodate 39,000 travellers daily;
- 1670ha of land is being formed which will add about 50% to the amount reclaimed in Hong Kong since 1945.

In addition, port expansion is part of PADS. Hong Kong's first seven container terminals were completed by 1988. The PADS strategy plans for five further terminals.

- Container Terminal 8, now operational, comprising three berths on reclaimed land abutting Stonecutter's Island;
- Container Terminal 9 on reclaimed land at the southeast of Tsing Yi Island;
- Terminals 10, 11 and 12 on the eastern side of Tsing Chau Tai Peninsula which will create the Lantau Port Peninsula.

Additionally:

- berths for handling river trade containers at Tuen Mun;
- a multi-purpose terminal in Kwun Tong.

The public sector is likely to remain the major client of the industry but it represents a myriad of faces as each project has to satisfy the specific needs of a particular sector of society. To cope with a workload of such complexity Government is moving to a position where it relies increasingly on the construction companies to manage whole projects on a design and construct basis. The contractor is responsible for designing and managing as well as constructing the facility so is in control of all designers and other professional contributors and will be free within constraints imposed by the client to innovate in solving design and construction problems.

3 Future scenarios

A recent study examined the future demand for construction in Hong Kong up to the year 2006[4]. The study generated alternative scenarios based on assumptions regarding:

- GDP growth rate, for which two alternatives were taken within the conservative framework employed; 6%, which is higher than levels achieved over recent years, and a more pessimistic view of a lower 2% growth rate;
- Total Consolidated Account Expenditure (TCAE) as a percentage of GDP. This has been running between about 16% and 20% in recent years. Four values (10%, 15%, 18% and 20%) were selected to illustrate a range of scenarios.

Further assumptions were made regarding the medium to long term future which was likely to be determined by four major political factors:

- continued modernization with controlled economic growth in China;
- China's commitment to the 'one country-two systems' concept;
- China's interpretation of the Basic Law;
- stability in China's political succession.

In considering future scenarios for Hong Kong's economy it was assumed that these four factors will be positive in favour of Hong Kong. Without such assumptions the outcome is likely to be one of reducing confidence and economic activity and the demise of Hong Kong as an international centre distinguishable from other parts of China. The more likely outcome is of China's economy expanding with Hong Kong taking full advantage to develop its own economy at a rate which is greater than in the recent past. Nevertheless a relatively conservative view was taken in developing the scenarios.

Judgment regarding which scenarios are most generally acceptable were made against criteria represented by the parameters within which Government decisions on expenditure had been made in recent years. The scenarios show that given future growth which is conservative relative to recent growth, the demands on the construction sector of the economy will continue to increase. The projected increases in expenditure on construction in the year 2006 range from 35% to 123% more than current expenditure (at current prices).

The scenarios which show better growth in construction all assume a 40%/60% split between public and private expenditure which reflects recent levels. Previous years' ratios have been about 33%/66% but the airport development has changed the ratio. Nevertheless confidence will need to be maintained in order to attract private investment. Increased public investment, as long as it is not too great, will contribute to increasing confidence. In this respect the management of the infrastructure development plans of over HK$200 billion e.g. the airport and container terminals, will continue to be significant in terms of the ratio of public to private funding. Only if public expenditure exceeds private expenditure do the demands for construction decline below the levels given above.

4 Market access

There is, in general, no formal restriction for entry to the contracting business in Hong Kong. However, contractors acting as main contractors undertaking any building works either public or private under the Building Ordinance must be registered. The Hong Kong Government maintains two lists of approved contractors qualified to tender for different sizes and categories of work. List I are local contractors and List II are foreign contractors. Firms first enter the lists on probation and membership is confirmed after satisfactory performance. Technically entry is relatively straightforward but, as Male [3] remarks, project size and complexity pose barriers to entry and this is particularly so for foreign contractors in Hong Kong who are only allowed to tender for contracts over HK$30 million.

A good indication of the attractiveness of Hong Kong to overseas contractors is given by a list of some of the contractors involved on the airport core projects :

Hong Kong
GEC (Hong Kong)
Dragages
Sung Foo Kee (Civil)
Leighton
Hip Hing
Shui On Civil Contractors
Hsin Chong
Wah Hin
Kumagai Gumi
Nishimatsu
Gammon

China
China Fujian
CSCEC

Japan
Maeda
Aoki
Penta Ocean

Britain
Kier
GPT (Exports)
Tarmac
GEC Alsthom
John Laing International
Balfour Beatty
Amec
Trafalgar House
Costain

Germany
AEG
Ed Zublin

France
GEC Alsthom
CNIM
GTM
CFE

Mitsiu

Spain
Entrecanaies
Cubiertas
CAF

New Zealand
Downer

5 The Contractors

Against this backcloth of a market in Hong Kong which whilst having the potential to increase will probably do so at a decreasing rate due to the completion of the airport, continuing increased competition from overseas contractors which have easy access to the market and are attracted to it due to limited opportunities elsewhere and the advent of 1997, Hong Kong's major contractors have to plan their business strategy for the next few years.[5]

The strategy of six of Hong Kong's major contractors, each of which has a quite different profile, is examined against this scenario:

- Shui On (Contractors) Ltd., a mainstay of Shui On Group Ltd., a home grown Hong Kong company;
- Hsin Chong Construction Group Ltd., part of the Hsin Chong Group but separately listed on the Hong Kong Stock Exchange, the longest standing of the major Hong Kong construction companies;
- Nishimatsu Construction Co. Ltd., the Hong Kong branch of a Japanese construction company but now classified as a local contractor;
- Leighton Contractors (Asia) Ltd., the Hong Kong arm of the Leighton Group, Australia's largest construction company whose major shareholder is Hochief, a diversified European construction company;
- China State Construction Engineering Corporation, the PRC construction giant, whose listed Hong Kong subsidiary is China Overseas Land and Investment Ltd.;
- Hip Hing Construction Co. Ltd., the construction arm of the famous Hong Kong developer, New World, but which also competes on the open market.

Table 1, which shows the 1995 workload of the six contractors, illustrates the current strategy adapted by each company. Whilst there are similarities, the distinctive nature of each company portfolio of current contracts is apparent.

Table Current contracts of six of Hong Kong's major contractors (1995).

	Shui On	Hsin Chong	Nishimatsu	Leighton	CSCEC	Hip Hing
Public housing	2083	240	-	208	2287	1076
Private housing	-	-	-	-	-	3008
Private non-residential	-	1180	191	181		1963
Airport related	1312	557	15301	3569	3759	-
Other civil engineering	532	202	5288	2166	2102	-
Public non-residential	1078	1872	-	198	1329	3676
Total	5005	4051	20780	6322	9477	9723

(HK$ million)

Shui On targets mainly public sector projects and is involved with the airport. Hsin Chong's profile is similar but with less civils and airport work (although it has won a large airport terminal contract for 1996) but has some private non-residential. Nishimatsu is quite distinctive focusing practically exclusively on civils work and having the largest portfolio with a massive exposure to the airport. Leighton similarly is heavily committed to civils work and the airport but is also involved in public housing and other sectors. China State focuses on the public sector including the airport for which it has obtained the major terminal building contract (in joint venture) which is not included in the table as it is not yet in progress. Hip Hing is again distinctive as it focuses particularly on the private sector (largely but not solely through its parent New World Developments Ltd). It undertakes no civils work except in connection with its building work.

It can be seen therefore that each company has a strategic focus which it will need to develop or adapt as the market changes as the airport related work reduces. The current strategies can be succinctly stated as:

Shui On - Public sector, building and civils
Hsin Chong - Both sectors, building and civils
Nishimatsu - Public sector, civils
Leighton - Both sectors, building and civils
CSCEC - Public sector, building and civils
Hip Hing - Private sector, building

The managing director (or chairman) of each of the companies was interviewed to ascertain their view of their company's major competitive advantages and of their company's future market and how their strategy was designed to match their company's strengths with their vision of the market. A summary of their view are as follows:

Competitive advantage:

Shui On
- Combination of discipline and endeavor traditionally found in local Hong Kong companies and a more creative people centered environment which has allowed the natural hard work of local people to be channeled into the development of initiatives and new ideas within a climate in which people seek responsibility.
- One of the earliest Hong Kong construction companies to begin operations in China (1985).

Hsin Chong
- Staff orientated innovation
  - retirement fund scheme (first in Hong Kong)
  - medical scheme
  - education fund
  - share option scheme
- Joint venture companies to undertake specialist work (e.g. building services)

- Developed expertise in design and build (particularly hospitals).

Nishimatsu
- Development and application of technology (supported by Japan headquarters).
- Build-operate-transfer infrastructure project experience.
- Lead in managing complex multi-discipline projects involving joint ventures.
- Financial strength to put together funding packages.
- Focus on design/build.

Leighton
- Focus is solely the effectiveness of the construction process ('Profits are made by completing projects efficiently'), not involved in development.
- Versatility, the company will bid for any project which it can cope with.
- Experience of design/build, re-design and alternative design approach.

CSCEC
- China connection extremely close, access to projects strong.
- Directly under the PRC State Council.
- Learning from its Hong Kong experience.

Hip Hing
- Pre-eminent in high quality construction for private sector developers.
- Has available substantial work from its parent company. (New World Developments).
- Design/build specialist with in-house design capability.

Future market:

Shui On
- Hong Kong is a mature market with limited opportunities for growth, more competitive.
- Will concentrate on selective tendering.
- Will approach design/build and other innovations with caution.
- China a growth area.

Hsin Chong
- Hong Kong market growth limited, more competitive.
- Will build on experience in China (project management and investment).
- Some expansion in Malaysia and Vietnam.
- Focus on design and build (valuable in above regions).
- Potential in Indonesia and Thailand.

Nishimatsu
- Hong Kong rate of growth will diminish and be more competitive.
- Hong Kong the 'engine room' of Southern China.
- Cross-border infrastructure projects.
- Chinese high-tech b-o-t projects.
- Create b-o-t projects for Hong Kong, China and other South East Asia countries. (A Big Projects Promotion Dept.).

Leighton	-	Hong Kong's market will continue to be attractive after 1997 but competitiveness will increase.
	-	Cautions diversification into other South-East Asia countries.
CSCEC	-	Hong Kong will become a more competitive market.
	-	Focus on Hong Kong and Southern China only.
Hip Hing	-	Maintain its position as a contractor for the high quality private sector of the market anywhere in the region.

6 Discussion

Whilst on the face of it the six companies considered here could be seen to be competitors, their individual orientation means that only some of them complete and even then only over a relatively small number of project types. In many cases the competitive advantage of individual companies give them a decisive edge.

An exception to this generalisation is Hip Hing which has a pre-eminent place in the niche (although a large niche) market of high quality private sector projects for which the company is often appointed by negotiation particularly by its parent company, New World. It will continue to focus on this market throughout South-East Asia. The remaining companies have some views which coincide and some which conflict.

Whilst all companies agree that Hong Kong market's will become more competitive (as contractors remain after the airport completion) they vary in their opinion of its rate of growth. Their views reflect the rang of predictions identified in the study referred to earlier. The more pessimistic are concerned about caution on the part of the SAR government immediately after 1997. Nevertheless, except for Nishimatsu and Hip Hing, they all intend to compete over a broad range of project types in Hong Kong.

The various companies' strategy in China is interesting. All of them recognise that China provides opportunities particularly from a construction management/management contracting base but also with some opportunities for a more entrepreneurial role. The out-and-out Hong Kong companies, Shui On and Hsin Chong, are bullish on China based on their long experience of working there. Nishimatsu and Leighton, whilst Hong Kong companies are not "Chinese" and are more cautions. Nishimatsu will aim to enter China on a b-o-t basis using its financial and technological strength rather than any Chinese "contacts". Leighton will only diversify into China cautiously as it has no special edge in that market. CSCEC, as a China State company, is working back in China out of Hong Kong and has ready access to projects.

Two elements are particular beneficial in penetrating the China market.. One is an expertise in design and build and b-o-t, the other is to be closely associated with a developer working in China. Shui On, Hsin Chong and CSCEC all have development arms and have already invested in developing projects in China. Hip Hing is also well placed in this respect. However Hsin Chong is the most experienced and committed to design and build. On the other hand Nishimatsu does not have a development arm but is the most experienced in b-o-t including, most importantly, providing financing and taking an equity interest. The opportunity for and attractiveness of joint ventures amongst such a group of contractors is clear. Leighton are the least likely to be able to offer such advantages to China projects but they do have strengths in design and construct civils projects and are focused solely on construction.

There is a contrasting approach to other South East Asia countries with Hsin Chong, Leighton and Nishimatsu keen on opportunities there but Shui On and CSCEC concentrating soley on China.

All those interviewed were fully aware of the environment within which their companies existed and the scope for strategic choices which theses presented. The way in which the companies were structured as a part of a wider organisation reflected the strategic contingency approach to organisation [6] aimed at providing organisations with a choice of strategies.

Construction companies are faced with complex business environments which included many elements other than those described here. In order to cope with increased risks, provision of finance, design and build contracts, scarcity of labour or increasing training, greater investment in plant and machinery etc. to tackle the larger high technology projects, it seems inevitable that only a relatively small number of companies in Hong Kong will be able to gear themselves up for such challenges. These will be impressive, well managed, technologically adept, people orientated companies with the critical mass of professionals and finance necessary to accept such risks.

Such a development is likely to favour multi-national companies with the capability to put project teams together anywhere in the world whilst retaining core functions such as designing and estimating at head office. A number of Hong Kong companies such as those described could develop in this manner to build increasingly in China and other Asian countries and even further afield. Such a development could see the amalgamation of companies to provide the strength required. Such mergers may not be solely between construction companies but could include financial and other organisations. The distinction between Hong Kong and overseas contractors in Hong Kong is becoming extremely blurred and such a development could blur the distinction even further. The success of the strategies adopted by the companies included in this paper will to a large extent determine how they will fit into such a scenario.

The companies' approach to defining their strategies is neither as clear nor paradoxically as complex as the theoretical literature on strategy implies. It appears that companies have broad visions of their strategy which in its implementation is strongly influential by opportunism as projects become available. Companies remain adaptable (which is the hallmark of Hong Kong) and ready to capitalise on any opportunity which the market provides.

7 References

1. Male, S.P. and Stocks, R.K. (1991) *Competitive Advantage in Construction*, Butterworth-Heinemann, Oxford.
2. Porter, M.E. (1980) *Competitive Strategy. Techniques for Analysing Competitors and Industries,* The Free Press, New York.
3. Male, S.P. and Stocks, R.K. op. cit.
4. Walker, A., Chau, K.W., Lai, L.W.C., (1995) *Hong Kong and China: Real Estate in the Economy*, Brooke Hillier Parker, Hong Kong.
5. Walker, A, (1995) *Hong Kong: The Contractor's Experience,* Hong Kong University Press, Hong Kong.
6. Scott, W.R. (1992) *Organisations: Rational, Natural and Open Systems,* Prentice-Hall, Englewood Cliffs, N.J.

DEVELOPMENT OF CONSTRUCTION MANAGEMENT BASED ON REGIONAL CULTURE

K. Baba
School of Civil Engineering, Asian Institute of Technology,
Bangkok, Thailand

Abstract
From the end of this century to the beginng of the next century, there will be a boom of construction in Asia. In order to implement these project efficiently, the method developed in the field of construction management shall be adopted.

However, there had been some cases when strong resistance against using these method and systems in the actual construction works. Conflicts mainly came from the culturaL differences between Western who developed these methods and systems and of Eastern people who wanted to employ these methods and systems.

In order to apply the methods and systems of construction management to the actual construction works in the region where culture are different development and adjustment of the existing methods and systems for the regional culture will be necessary.
Keywords: Asia, construction management, culture. region.
Sommaire
D'ici à la fin de ce siècle et au début du prochain, l'Asie connaîtra un boom dans le secteur du bâtiment et des travaux publics. Afin de mettre des travaux en place efficacement, les méthodes développées dans le domaine de la gestion des travaux devraient être adoptées.

Cependant, on a pu assister dans certains cas à de fortes résistances contre l'utilisation de tels systèmes et méthodes dans les projets mêmes. Les conflits ont été en général générés par les différences culturelles, différences entre les occidentaux qui développaient ces systèmes et méthodes et les orientaux qui allaient les utiliser.

C'est pourquoi il est indispensable que les dits systèmes et méthodes soient développés et ajustés à la culture et à l'environnement dans lesquels ils vont être utilisés.

1 Introduction

The objective of "Construction Management" is to accomplish a construction project effectively to meet the requirements specified for that project. For this purpose, many practical methods and systems have been developed in advanced countries such as in U.S.A.,United kingdom and other countries as well. In transfering and utilizing these methods and systems to suit the other countries'needs, sometimes conflicts and resistance will come across due to the differences of principles between the source of those

The Organization and Management of Construction: Shaping theory and practice (Volume One).
Edited by D.A. Langford and A. Retik. Published in 1996 by E & FN Spon. ISBN 0 419 22230 8.

methods and systems and the user, those people who are trying to implement these methods. These conflicts mainly come from the differences in cultures.

To explain these facts more clearly, it may be helpful to cite experiences in Japan concerning conflicts caused by differences of culture in construction management.

These conflicts in the construction management were not so important when major part of the world construction was carried out in the region where these construction management methods and systems were developed and established. However, recent rapid development and growth of the economy in Asia, considerable part of the world construction are carried out in Asia where the methods and systems of the construction management are transfered from U.K and U.S.A.

These transefered methods and systems are sometimes quite strange to Asian people and cause some conflicts. Due to the increase of the volume, it is impossible to neglect the existence of the conflicts and it is necessary to find out the a countermeasure.

Differences in cultures will cause these conflicts,but, they can be more clearly classified into three categories:(a) traditional organization structure,(b)managerial differences,and (c)differences in fundamental concept and philosophy which contracts and laws are based upon.

For this purpose,we must start with differnces in traditional organization sturacture in comparison with that of U.S.A and Japan. Also,influences of managerial differences,and differences of fundamental concept and philosophy which contracts and laws are based on are also considered.

2 Differences in construction management between U.S.A and Japan

The differences in management practices related to construction projects between the United States and Japan consist mainly in three categories :

(1)Differences in traditional organization structure for execution of construction project.

(2)Managerial differences in executing the work.

(3)Differences in fundamental concept and philosophy which contracts and laws are based upon.

These differences for each item shall be discussed and describrd hereinafter.

2.1 Organizational Structure for Execution of Construction.

One of the biggest differences in executing construction systems is the organizational structure for executing construction projects as illustrated in Fig.1. In the American system, three parties are involved in project proceedings, while that of the Japanese system, only two parties are mainly involved.Consultants act as in-house engineers of owners.

The triangle organizational structure adapted in the U.S.A.,so to speak,is one of the commonly used system in the countries where present construction management methods and systems are developed and accepted as the most

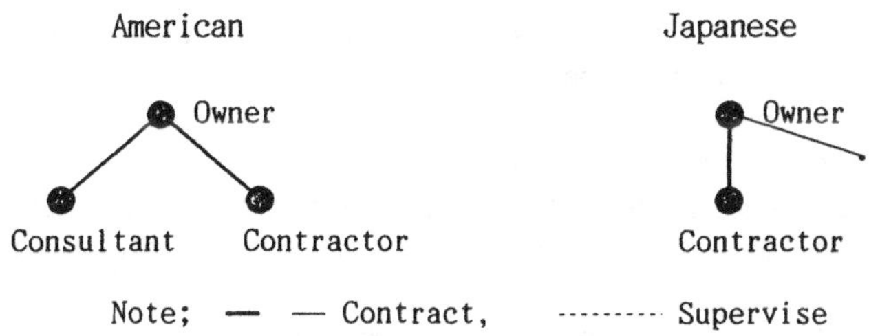

Fig.1. Organizational structure for construction

reasonable structure for executing construction project. However, if two systems were analyzed by using communication theory, there are many advantages in the Japanese system over the American system as shown in Table 1.

The disadvantage of Japanese system is in its nature of organization structure. That is, Japanese system tends to be close and unilateral.

Table 1. Nature of organizational structure

Items		American	Japanese
Nature Tends to be to be		Open, Bilateral	Close, Unilateral
Comunication:	Effectiveness	Low	Moderate
	Centralization	Low	Moderate
	Leadership	Low	Moderate
	Satisfaction	Moderate	Moderate

Therefore, to evaluate Japanese system, it is necessary to compromise between efficiency of communication and unilateral nature.

Also, the difference in the organization structure of execution of the construction projects leads to the differences of the role of each parties as illustrated in Fig. 2.

A big difference is found in the role of the consultants. In the Japanese system, consultants are some kind of assistants to owner and are not completely independent in the organization structure. Their role is limited while that of the owner is quite big.

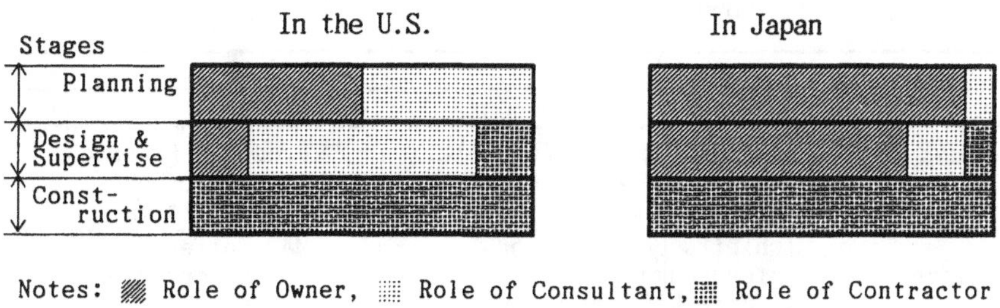

Fig.2. Role of each parties

In the American system, consultants take more active role and roles of each party involved are some what balanced.

2.2 Nature and method of management in Practice

There are many differences in the actual practice of management. Differences in management in practice of both systems in general are listed in Table 2.

Table 2. Nature of American and Japanese management

Items	American	Japanese
1. Decision Making	by Individual	by Group
2. Responsibility	by Individual	by Group
3. Promotion	Rapid	Slow
4. Control Mechanism	Explicit	Implicit
5. Career Path	Specialized	Nonspecialized
6. Employment Term	Short Term	Lifetime
7. Object of Employment	Function	Person

As typical examples of differences in form of control and communication are listed in Table 3 and 4 respectively.

Table 3. Forms of control

	In the U.S.	In Japan
Kind of Control	Bureaucratic Control	Clan Control
Goal	Employee Compliance	Employee Commitment
Formality	Strict Rules,Formal Control, Rigid Hierarchy	Group Norms, Culture, Self-control
Performance Focus	Direct toward Minimum Level Acceptable of Performance	Direct toward Enhanced Performance above/beyond the Minimum
Organization Design	Tall Structure, Top Down Influence	Flat Structure, Shared (Bottom Up) Influence
Reward System	Directed at Individual Performance	Directed at Group Performance
Participation	Limited and Formal	Extended and Informal

Table 4. Forms of communication

	in the U.S.	in Japan
Form	Written	Oral
Advantages	Tends to be More Accurate. Provides a Record of the Communication.	Promotes Feedback and Interchange Is Esay to Use.
Dis-advantages	Inhibits Feedback and Interchange. Is More Difficult and Time Consuming.	May Suffer from Inaccuracies. Leaves no Permanent Record

2.3 Conformity with respect to contracts

Another difference is the comformity of the contract between both countries as listed in Table 5.

In general, Japanese system reads between the lines of the contract, while the American system reads the lines of the contracts. It should be added that in the Japanese system, only after making a good relationship between representative from owner and contractors,contract issues will be discussed.

Table 5. Meaning of conformity of contract

	In the U.S.	In Japan
Contract Attitude	Highly Esteemed	Highly Esteemed
Process of Conformity	Each Clause and Line of the Contract is Carefully scrutinized	Put the stress on the Sense of contract as a Whole
Interpretation	Judgement by Rationality Rationality	Judgement by Humanity
Solving Dispute	By Legal Proceedings, By Lawyer	By Negotiation, By Engineers

2.4 Social background

These differences in construction management of both systems are backed and originated by social systems of both countries.

One of the examples in the business is the number of lawyers as shown in table 6.

Table 6. Number of new lawyer

Items	U.S	Japan
Number of New Lawyers / year	approx.25,000	approx. 500
Population (Ratio):	2	1
Number of New Lawyers /year(ratio)	2 5	1

Table 6 shows the fact that U.S.A.produces lawyers 25 times as much as in Japan. Too many lawyers sometimes result in existing of an " Ambulance Chaser " and / or a " Philadelphia Lawyer." Also, because of this, the recent trend would be to have a question to solve almost every dispute in construction project by lawsuit.

3 Different countries have different cultures

There has been quite a bit of discussions concerning comparison between managerial systems of U.S.A.and Japan. For example, Peter F.Drucker points out many aspects of the differences between American management and Japanese management in his paper.

Also, a simple but systematic analysis of the differences in the two style of management is discussed and presented in the form of comparison table in the book written by H. Koontz. These facts show that American management has been built on logic and reason, while Japanese management is based on harmony and humanity.

Throughout the analysis, the differences in the principles of construction management between U.S.A. and Japan come from the differences of the history and culture of both countries.

Japanese system is on the premise that everything should be preliminarily controlled. This system is established based on the fact that Japan consists of several islands. In this case, the best way of defending itself is to fight against invaders at the shoreline.

This kind of process becomes a traditional way of control and is applied in many aspects of management. Similarly not only in the Japanese system, but also in the American system, management system is completely dependent on its history, culture and given background.

Some outstanding characters of both countries, U.S.A. and Japan, which influence most the process of construction management are listed in Table 6. Facts listed in Table 7 explain the origins of the differences in construction management between U.S.A. and Japan more clearly than explained by any other theory.

Table 7. Outstanding character of U.S.A and Japan

	U.S	Japan
Country	Continent	Island
Border	River	Sea
Race	Multiple	Uni
Culture	European Heritage	Hybrid
Constructed by	Immigrant	Aboriginal
Movement of People	Come in	Defense at shoreline
Language	International	Inherent
Tends to be	Open	Close

4 Compromising the differences in the age of internationalization

Recent rapid changes of the world economy due to globalization and diminishing borders, have made it impossible for any country to remain as a closed society. Therefore, every country should open its market to the rest of the world. This should be done for the benefit of both parties.

However, the globalization of the world economy, in general, will bring much more chance to cause conflicts by the differences in cultures.

Actually, the above mentioned differences in construction management, have been discussed by the U.S.and Japanese goverment to open the Japanese construction market to U.S. contractors.

Actual solution for these conflicts shall be found only in compromising the differences. Also, it can be said that in the process of finding the solutions for the conflicts, new methods and systems of construction management will be disclosed.

5 Expected rapid increase of construction projects in Asia

5.1 Growth of economy in Asia

In order to forecast the possibility of executing many projects in Asia towards the end of this century, it is useful to review the recent world economy and forcast its future.

Fig.3 shows the comparison of three major economic regions in 1991: EAEC ; Asia, NAFT ; North America, EEA ; Europe.

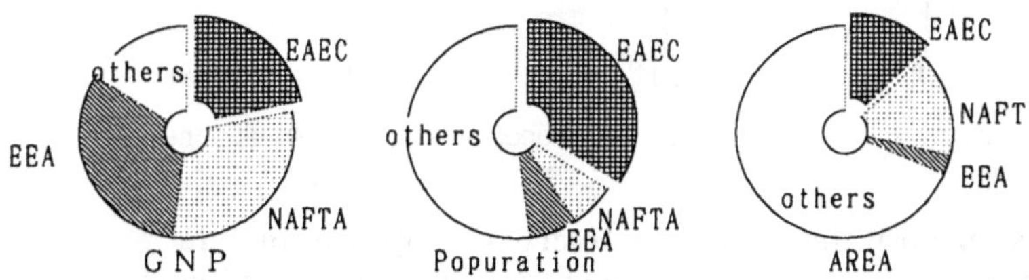

Source : The World Bank. The World Bank Atlas. 1991
Fig. 3 Comparison of three major economic regions (1991)

Also, Fig.4 shows annual averaged growth rate of economy in these region in '90s. Due to the rapid decrease of ratio of people in poverty in the countries in Asia as shown in Fig. 5 , growth rate of economyu will be maintained in considerable high level for several years in the future.

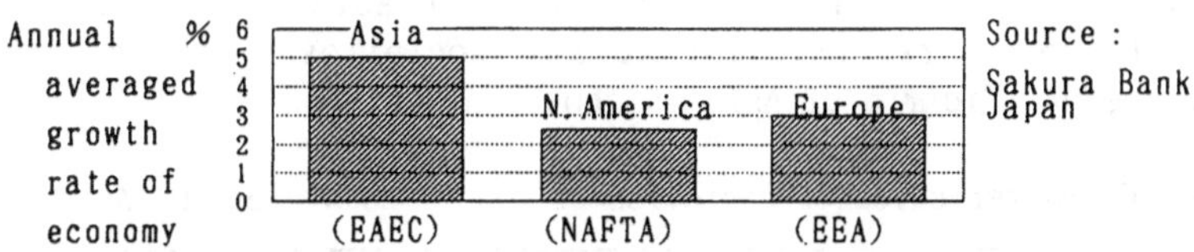

Fig. 4. Growth of economy in the three regions

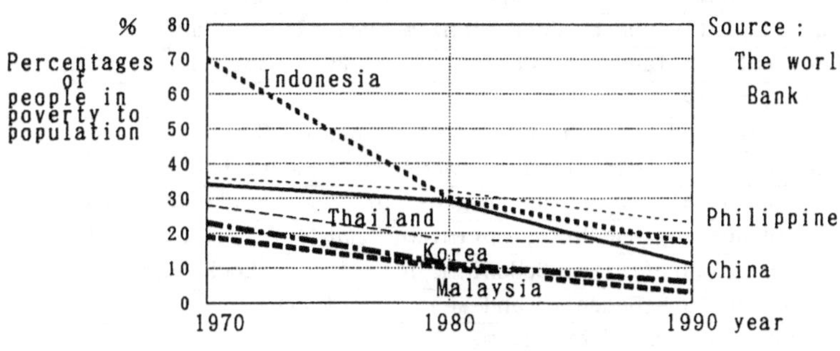

Fig.5. Decrease of ratio of people in poverty in Asia

Under the assumption that same growth ratios of the economy of these three regions will be continued in the future, the amount of GNP in these regions will be as shown in Fig. 6. At the beginning of the next century, economic regions of Asia will be the largest among the three big economic regions in the world. This rapid growth of economy in Asia will provide and require the enormous amount of construction investment in this region.

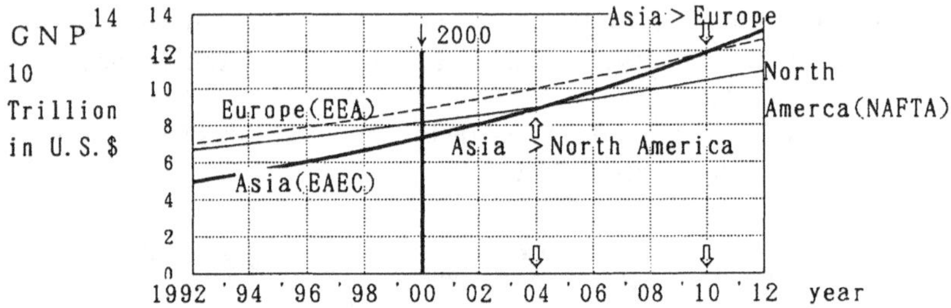

Fig. 6. G.N.P. of three economic regions in future

5.2 Cross cultural issues in construction management in Asia

There are many discussions concerning the differences of cultures of East and West. Of course, this is one of the subjects to be studied. However, culture of East itself is not monolith. As anyone can find, there are many varieties such as different languages, even different letters, religions, etc., in the region of Asia. Therefore, cross cultural issues are much more important in Asia.

5.3 Each country has it's own culture

Compared with European countries, countries in Asia have much more varieties. Outline of the characteristic of societies of the several countries of the pacific rim of Asia is shown in Table 8.

Table 8. Characteristics of societies of major countries in Asia

Country	
People's Rep. of China	*Unidentified* Under communism but remarkable change, Changing
Korea	Strong influence by Confucianism, Moderate by Buddhism
Japan	Moderate influence by Confucianism, Buddhism, "Harmony"ony"
Taiwan	*Unidentified*
Hong kong	*Unidentified* " Window to main land of China "
Philipine	Influenced by Christianity, American & Spanish culture re
Vietnam	Under socialism, In the progress of " *Doi Moi*", Changing
Thailand	Influenced by Buddhism,(optimism," *Sakdi Na*", Individualism)
Malaysia	Mixed culture(Islam, Buddhism, Hinduism) "Look east"
Singapore	Mixed culture(Buddhism and western culture)
Indonesia	Strong influenced by Islam,

Source; some come from *Kokusai Hikaku Keieiron* of reference 7

5.4 Today's role of Japanese engineers

As stated previously, Japanese management in the construction industry has many unique characteristics. This had caused some conflicts with the American management system in the construction management which have been discussed by both governments to find a solution. Since Japan was isolated from the rest of the world by sea, it has been able to develop its own culture. But because of the globalization of the world economy, Japan can not stand as a closed society.

Other Asian countries will experience the same kind of situation in the near future which Japan had experienced in past several decades.

In fact, Japan was a receiver of technical know-how from the West. However, Japan is now in a position to transfer this knowledge to the rest of the Asian countries. Since Japan has experienced two different and opposite situations, Japan might be an example to other Asian countries.

Today's role of Japanese engineers might be to listen to people of the other countries and to give adequate advice in executing construction projects in the Era of " golden age " to come giving due consideration to cultural differences and compromising conflicts.

6 Conclusion —" Let's challenge to new era of world construction "

We are facing the age of paradigm shift in the world of construction. The main stream of world construction will be shifted from the region of Western culture to the Eastern culture, Asia. In this transfer, conflicts will appear due to the differences in cultures of West and East.

But eastern culture itself has many varieties. Therefore, it will be difficult to find solutions if there arises a conflict. However, efforts to solve these conflicts lead to development of new methods in construction management. Moreover, there is a possibility that the future system of management in construction world, could be developed from a combination of Eastern and Western culture, but should also take regional culture into consideration.

This new form of construction management would comprise the characteristics of pragmatism and humanism along with a sense of many differnt religions.

It can be said that America and Japan are both extreme examples of construction management. America is a rather new country consists of immigrants mainly from European countries. Japan has a long history of isolation from the rest of the world by the ocean. Therefore, there are many differences not only in construction management, but also everywhere in social system. However, it is quite useful to compare two completely different systems to find out a new and better system for the future.

In the way of comparison, many advantages and disadvantages over each other are found. This analysis is bound to produce and establish a new system which will be used in future construction management system, to lesser cross cultural problems, which will be one of the most important issues in the next century.

7 References

1. Baba, K.(1990) Principal Nature of the Management in Japanese Construction Industry. *Jounal of Construction Engineering and Management,* ASCE, Vol.116,No.2, N.Y. pp.351-364.
2. Baba, K.(1989) *Management by the Unconscious*— Reason of the Strength of Japanese Management - (in Japanese), Chuoh Keizaisha Tokyo, Japan.
3. Drucker,P.F.,(1971). What We Can Learn from Japanese Management. *Harvard Business Review,* (March-April) pp.119-122.
4. Koontz,H.,O'Donnell,C. and Weihrich,H.,(1984) *Management,* McGraw-Hill Book company, New York .
5. Dunham,C. W.,Young,R. D.(1971) *Contracts,Specifications,And Law For Engineers,* McGraw-Hill Book Company, New York.
6. Rose,H.F., Barrow,M.H.,(1957) *Project Engineering of Process Plant,* John Wiley & Sons, Inc., New York.
7. Tanno,Isao (1994), *Kokusai Hikaku Keieiron,* (International Comparative Management, in Japanese), Dohbunkan,Tokyo.
8. Hellriegal, D., Slocum, J.J., and Woodman R. W. (1983) *Organization Behavior* (third Ed.), West Publishing Company, New York,
9. Griffin,R. W. (1990) *Management* (3rd Ed.), Houghton Mifflin Company, Boston Mass.
10. U.S.Department of Commerce (1990) *Statistical Abstruct of the United State 109th Ed.,* U.S. Government of Printing Office,Washington DC.
11. Engineering News Record.(1991) *Quiet Revolution Brew for Settling Dispute,* August 26, McGrow-Hill Book Company, New York.

THE DEVELOPMENT OF AN INDUSTRY FUTURES BULLETIN

J.R. Brightman, C. Eden and K. van der Heijden
Department of Management Science, University of Strathclyde, Glasgow, UK
D.A. Langford
Department of Civil Engineering, University of Strathclyde, Glasgow, UK

Abstract
The purpose of this study was to devise a means of influencing the strategic thinking of the managers within construction organisations, of supporting their decision making and helping them to make sense of the vast amount of data about the business environment which could potentially be used to inform strategic decisions. Two tools were used to assist with the design of the decision support system, namely, scenario development and causal mapping software, known as Graphics COPE. Scenarios have been used successfully in other industries as part of the strategy development process. In this instance the scenario methodology developed by Royal Dutch/Shell was used and adapted to the needs of construction contractors. The second tool, Graphics COPE, has been used to support strategy formulation in a number of different industrial and public sector contexts. In this study it was adapted and developed for use as the vehicle for presenting and structuring data which managers can then use to explore multiple possible futures. The development was aimed at facilitating the manipulation of generic data in order to generate company specific scenarios relating to their future business environment.
Keywords: construction contractors, decision support, information technology, multiple possible futures, scenarios, strategic management.

Sommaire
Cette étude repose sur un tryptique d'objectifs: influencer les entrepreneurs dans leur façon de penser les stratégies, leur apporter des outils à la prise de décision, les aider à reconnaître, parmi les nombreuses données de leur environnement sectorial, celles qui pourraient contribuer à l'élaboration des stratégies. Deux outils ont été utilisés en phase de conception du système d'aide à la décision: un logiciel de développement de scénario d'une part, un logiciel de repérage causal, d'autre part, connu sous le nom de Graphics COPE. Ces mêmes simulations ont été appliquées avec succès dans d'autres secteurs pour le développement des stratégies. Ici, c'est la méthodologie de développement de

The Organization and Management of Construction: Shaping theory and practice (Volume One). Edited by D.A. Langford and A. Retik. Published in 1996 by E & FN Spon. ISBN 0 419 22230 8.

Royal Dutch/Shell qui a été utilisée, après avoir été adaptée aux besoins du bâtiment et des travaux publics. Le second outil, Graphics COPE, a été utilisé pour la formulation des stratégies dans un certain nombre de contextes de l'industrie privée et des travaux publics. Pour cette étude, il a été adapté et développé afin de servir d'instrument de présentation et de structuration des données qui s'avèreront utiles pour l'exploration de différents scénarios prospectifs. Le développement a été fait dans le but de faciliter la manipulation de données génériques qui puissent aider à l'élaboration de scénarios spécifiques à l'entreprise dans son traitement leur environnement d'applications professionnelles futures.

1 Introduction

This paper reports on work carried out over the past three years and follows on from the paper Scenario planning for construction [1], published in the proceeding of CIB W-65 1993. The intention of this project was to combine the principles of scenario development [2,3,4] and the use of interactive causal mapping software [5], in order to structure and explore the information which managers in the construction industry use to support their strategic decisions. Construction organisations operate in a particularly complex and turbulent business environment and there is a vast amount of potentially useful information which managers could use to support their strategic thinking. Our aim was to find a means of filtering this information, of focusing effort, in order to facilitate more appropriate opportunistic strategic decision making.

2 Foundation research

At the start of the project it seemed appropriate to devote a little energy to validating those assumptions which were key to the project's success. The outcome of this foundation research was the precursor to designing the overall research methodology for the project and the project plan was established after its completion. The purpose of this initial research was to:

* test our assertions about the nature of strategic management within the industry and in particular to establish the level of sophistication of the strategic management processes used.
* establish whether there was indeed an unarticulated need for assistance to resolve the problem of assimilating the large volumes of strategically important business data available and to establish the nature of the most widely used data sources.
* assess the extent of the managerial use of information technology in the industry, both as devices to access information and as decision support tools.

The basis for the foundation research was interviews with senior managers in a range of construction industry organisations and a review of literature. This research proved to be significant in shaping the future of the project.

2.1 Interviews

Sixteen organisations throughout the UK were approached to be interviewed for this part of the study and ultimately eleven individuals from eight organisations took part. The basis for choosing this sample was a mix of pragmatism and breadth of population. Organisations which might be amenable to acting as collaborators later in the project were chosen, as well as selecting those that would be representative of different sizes of organisation, location and assumed level of management sophistication. Very small organisations, unlikely to have a great interest in formalised strategic management, were excluded from the sample, although cognisance of the fact that the construction industry is characterised by a large number of small organisations was taken in the research, so that the resulting system has been designed to be flexible enough to be used by a wide variety of organisation types and sizes.

2.2 Foundation research outcomes

Managers tend to rely on soft sources of information for their business intelligence and they take an intuitive approach to strategic decision making. Contractors operate with an outward focus, reacting to market needs. Managers' visions are resource based - 'if we have X,Y and Z, then we can play it as it happens'. This is typified by the comment, from one of the interviewees, that "if it's bridges next year, we'll be in bridges". The antithesis of this approach is a predict and control view of managing, which is seen by managers as inappropriate in the construction industry. Rather than having explicit strategy formulation processes it appeared that strategic growth comes from a series of step moves. These strategic moves are a response to triggers, such as reports that "design and build is on the increase", not the result of what might be called 'traditional' strategic planning. The reactive nature of construction organisations and project life cycles mean that planning horizons are generally limited to between three and five years.

Interviewees responses to questioning about the type of information used to support strategic decisions led initially to the identification of market or trade data, that is information about volumes and prices. Trade data gives pointers about how to act in current and immediately imminent commercial activities, for example, how to tender. It provides information about what is going on now and in the very recent past. Three types of data are used for planning: trade data, trade rumours (planning lead type information) and data which has long term implications. The last of these types informs strategic thinking; this planning data tends to come from interpretive sources, digests of information and commentaries, and includes sources such as newspaper and trade press reports. This is in contrast to the use of large, government or industry wide, databases such as the housing and construction statistics. The interviews demonstrated a consistent pattern of information sources used to inform strategic thinking.

The contractors interviewed did not use statistical information directly. They said that they do not "clinically analyse statistics" but rather they will cast an eye over them and use them to confirm the validity of previous decisions. Strategic decisions are based on anecdotal evidence, which precedes statistical evidence. Analysing statistics was taken to be different from using them to inform their strategic thinking, in that managers may be powerfully influenced by statistics even though they do not analyse them. Although organisations may vary in terms of size and their degree of management sophistication, they still tend to scan the same data.

The information technology (IT) capabilities of the managers interviewed were universally poor. It was clear that there would be general antipathy towards using a computerised Group or Individual Decision Support system, although individual champions exist within organisations and such an individual might still use a computer based system.

2.3 Implications for the design and prosecution of the research.
The level of sophistication of strategic management implies that the impact of scenarios can only be achieved indirectly. Rather than attempting to persuade the industry to change their approach to strategic planning it was more appropriate for this project to focus on finding ways of influencing their thinking about the future. Any attempt to influence thinking in the industry would have to reflect the need to intervene in a fluid and less formal planning process.

While the foundation research suggested that a generic scenario may not provide organisations with a useful means of thinking about their own future, there was support for the project working with generic data sets; information which 'everybody collects and uses' and which it is assumed, by managers, that everybody else has. A shift away from the typical scenario methodology suggested the creation of an adapted scenario approach or, what we then called, an Industry Futures Bulletin (IFB), based on generic data and encompassing multiple futures. The structuring of data into a scenario type format would provide the screening process through which large amounts of data could be utilised and interpreted and which would support strategic thinking.

The most powerful data used for strategic thinking is of a qualitative nature, thus the project concentrated on selecting, organising and structuring qualitative data to suggest the type of multiple futures which are fundamental to a scenario aspect of strategic planning. The contractor interviews provided the portfolio of data sources which the project should explore for the structuring of generic data sets.

The attitudes to the use of IT suggested a potential lack of interest in, or opposition to, the originally conceived system. Managers who are not themselves familiar with IT will be more likely to use a system which is run for them by a facilitator or chauffeur. To overcome this problem the methods devised for interacting with the IFB had to be kept simple and the system design such that the user feels that they are gaining added value, new insights into their business environment, quickly. Contractors would also require a decision support system which fits in with their flexible approach to business.

3 The revised project plan

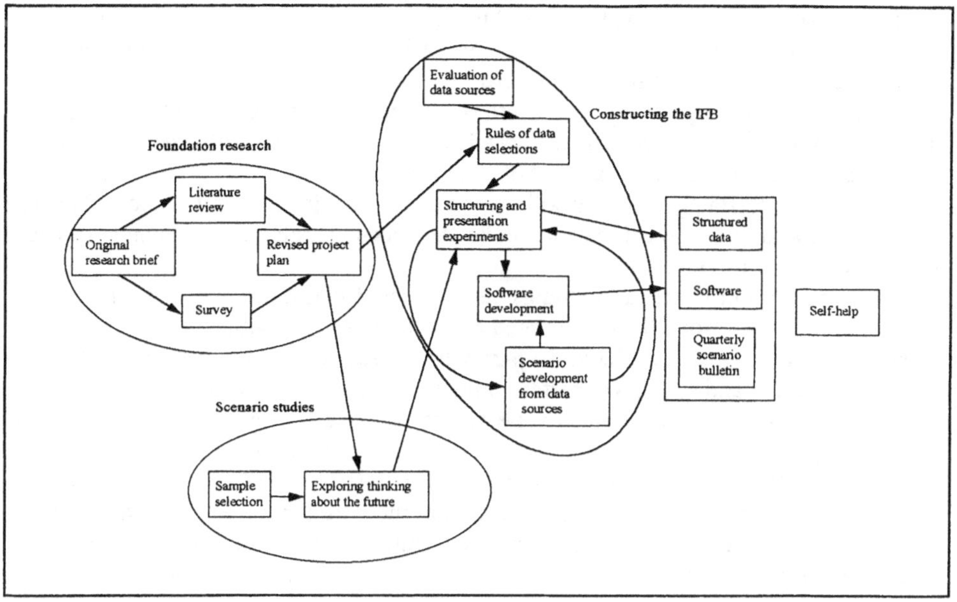

Figure 1. Revised project framework.

Figure 1 shows the revised framework used to guide the project. It depicts two parallel research activities which follow on from the foundation research: i) scenario studies, undertaken to establish the differences between what can be provided through an IFB distributed to all construction companies and the individualistic demands of companies; ii) methods for constructing the IFB - what data to extract from industry data sources, how to translate the data into an IFB and the adaptation of computer software to depict graphically the IFB in an accessible form. On completion of these parallel activities the final stage was to test the IFB with the managers of a construction firm.

4 Exploring "thinking about the future" in the collaborating organisations

4.1 Scenario studies

The scenario studies consisted of interviews and workshops, the aim of which was to make explicit the way managers envision their future. By going through a traditional scenario development exercise visions of the future are shared and explored by the management team. This process of explication was vital to collecting cogent data about the type of scenarios that would be of interest to managers in the construction industry - the research process followed the action research method [6]. This part of the research was conducted with directors of each of two collaborating organisations.

As a result of the above process areas of strategic concern were identified. From these areas a smaller number, three or four areas considered to 'really make a difference' to the business, were identified. These areas constituted a scenario agenda. This agenda was used as the basis of research and to collect further expert knowledge which would inform thinking about the future. The information derived from experts was then used to contribute to a number of alternative scenarios.

The most significant outcome of this exercise was absolute reinforcement of one of our initial conclusions from the foundation research, namely that the role of generic scenarios, which can service the industry as a whole, is problematic and that any 'large database' designed to inform strategic thinking will only be used by the industry if it is accessible and presented in a form which can be adapted to the specific strategic environment of each individual organisation.

Experience from the scenario studies also indicated that the process of introducing alternative futures, beyond those of immediate concern and set within managers' current mental framework, is very difficult. In particular it became clear that the unique history of an organisation and its growth pattern influences the extent to which new futures are considered. Possible futures which appeared to be close to negative past experiences would be rejected. This raised important issues for the success of introducing a 'generic' IFB. Thus the IFB needed to be designed in such a way that initial exploration reflected the particular needs of the organisation, and so that the introduction to alternative futures was gentle, after a degree of comfort about futures of obvious interest had been established. Access to the IFB needed to be flexible and contingent and users would need to be encouraged to add company specific data, to customise and change the complexion of the 'scenarios'.

4.2 Building the IFB

The foundation research identified the sort of information which should be drawn on for computer modelling and which would be used for scenario development. The first modelling experiment involved extracting headline type information from eighteen trade press publications, from five sources. Some of the information selected was descriptive, about what was already known, and some of it was speculative. Information was presented as short phrases (concepts) which were linked on a cause and effect basis to form maps. The concepts were expressed in terms of events, that is a statement such as 'Labour say that they will introduce a minimum wage' would be translated into 'Labour introduce a minimum wage'. Information supporting this statement would be added and the consequences of this event would be explored and linked to a number of other events. In this way the interrelationship between events could be examined. Thus the IFB presents 'narratives of causality'. An example of an IFB map is given in figure 2.

This exercise raised two significant issues, those of sourcing and inter-coder reliability for data selection. Sourcing the IFB data was an important issue because of the credibility which users place on different sources, for situations where sources contradict one another and to allow users to make a judgement about the bias of a source. The sourcing issue was dealt with by the development of a 'memo card' facility within the software. A memo card sits behind a concept, storing the source reference and additional information relating to the subject of the concept, allowing the user to get a fuller picture, beyond the headline information.

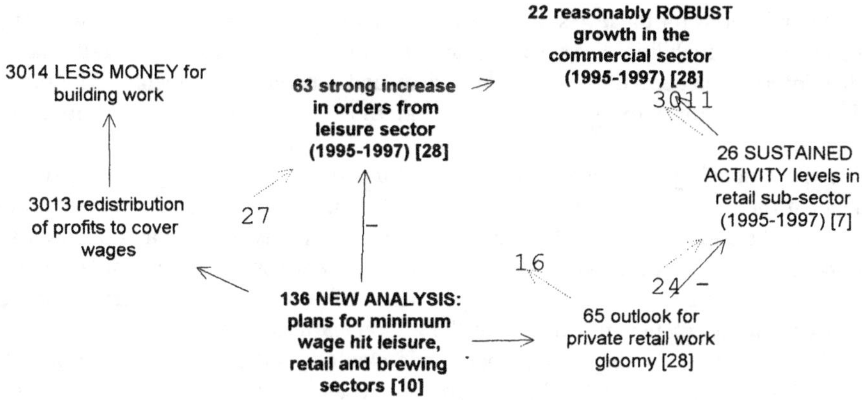

Figure 2. IFB sample map.

To establish a basis for reliability in the selection of data a set of rules was needed to guide the extraction of scenario data from the published sources. An initial set of guidelines for data extractions was developed by J.R. Brightman, based on knowledge of and experience within the industry and insights gained from the contractor interviews. Through the evaluation of data from a number of sources, C. Eden and K. van der Heijden, who have extensive experience of scenario development, characterised the elements which constitute scenario data and developed the initial guidelines into a number of interconnecting rules for the extraction of data to be included in the IFB. Some of these rules are shown in figure 3.

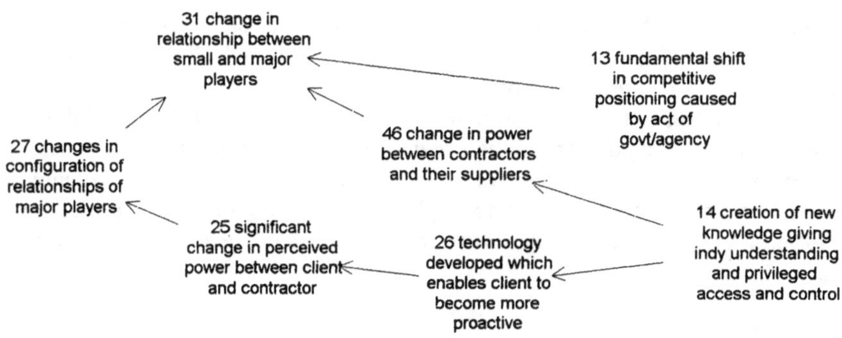

Figure 3. Examples of rules for data extraction and inclusion in the IFB.

Several experiments testing different forms of presentation and structuring of the data were carried out. Colour and font variations (styles) were used to distinguish between information categories. Initially three styles were used to categorise information; that which had implications for the levels of construction activity , that which had implications for organisational capabilities (either to be acquired or maintained) and 'the rest', that is supporting information needed to explain the activity and capability statements. The maps were hierarchical with activity and capability at the top of the map and supporting statements below.

One of the modelling experiments involved the use of generic activity and capability statements at the top of the model. This was an attempt to move away from presenting information in the commonly used classifications within the industry. The aim of this was to encourage an alternative and wider view of the business environment, drawing attention to other areas which might, at first glance, appear to be of little or no consequence for the business. This approach ties in with one of the lesson learnt from scenario development which is that it may be secondary issues that herald changes in the business environment. However, this structure proved to be unworkable because it was not possible to find a useful generic capability statement and generic activity statements of the type 'more building work' or 'less building work' are too bland and unspecific and tend to condition thinking so that only two, not very radical, or necessarily helpful, possibilities are dealt with.

4.3 Comparison of scenario studies and IFB development

In theory, if the data extracted from the sources identified by foundation research was accurate and this data was truly of relevance and interest to the managers involved in the project, then there would be a significant overlap between the data contained in the IFB and that gathered from the scenario studies. With company 1 the scenario study was completed using a 'traditional' scenario development workshop. No computer software was used during this session. The workshop involved discussion of the areas on the scenario agenda, with the participants' knowledge being supplemented with feedback from research and collected expert knowledge.

It was clear that there were areas of overlap between the IFB data and the data from the scenario studies, but that what might be termed as 'blind spots' existed. These blind spots covered issues where information was contained in the IFB but which the managers did not appear to want to consider. For example, it was extremely difficult to elicit information about a 'bad future', in relation to those issues which had been identified, in scenario terms, as most important for the organisation and least predictable in nature. This had implications for the final design of the IFB, in that it would not be fulfilling its function if it did not encourage users to explore these 'blind spots' as part of their consideration of a number of possible futures and to assess if it made sense to dismiss these areas of business.

4.4 Modified IFB

After the initial IFB structuring and presentation experiments and the scenario studies, the next stage was to produce an IFB which was much nearer to a 'commercial' version of the IFB. Over the period of a month approximately 100 data sources were scanned for information which could be of strategic importance to construction organisations and which would form the basis of a number of scenarios. By this stage the software development for referencing the data sources and providing additional information 'behind' concepts had been completed and the rules for structuring the maps had been developed.

Ultimately these modelling experiments and comparison with the scenario studies led to the use of five categories for the classification and presentation of the IFB data: activity, capability, 'so what', commentary and past event. 'Past event' represents historical information needed to put today's events into context. The need for these historical references came through very strongly from the scenario work and so more history was added to the model. 'So what' is explanation added at the top of the model where a line of argument does not end in an activity or capability statement drawn from the literature (i.e where the user or model builder adds an outcome). Commentary (formerly the 'the rest' category) was renamed to suit its purpose, providing supporting information for the activity, capability and 'so what' statements. Two further categories, theme and seen, were added to facilitate exploration of the model. The theme category contained concepts which, from analysis of the map, were the most highly elaborated concepts, that is they had the largest number of links going into and out of them. This was taken as an indication of the importance of these events as the 'most talked about' issues in the data sources. The 'seen' category was for ticking off theme items which had been explored.

4.5 Testing the modified IFB.

With company 2 the scenario study was completed using the IFB. This workshop was run with two of the researchers acting as facilitators. The session was filmed, recording screen displays and participants' reactions so that responses to the exercise could be analysed. A structured approach to the exploration of the model was devised, whereby the managers were presented with a list of theme concepts (the 'most talked about' issues in the business environment) and invited to identify which of these themes was of most interest to them. The maps were then explored in small sections, building up a picture of the issues surrounding the chosen theme event. When a theme had been explored it was 'ticked off', changing its display style to 'seen' and another theme chosen and explored.

When faced with the IFB the managers' first question was "where has this information come from", confirming the importance of sourcing and the facility to easily access back up information. The development of the memo card had solved this problem, however discussions in this session also focused on the question of exactly what should be contained in the memos. In cases where quantitative data existed behind the qualitative judgement the managers wanted to be made aware of that, so that they could dig deeper if that was deemed to be necessary.

In contrast to the workshop with company 1, where 'blind spots' had been encountered, with company 2, where statements in the IFB went against the managers' beliefs they followed contradictory lines of argument, to evaluate whether or not there was any substance in the argument, any information which they might not have been aware of and which might influence their decisions.

No client specific information was added to the model but as the participants had been interviewed during the first part of the scenario development process their scenario agenda was known. The model contained generic data and had approximately 450 concepts. During a five hour session only approximately a quarter of the model was explored and some issues previously identified as being on the scenario agenda were not explored. This rate of model exploration was mainly attributable to dealing with and learning about a new form of information presentation. As managers become more familiar with the structure and presentation of data in the IFB exploration will become quicker.

The participants were very positive about what they had seen but the leap between the data presented and the development of a number of clearly defined, company specific scenarios was not adequately achieved during this session. However the managers' thinking was most definitely influenced by what they had seen, as acknowledged in subsequent correspondence.

5 Conclusions

The positive response of the collaborating firms suggests that the product of this research does have potential industrial use. This research has sought to bridge the gap between management theory and construction management practice, by devising a usable and flexible means of influencing managers' thinking about future business environments and encouraging thinking about multiple possible futures.

The research has produced clear evidence that it is possible to enhance futures thinking amongst managers in the construction industry. By identifying activity and capability issues the method focuses on the building up of distinctive competencies and capabilities in anticipation of possible and possibly new demands from the market. Decision support systems of the type which we have been experimenting with can be extremely successful in assisting with this process, but so far this requires facilitation through strategy development workshops.

We are currently undertaking further work to refine the method of entering and exploring the model and to add a help system so that independent users are guided through the model and prompted to add their own data. The use of memo cards is also being more clearly defined so that the additional information which is given in them is 'what the managers want'. We are extremely encouraged by our results todate and are continuing this work with the ultimate aim of developing a stand-alone decision support system which could be used by organisations without expert facilitation.

References

1. Langford, D, Eden, C and van der Heijden, K. (1993) *Scenario planning for construction*, CIB W-65
2. Wack, P. (1985) *Scenarios uncharted waters ahead*, Harvard Business Review. Sept/Oct. pp73-89.
3. Wack, P. (1985) *Scenarios: shooting the rapids*, Harvard Business Review. Nov/Dec. pp139-150.
4. van der Heijden, K. (1995) Scenario thinking about the future, in Rosell, SA et al, *Changing maps*, Carleton University Press, Ottawa.
5. Eden, C. (1990). *Strategic thinking with computers*. Long Range Planning, Vol 23, No. 6, pp55-43.
6. Eden, C. and Huxham, C. (1996). Action Research for the study of organisations, in S. Clegg, C. Hardy and W. Nord (eds), *Handbook of Organisation Studies*. Beverly Hills: Sage.

ORGANISATIONAL STRUCTURE, CORPORATE STRATEGY AND SURVIVABILITY IN THE SCOTTISH CONSTRUCTION INDUSTRY

J. Docherty
Barclays Bank, Newcastle, UK
D.A. Langford
Department of Civil Engineering, University of Strathclyde, Glasgow, UK

Abstract

This paper examines the organisational structure, corporate strategy and survivability of medium to large sized Scottish construction firms. The study looked at ten of the largest privately owned construction firms, which in 1992 accounted for almost nineteen percent of the total construction output in Scotland.

The paper explains why the construction industry behaves differently from other industries and shows why these differences combine to make it particularly vulnerable during downturns in an economy.

Structure, strategy and survivability are discussed in turn, drawing on some of the broad range of literature available, before developing a research model, which defines the relationship between variables used in the study.

Organisational structure is the basis by which a firm is physically organised to carry out the tasks it has set. The structural variables which are used in the study are; Functional, Holding Company, Divisional and Functional with Subsidiaries.

Strategy is defined as encompassing the organisational behaviour of an organisation. The strategic variables used in the study are Competitive, Co-operative, Innovative and Retrenchment.

Survivability is the term used to describe a firms likelihood of remaining solvent in the future, based on its past and current trading performance. The method used to determine this is the Multi-variate Discriminate Analysis technique, also known as the Z model technique.

The organisational structure a company adopted was found to affect its survivability. A Divisional structure gives the best survivability rating. Similarly, strategy was shown to have an affect on survivability, with a Competitive strategy yielding the best results. Both a Divisional structure and a competitive strategy were found to be more prevalent in larger firms.

Keywords: Construction; strategy; organisational structuring; financial performance.

The Organization and Management of Construction: Shaping theory and practice (Volume One).
Edited by D.A. Langford and A. Retik. Published in 1996 by E & FN Spon. ISBN 0 419 22230 8.

Sommaire

Cet article examine la structure organisationnelle, la stratégie des affaires et la capacité de survivre des moyennes et des grandes entreprises de batiment écossais. L'étude a consideré dix des plus grandes entreprises de bâtiment privées, qui en 1992 ont representé presque dix-neuf pourcent du rendement total de bâtiment en Ecosse.

L'article explique comment le bâtiment se comporte différemment des autres industries et montre comment ces différemment des autres industries et montre comment ces différences se combinent pour le faire particuliérement vulnerable pendant les tendances à la baisse dans une économie.

On discute la structure, la stratégie et la capacitéde survivre à tour de rôle, on a tiré des informations du grand choix de littérature a notre disposition, avan de développer un modéle de recherche, qui détermine le rapport entre les variables utilisées dans l'étude.

La structure organisationnelle est le fond autour duquel une entreprise s'organise physiquement, pour realiser les taches qu'elle s'est fixée. Les variables structurelles qui sont utilisées dans l'étude, ce sont des structures "Foncitionnelles" (Functional), "Holding" (Holding Company), "Divisionnelles" (Divisional) et "Fonctionnelles avec Filiales" (Functional with Subsidiaries).

On définit la stratégie comme etant le comportement organisationnel d'une organisation. Les variables stratégiques utilisées dans l'étude, ce sont des stratégies "Competitifs" (Competitive), "Coopératives" (Cooperative), "Innovatives" (Innovative et "Réductives" (Retrenchment).

Le terme "la capacité de survivre" dé crit la probabilité pour qu' une entreprise reste solvable dans l'avenir, ce qui est basé sur ses resultats commerciaux actuels et du passe. Cette proabilitéest determinée par la technique "Multi-variate Discriminate Analysis", aussi appelée la technique du modéle Z (the Z model technique).

On a trouvé que las structure organisationnelle choisie par une entreprise, influe sur sa capacité de survivre. Une structure divisionelle donne la meilleure évaluation pour la probabilité de survivre. De la meme façon, la stratégie influe sur la capacité de survivre mais une stratégie competitive donne les meilleurs résultats. On a trouvé que la structure divisionnelle et la strategie competitive étaient toutes les deux plus courantes dans les grandes entreprises.

Introduction

The well-being of the construction industry is of major economic importance to the United Kingdom.

In the year to December 1992, it accounted for £30,126 million, or 6% of the Gross Domestic Product (Joint Forecasting Committee for the Construction Industries, 1993). In the winter of 1992/93 it employed a total of 1,586,000 people , or 6.56% of the total workforce (Government Statistical Service, 1993). Given the size and its power to generate employment an understanding into how the construction industry works is an essential component of industrial policy. The argument concerning how governments have used construction as an economic lever are too well known to be rehearsed here but the evidence is that construction has had a very difficult time during the recession.

The deep recession experienced by the construction industry has focused thinking concerning strategic management and how to survive. The depth of the recession can be charted from the second quarter of September 1990 to December 1992, when the UK. economy, as measured by Gross Domestic Product, fell by 3.7%. At the same time construction output fell by 15% (Source: Central Statistical Office, 1993). Whilst employment in the UK. as a whole dropped from 25.726 millions in the spring of 1990 to 24.176 million in the winter of 1992, or some 6%, those employed in the construction industry fell from 2.026 million to 1.586 million, a fall of almost 22%.

These figures demonstrate the importance of understanding the financial implications of structure and strategy within the construction industry.

Scottish Construction

The research focused upon the Scottish construction industry. The reasons for limiting the study to Scotland are laid out below.

The construction industry in Scotland behaves differently from that of the UK. as a whole. For example between 1985 and 1990 construction output soared in England amid an economic boom. At constant factor costs (1985 = 100) construction output rose from 100 in 1985 to 134.3 in 1990 (Central Statistical Office, 1993). In Scotland the rise was more modest, from 100 to just 107 (Scottish Economic Bulletin, Summer 1992). The big, it may be averred, bang was not heard in Scotland.

Even the very nature of the spending on construction is different in Scotland. In 1981 the public sector in England accounted for 38.5% of new orders obtained by contractors, falling to 25.1% in 1990 (Government Statistical Service, 1992). At the same time the figures in Scotland were 53.8% and 29.1% respectively. The nature of home ownership in Scotland differs greatly from that in England, such that while government spending on housing fell in England between 1981 and 1990 by 5%, it actually rose in Scotland by 6% (Government Statistical Service 1992).

Not surprisingly, employment levels differ, in Scotland the percentage of the workforce employed in construction being 4%, as opposed to over 6.5% in the UK. as a whole (Source: Scottish Economic Bulletin, Winter 1992). Even so, while construction output accounted for 5.9% of the Scottish Economy, the industry shed over 28% of all those made unemployed in Scotland in the [last] recession.

Clearly, with growth, expenditure and employment patterns different in Scotland an investigation into the construction industry there is merited.

These structural issues overlay the individuation of Scotland, a country with its own customs, language, tradition and character. These factors alone would be cited by many as reasons enough for a special study of the Scottish construction industry.

STRUCTURE, STRATEGY AND SURVIVABILITY

There has been a great deal of work carried out in each of the three fields investigated. It has been possible to distil the mass of opinions in to variables used in the study.

Structure

Studies into organisational structure tend to take on one of two forms; those concerned with organisational behaviour and environment, and those which focus on physical organisational structure. While both were examined in the course of the research, it is inevitable that given the nature of the data collection, an emphasis would be placed on the descriptive, i.e. on physical organisational structure.

However, there are areas which are of relevance to this study in terms of providing a fuller understanding of organisational behaviour as a whole. There is no universal agreement on which is the correct theory on organisational structure but the Contingency Theory, defined by Woodward (1965) and built upon by others, is the nearest to orthodoxy anyone has yet come to. Her acknowledgement of the social, economic and political factors which have an influence on organisational structure was to open the doors for further research in to the broader environment. Put simply, the contingency theory suggests that organisational structures are dependant on the parameters of the organisations environment. Or, as Mansfield (1986) put it "the performance of the organisation is contingent upon the fit between its structure and the organisational context and environment".

The concept of the environment within which an organisation operates has provided much scope for debate. Its importance is almost universally recognised. Companies which fail to take an adequate reconnaissance of their environment will suffer. Closed system organisations (Newcombe, Langford and Fellows, 1990) which fail to recognise influences outside their own offices tighten internal functions when under stress. One assumes that this makes the organisation more rigid and inflexible at precisely the time when flexibility may be called for.

Such inflexibility is often attributed to bureaucratic forms of organisation. The basis for a bureaucratic form comes because the needs are expressly designed to achieve certain goals. Weber (1930), whose view this was, believed the bureaucratic organisation was technically the most efficient organisation possible. Of it he stated "precision, speed, unambiguity, knowledge of files, continuity, discretion, unity, strict subordination, reduction of friction and of material and personnel costs - these are raised to the optimum point in the strictly bureaucratic administration". Weber was, of course, a philosopher, much of whose work was concerned with religion and morality. Religion and morality, while important, have not been the subject of empirical study in relation to the construction industry, but some of his views on bureaucracy survive and were echoed by Woodward who believed that mechanistic systems of management, where there is a recognised hierarchy and communications take the form of orders and requests, were made suitable in stable conditions.

But bureaucracy is only one of the more recognisable definitions of a style of organisational structure, and is often attributed to the older and larger organisations. Indeed size has been cited by Harris and McCaffer (1989), Mansfield (op cit.) and Newcombe, Langford and Fellows (op cit.) as being of particular importance to organisational structure. Indeed the Aston Group (Pugh and Hickson, 1976) found that the larger an organisation became the more likely its employees were to work in very specialised functions, following standardised procedures, or in other words a bureaucracy. The Aston group defined four types of bureaucratic organisation, but there are other descriptions of organisational structure. One of those is Shittu (1988), who drew on the works of both Channon(1978) and Grinyer (1980) to produce four categories of organisation especially suited to the construction industry. These are;

1. Functional Structure:- The firms activities are subdivided into a series of specialised functions with the Chief Executive Officer co-ordinating all these specialised functions.
2. Holding Company Structure:- Where the firms are composed of semi-autonomous units, loosely held together as a corporate legal entity.
3. Multi Divisional Structure:- These are firms subdivided into a series of autonomous multi function divisions responsible for operations, but with a central office which serves and monitors their function.
4. Functional with Subsidiaries:- These are firms which are basically functional in structure, but also have one or more separate functions (not necessarily true subsidiaries) reporting to top management or, in some cases, to one of the functional managers (i.e. fully centralised and outside the operating units).

For the reasons of simplicity, relevance to the construction industry and the method of data collection, the four variables above, defined by Shittu, were the organisational variables used in this study.

Strategy

A key question is: What is strategy ?. As stated, the dictionary definition is "a particular long term plan for success, especially in business". The question most asked by academics is what plans do strategy encompass. The choice has perhaps been best posed by Mansfield (op cit.) who asks whether the concept of strategy includes both the objectives an organisation wishes to achieve and the means that will be used to achieve them, or whether it includes only the means to the objectives. Mansfield answers this question by defining strategy thus: "Basically, the strategy of an organisation may be regarded as the more or less planned collective predisposition of senior management to make particular decisions regarding the task of the organisation and its relationship with the environment". This view is at odds with that of Rumelt (1986), who cites Chandlers' (1967) view, which states that strategy encompasses "the determination of the basic long term goals and objectives of an enterprise and the adoption of courses of action and the allocation of resources necessary for carrying out these goals". For the purposes of this study, strategy broadly encompasses the objectives of an organisation. How these are to be carried out can be defined as tactics, which themselves may be flexible.

A number of factors have to be taken into account when deciding on strategy, not least the fear of being left behind.(Lansley 1987) (Rumelt 1986). The need to create flexibility is perhaps more explicit. The drop in construction output in the 1970s highlighted the need for flexibility, and the ability to realign existing resources with a change in demand.. Indeed Lansley stated "in an uncertain environment... the style of management needed to reflect a long term vision of the future, to provide motivation to overcome the inevitable role stresses caused by change and to encourage the individual to develop the skills necessary to fulfil a new role. Thus a corporate and people oriented style became more important than the task oriented style which had characterised the 1960s". Managing in such turbulent environments demands rapid and creative responses by managers blessed with flair. The organisation structure should be tolerant of such flair with organic, rather than mechanistic, structures.

Consequently the skills and motivation of decision makers involved in a business is crucial to a successful strategy. The view that successful strategies require creative and innovative people is well founded and has specific implications for construction. This point has been emphasised by Kinsman (1987) who, like many futurists emphasises the increasing demand for a deeper human dimension in business. Rambo style management is out. A need to understand more deeply the motivation of individuals is in. Kinsman believes "the intuitive capacity for assessing situations... will come to represent the very essence of tomorrow's management skills".

In this research firms were categorised in to one of four types of strategy. Drawing upon the work of Mansfield the strategic typologies used are:- Competitive strategy, Co-operative strategy, Innovative strategy and Retrenchment strategy. These are briefly explained.

The purely Competitive business strategy involves an acceptance that there will be other firms carrying on very similar activities. To be successful the implementation of such a strategy is dependant upon being able to carry out those activities in some way better than competing enterprises.

The essence of the Co-operative strategy is an avoidance of extreme forms of competition. It may involve direct collaboration with competitors, but more normally it is based on a general attitude of co-operativeness- without any actual collusion. The general strategy would seem to be particularly prevalent in oligopolistic market situations. The essential environmental orientation of all firms involved in such a situation is defensive.

An Innovative strategy involves the invention of new products or methods of working which allow the elaboration of the interlocking cycles of production and consumption in which the organisation is involved. The kinds of innovation included, whether technological or social, are those that go further than the replacement old products or systems. This sort of strategy, if successfully implemented, is one that positively adds to the wealth of the parties to the particular cycles of production and consumption.

The Retrenchment strategy also involves the revision of input and output exchange relationships and the modification of the pattern of production consumption cycles in which the firm is involved. However, in this case, far from elaborating the pattern, the move is towards curtailment and simplification, normally involving a reduction of both internal and external differentiation. Usually this strategy is based on an attempt to achieve survival by cost cutting and focusing on core activities.

Survivability

It would not be possible to link strategy to survivability were there no method of comparing companies financial health. Predicting a companies financial health is like predicting the winning horse in the Grand National, with the exception that, rather than be given the Jockeys weight, you are given an accountants guess as to what the jockey would like to weigh.

When making comparisons, it would be simple to extract the figure for profit from the annual accounts and then make a direct comparison with other organisations. However such a comparison would be largely irrelevant, given the variety of accounting techniques which can make comparisons between organisations misleading.

The absence of a coherent set of legal definitions in accountancy has been recognised as a draw back by, among others Jones (1991), Freear (1980) and Tamari (1978). Even so two methods have historically been used for forecasting company financial performance, Financial Ratios and Multi variate Discriminate Analysis.

Views vary on the applicability of ratios for prediction purposes. Jones stated that for any real conclusions to be drawn at least five years data would have to be used. Freear cited Beaver (1968) who found perhaps not surprisingly, that longer term ratios performed better than shorter term ratios, while Altman (1968) noted that "the use of financial ratios for analytical purposes is usually tempered by qualifications as to the applicability in the deductive reasoning process". However Langford, Iyagba and Kamba (1992) found that the performance of an organisation could be measured using ratios. They stated that there would be an industry average and that over a long period of time a trend for a particular organisation would emerge. It is not possible to predict what the managers, shareholders, or creditors of the company will do to improve the situation.

All in all, ratios can only produce a snapshot of a companies financial position, and cannot provide the predictive capacity required for the research. A more accurate predictor is based on the Multi variate Discriminate Analysis, of which Altman states "the use of a multi variate analysis (is) appropriate...to arrive at an over all index of a companies position".

The technique used in the research replicated Mason and Harris (1979) and is known as the Z model technique. It provided impressive results and, as importantly, used data which is readily available in each companies annual accounts.

The method, designed specifically for application to construction firms, uses published financial accounts to create a solvency profile, and that is compared to the solvency profiles of known financially stable/unstable companies. The solvency profile is summarised in a single score.

To produce the score two sets of financial ratios were used in an equation designed to give the final solvency indicator. The first set was derived from the final years accounts of twenty construction companies which had failed since 1969 and the other from the 1976 to 1979 accounts of twenty construction companies that were thought to be particularly sound on a traditional financial ratio analysis basis.

The authors then considered twenty eight discriminating variables and with the use of computers they managed to reduce these to the six with the most discriminating powers. Although the more variables used the better, or more accurate the model, for it to be of value it had to be easy and quick to apply. They then produced a function of the form

$$Z = C(0) + C(1)R(1) + C(2)R(2) + \ldots + C(N)R(N),$$

where

Z = Z score, C(0) is a constant C(1) to C(N) are the ratio weights and R(1) to R(N) are the discriminating financial ratios. The final model was of the form

$$Z = 25.4 - 51.2R(1) + 87.8R(2) - 4.8R(3) - 14.5R(4) - 9.1R(5) - 4.1R(6)$$

where

R(1) = Profit Before Interest and Tax / Opening Balance Sheet Net Assets

R(2) = Profit Before Interest and Tax / Opening Balance Sheet Net Capital Employed

R(3) = Debtors / Creditors

R(4) = Current Liabilities / Current Assets

R(5) = Log 10 days Debtors for Year

R(6) = Creditors Trend Measurements

and Days Debtors for Year $N = \frac{D(N-1) + D(N) \times 365/2}{\text{Turnover}}$

and Trend Creditors for Year $N = \frac{C(N) + C(N-1)}{2C(N-2)}$

C(N-1) = creditors for year N-1, C(N) = creditors for year N etc.

On completing their research the authors found that companies with positive Z Scores (i.e. above zero) had all remained solvent, while companies with negative Z scores had, without exception, failed. The authors do, however, qualify their findings. They admit that the Z score cannot accurately predict failure. Unhealthy firms can recover from having a "bankrupt" profile by the actions of debenture holders, bankers and creditors; "The Z score alone cannot predict failure, it only indicates that a company has a profile very similar to failed companies, and therefore has a high probability of failure".

Research Model

Organisational structure can be assumed to have an effect on the success of an organisation. Similarly there is an established link between an organisations strategy and its subsequent financial performance. It has been noted also that company strategy will influence organisational structure and vice versa. Using the variables already detailed above, these relationships can be illustrated diagrammatically below.

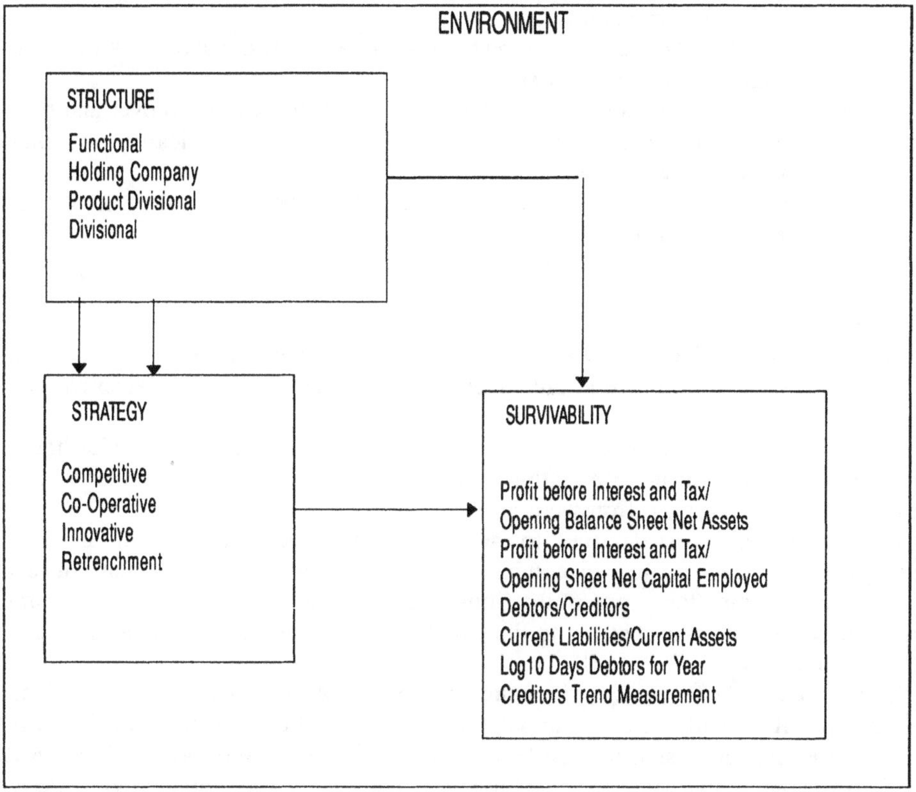

Fig 1. Research Model

The Hypothesis

For the results to be of any value they must be tested against a set of hypotheses. It can be seen from the research model, drawn for our earlier observations, that strategy and structure are to a degree dependent on each other. It has also been noted that they both affect a companies financial performance and therein its survivability. From this three major hypothesis are derived:-

1. An organisation's structure is linked to its subsequent financial performance, and therein its survivability.
2. An organisation's strategy is linked to its subsequent financial performance, and therein its survivability.
3. An organisation's structure combines with strategy to provide the best financial performance, and therein its survivability.
   As an aid to the research the link between an organisation's turnover and both its structure and survivability was examined. More specifically, two sub-hypothesis were examined:
4. An organisations turnover is linked to its subsequent financial performance, and therein its survivability.
5. An organisations structure is linked to its turnover.

The Data Collection

The data used in the study was retrieved from three sources; annual accounts, a piloted questionnaire and face to face interviews with the most senior executive in a number of firms.

The most recent available annual accounts for each firm filed at companies house were examined for relevant information. The Z model adopted meant that at least three years of accounts for each firm were utilised.

The questionnaire was designed to identify which of the four structural and four strategic variables applies to each company. A pilot questionnaire was written and sent to the managing director of one of the firms taking part in the survey. Upon its return an amended version was sent out to the twenty largest privately owned construction firms in Scotland.

It is accepted that the questionnaire method of data collection is not ideal. Organisational structure, not commonly held to be confidential or confusing, can indeed be established using this method. That being so, no matter how well planned the questionnaire it cannot be expected to record the finer nuances of strategy. Indeed, assuming that senior executives are aware of their own corporate strategy, commercial confidentiality makes it uncertain as to whether they shall wish to share that information. However, given the time available to complete the data collection and the extensive geographical spread of the companies involved, a piloted questionnaire was included.

The purpose of the interviews was to build upon the information provided in the questionnaires and to add any information which can only be provided through conversation. This was intended to corroborate / modify any findings from the questionnaires. The ten firms used in the sample comprised 19% of the turnover of the Scottish construction industry.

Method of Analysis

The first step in discerning what conclusions should be drawn is to create a null hypothesis against which to test the propositions derived from the research model. The method used to test this null hypothesis was significance testing. Given the size of the sample the method used to determine any links between the variables was the Chi Square distribution.

To determine what, if any, link that exists between structure and strategy and survivability (i.e. major hypothesis number three) it is proposed to use the Two Way Analysis of Variance with Interaction Technique.

The results of the interviews will be added to the findings of the quantitative evidence. The data provided by the managing directors will provide an accurate quantitative assessment of the relationship between strategy, structure and survivability in construction companies. The data collected has been analysed thematically and so enriches the understanding of any relationship that exists between the research variables.

Results

The finding of the questionnaire revealed that only two of the four strategic variables were being used; competitive and retrenchment. A broader range of organisational structures were given, covering Holding Company, Functional, Divisional and Functional with Subsidiaries.. The accounts revealed a range of Z scores, from -24.0 to +9.9. The exact combination of strategy, structure and survivability ratings can be seen below.

Company	Strategy	Structure	Z Score
1	Competitive	Divisional	+9.9
2	Competitive	Divisional	+6.7
3	Competitive	Divisional	+0.2
4	Competitive	Divisional	-1.9
5	Competitive	Holding Company	-2.5
6	Competitive	Functional	-3.2
7	Retrenchment	Functional	-4.7
8	Retrenchment	Holding Company	-13.0
9	Retrenchment	Functional	-13.2
10	Retrenchment	Functional with Subsidiaries	-24.0

At first glance it would be apparent that a Divisional Structure is reflected in a positive Z score and that a competitive strategy will also reflect favourably on a firms chances of survival.

The Chi square tests prove statistically that strategy has an effect on survivability. Accordingly, the first hypothesis, that an organisations structure is linked to its financial performance, and therein it survivability, is supported. More specifically that;

A divisional structure improves a firms financial performance, and therein its survivability.

Following this a test was conducted to explore the relationship between an organisations strategy and its subsequent survivability. Using a Chi square test it was possible to establish a link and, in particular, support the second hypothesis, demonstrating that;

An organisations strategy is linked to its subsequent financial performance, and therein its survivability

The next test was to try and establish statistically whether a competitive or retrenchment strategy combined with one of the four structural variables to provide the optimal survivability. The test was unable to prove any relationship, although it would appear, and only appear that a competitive strategy and divisional structure combine to provide survivability.

While it was not possible to link statistically an organisations structure, strategy and survivability it was possible to show that structure was linked to company turnover and that company turnover was linked to survivability. More specifically that;

A divisional structure is linked to a larger turnover (a turnover of more than forty million pounds was used as a break point).

and

A large turnover will lead to positive Z score and accordingly is more likely to maintain solvency.

Conclusions

The research has supported two of the main hypothesis, whilst the third cannot be validated. Furthermore the relationship between company size and both structure and survivability was upheld. It is necessary, however, to look at the choice of variables given in response to the questionnaire and relate them to the current literature studied and the views given in the in face to face interviews.

Only two of the four strategic variables were identified by the questionnaire respondents as being relevant to their firm, competitive and retrenchment. Those adopting a competitive strategy sought to pick up work across a wide range of project types and geographical location - risk spreading. The group adopting a retrenchment had an approach. They sought to adopt a focus strategy. These two strategies are typical of firms under pressure and may be caricatured as "get more work - buy it if necessary" or "lets downsize, get lean and mean and do what we know we do well". The lack of technical sophistication in the industry would mean the absence of an innovative strategy, involving "the invention of new products or methods of working" (Mansfield, 1986) would not be surprising. It is worth noting that in the interviews the only firm which said it would take on work for which it had no previous technical experience had the second lowest financial rating of the ten firms surveyed.

In construction industries competitiveness is matched by its extreme volatility, illustrated by the Government figures for output and employment. The interviews showed that the two firms which most recognised the instability in the industry were among those with the healthiest survivability ratings. This is supported by Lansleys assertion that the business environment is of "fundamental" importance to corporate strategy.

The questionnaire data demonstrated a link between the size of an organisation and the organisational structure of the firm. The simplest structural variable, functional, was not identified in the largest of the firms sampled, which would lend weight to the view that the larger the firm, the more complicated the structure, as voiced by Mansfield (op. cit.).

The link between a divisional structure and large firms was upheld and reflects the views of Harris and McCaffer that larger firms will be more divisionalised. That the link between structure and survivability was upheld is reinforced by the findings of the Aston Group (op. cit.), that organisational size matters. Of the interviews, the two firms which had implemented a specific restructuring had done so in response to a crisis, and this concurs with the view of Woodward (op. cit.) who found reorganisation to be the "result of crisis".

The study concentrated on private construction firms, public companies are bound by different accounting rules and therefore comparison would not be possible. It can be seen from the public accounts of the ten firms surveyed, eight were substantially owned by members of the board, the lowest figure being fifteen percent and that owned by the Managing Director. This may have a bearing on the attitudes and thinking of the board on matters of management and certainly reinforces the findings of Hillebrandt (1985)

The interviews revealed that almost all the managing directors had specific construction backgrounds and qualifications. This demonstrates the lack of skills transferability in construction managers which will have a direct effect on not only the "way of life" attitude to managers but on strategy, as cited by Rumelt (op. cit.).

The paper has identified the difficulties faced by construction firms during a recession. This was confirmed by the findings; of the ten firms surveyed only two are healthily of solvent, with a third firm just showing a positive score. The other seven all showed a financial profile similar to those firms which previous research identified as being at risk.

Limitations of the Research

This work has established a clear link between strategy, structure and survivability in the Scottish construction industry. Even so there are limits to the findings of the research which should be noted.

One drawback is the sample size. This has meant that because of the requirement that at least five units of data be contained in each cell, the Chi square test is rendered indicative rather than precise. Consequently with only ten samples the conclusions drawn are indicative rather than conclusive.

However, the sample size is, in part, a reflection of the industry in Scotland. There are only eight publicly quoted construction firms in Scotland (Scotland's Top 2000 Companies). The ten private firms which took part in the survey accounted for 609 million pounds of the 3073 million pound construction market in Scotland. (Source: Scottish Economic Bulletin, 1993/4, op. cit.) or some 19%. A further nine firms, representing over 7%, declined to participate in the research. To extend the search for firms in Scotland any further would have meant having to contact small firms, which may have made an accurate comparative assessment of the strategies and structures misleading. The sample was restricted to Scotland to minimise the impact of the variability of the construction environment, so important to how a firm behaves in a recession (Lansley, 1994). Given this constraint it would certainly have been preferable to interview every managing director who took part in the survey, but while all were approached, only four agreed to an interview.

Further Research

Further research would include a comprehensive analysis of construction firms. Given the limited number of firms with a substantial turnover in Scotland, the scope for the research would have to be extended. It has already been shown that Scotland's economy behaves differently to that of the UK. as a whole where the construction industry is concerned. Therefore to increase the sample size, without affecting the validity of data collected, the research should be extended to the north of England only, where the economic differences with Scotland are at a minimum. By doing so a fuller understanding of the structure, strategy and survivability in the Scottish construction industry would result.

REFERENCES

1. Altman, E., "Financial Ratios; Discrimination Analysis and The Prediction of Corporate Bankruptcy. Journal of Finance 23. 1968.
2. Beaver, W.H., "Market Prices, Financial Ratios and the Prediction of Failure", Journal of Accounting Research, 6(2), 1968.
3. Central Statistical Office, "Housing and Construction Statistics, G.B., 1981-1991", H.M.S.O., 1992
4. Central Statistical Office, "Annual Abstract of Statistics", H.M.S.O., 1993
5. Chandler,A.D., "Strategy and Structure", Boston M.I.T. Press, 1962.
6. Channon, D.F., "The Service Industries; Strategy, Structure and Financial Performance", MacMillan, 1978.
7. Collins, "Collins English Dictionary", Collins., 1991.
8. Freear, J., "The Management of Business Finance", Pitman, 1980.
9. Government Statistical Service, "Labour Force Survey Quarterly Bulletin", H.M.S.O., June, 1993.
10. Grinyer, P.H., "Systematic Strategic Planning for Construction Firms", Building Technology and Management, February, 1972.
11. Harris,F. and McCaffer,R., "Modern Construction Management", B.S.P. Professional, 1989.
12. Hillebrandt, P., "Economic Theory and the Construction Industry", MacMillan, 1985.
13. Joint Forecasting Committee for the Construction Industries, "Construction Forecasts 1993-1994-1995", Construction Forecasting and Research Ltd., Summer 1993.
14. Jones, E., "Understanding Business Finance", Pitman, 1991
15. Kinsman, F., "Managing the Future", Management Today, April, 1987.
16. Langford, D., Iyagba, R., and Kamba D., "Prediction of Solvency in Construction Companies. Journal of Construction Management and ##### 11, 1993.
17. Lansley, P.R., "Corporate Strategy and Survival in the U.K. Construction Industry", Journal of Construction Management and Economics, May, 1987.
18. Mansfield, R., "Company Strategy and Organisational Design", Croom Helm, 1986.
19. Mason, R.J. and Harris, F.C., "Predicting Company Failure in the Construction Industry", Proceedings of the Institute of Civil Engineers, May, 1979.
20. Newcombe, R., Langford, D. and Fellows, R., "Construction Management One: Organisation Systems", Mitchell,1990.
21. Pugh, D.S., and Hichson, D.J., "Organisational Structure in its Context: The Aston Programme 1", Gower Publishing, 1976.
22. Rumelt, R.P., "Strategy, Structure and Economic Performance", Harvard Business School Press, 1986.
23. Scottish Office Industry Department, "Scottish Economic Bulletin", H.M.S.O., Number 45 (Summer), 1992/3
24. Scottish Office Industry Department, "Scottish Economic Bulletin", H.M.S.O., Number 46 (Winter), 1992/3

25. Shittu, O., "Corporate Strategy, Organisation Structure, and the Concept of Fit in the Construction Industry", Unpublished Thesis, University of Bath, 1988.
26. Tamari, M., "Financial Ratios: Analysis and Prediction", Elek, 1978.
27. Weber, M., "The Protestant Ethic and the Spirit of Capitalism", Allen and Unwin, 1930.
28. Woodward,J.,"Industrial Organisation: Theory and Practice", Oxford University Press, 1965.

Part II
The firm, its market and competitors

1 The firm and its markets

MARKET SHARE OF CONTRACTORS ACCORDING TO TYPE AND SIZE OF CONTRACT

D.S. Drew and K.T. Cheung
Department of Building and Real Estate,
Hong Kong Polytechnic University, Hong Kong

Summary
This paper examines market share of contractors in Hong Kong according to different types and sizes of government civil engineering work. Successful companies are often identified by market share. Although economists commonly view markets and market share according to product type and location, definitions of the construction market indicate that it exists according to the three dimensions of (1) contract size and complexity, (2) contract type and (3) location. Contract size and complexity is regarded as an important additional dimension because of the wide range of contract sizes that exists in the construction market. It is found that Hong Kong's four civil engineering market sectors vary enormously in terms of market value and number of contracts awarded. However, such variations are not reflected in the numbers of approved contractors for each market sector. Under such conditions, in which open tendering is used, contract awards are more evenly distributed amongst competitors in those sectors which have fewer contract awards. The disparity of civil engineering contract sizes are such that, in addition to the number of contract awards, the relative size of contract awards is a major factor influencing the overall market share of contractors. Although market share can be used to some extent in determining successful contractors within the various construction market sectors, it seems limited in the sense that it does not reflect the contract size distribution on which the market share is based.
Keywords: Contract size, contract type, international contracting, market share, project size, project type

Résumé
Ce rapport examine la part de marché des entrepreneurs de Hong Kong selon les différentes catégories et tailles des travaux publics. Les sociétés qui réussissent sont souvent identifiées par leur part de marché. Bien que les économistes considèrent communément les marchés et la part de marché suivant la catégorie et la situation

The Organization and Management of Construction: Shaping theory and practice (Volume One).
Edited by D.A. Langford and A. Retik. Published in 1996 by E & FN Spon. ISBN 0 419 22230 8.

géographique du produit, les définitions du marché du bâtiment indiquent qu'il existe par rapport aux trois dimensions de (1) la complexité et la taille du contrat, (2) la catégorie du contrat et (3) sa situation géographique. La complexité et la taille d'un contrat sont considérées comme des dimensions supplémentaires importantes en raison de la grande fourchette de tailles de contrats qui existent sur le marché du bâtiment. On a trouvé que les quatre secteurs du marché du génie civil de Hong Kong varient énormément en termes de valeur marchande et de nombre de contrats signés. Cependant de telles variations ne se reflètent pas dans le nombre d'entrepreneurs agréés dans chaque secteur du marché. Dans ces conditions, où les appels d'offres sont publiques, les contrats sont distribués de manière plus uniforme entre les concurrents des secteurs qui emportent moins de contrats. Les disparités dans la taille des contrats de génie civil sont telles que en plus du nombre de contrats signés, la taille relative des contrats signés est un facteur principal qui influence l'ensemble de la part de marché des entrepreneurs. Bien que la part de marché puisse servir dans une certaine mesure à déterminer les entrepreneurs qui réussissent dans les divers secteurs du marché de la construction, elle paraît limitée dans le sens qu'elle ne reflète pas la distribution de la taille des contrats sur laquelle est basée la part de marché.

Mots-clés: taille du contrat, catégorie du contrat, travaux publics internationaux, part de marché, taille du projet, catégorie du projet.

1 Introduction

Construction work is commonly allocated to willing contractors through a process of contract bidding in which 'all the bidders supply the client with their terms and conditions in sealed envelopes which are opened at a fixed date' [1]. Bidders 'must choose a price high enough to provide sufficient contribution to overheads and profit, yet low enough to ensure that a sufficient volume of work is actually obtained ... in an environment of considerable uncertainty' [2]. This is exacerbated, in many instances, by large numbers of contractors being encouraged to bid in order to ensure a fully competitive competition [3]. In addition, the construction market within which this takes place is highly fragmented with the dominant firm being the small contractor [4].

In such a fragmented, uncertain and highly competitive environment, it is crucial for contractors to monitor their success relative to their competitors and to be aware who the market leaders are within the different sectors of the construction market. Likewise, particularly for the purposes of contractor selection, it is useful for construction clients to be able to identify the successful contractors within different market sectors. Success can be measured by determining each contractor's share of the construction market. Competition within the market place determines which competitor obtains a share of the market value. A superior competitor will obtain a larger market share while competitors with essentially equivalent products and resources should receive approximately equal shares [5]. Market share is a measure of success and is regarded [6] as being the most important indicator of the firm's degree of monopoly power.

Construction contract bidding, like all other forms of pricing, is essentially about contractors making strategic decisions in respect of which contracts to bid for and the bid levels necessary to secure them [7]. Since 'tendering strategy is, in general, affected by the type of project and value range' [8], one approach to determining the relative success of contractors is to analyse contractors' market share according to the size and type of contract. Such an analysis is undertaken in this paper, in which the relative market share of local contractors, foreign contractors and joint ventures are determined, according to contract size and type, for Hong Kong Government civil engineering work.

2 The construction market

2.1 Construction market dimensions

Hillebrandt [9] suggests that 'markets in the construction industry should ... be defined in terms of the total demand for a particular service of a certain degree of complexity and size of contract and in a geographical area which may be covered, without undue increases in costs, by firms likely to be capable of undertaking work of that type. The total number of firms interested in work of this type will be referred to as being in a particular market'. Male [4] develops this definition further by introducing the notion of a project-based vertical market defined by project size and complexity. He considers 'construction in terms of a geographically dispersed project-based vertical market that operates world wide from local to international arena, and as we go up the vertical market, defined by project size and complexity, there are fewer and fewer companies able to undertake particular types of project and fragmentation tends to decrease as the industry is segmented by overlapping project based market structures' [4]. He also suggests that it is hierarchically structured in terms of company size with fragmentation being high at the smaller end of the vertical market structure in repair and maintenance work while for new build work fragmentation decreases according to project characteristics.

Common threads in both Hillebrandt and Male's definitions of the construction market appear to be (1) contract size and complexity, (2) contract type and (3) location. With the exception of contract size and complexity these definitions appear to conform with the economists' view of the market which is seen as existing in two main dimensions: (1) product type and (2) geographic area [6]. It seems that contract size and complexity is regarded as an additional important dimension in the construction market because of the wide range of contract sizes that exists within the construction market and it is this which is the major determinant of the number of firms able to undertake the work [9]. This is supported by Male [4] who points out that 'as project size and complexity increases there are fewer companies about to undertake particular types of project - through managerial capability and access to finance. He also states 'these pose barriers to entry for particular types of project and hence arrange contracting into a project based vertical market defined by project size and complexity'.

The relative influence of the construction market dimensions is likely to vary from market to market. For example, in the Hong Kong construction market the influence of geographic area appears to be minimal. Most contractors tend to operate territory wide with the exception of undertaking work on some of Hong

Kong's more remote islands [10]. Hong Kong's construction market, therefore appears to exist according to two main market dimensions, that of contract size and type.

2.2 Market sectors within the construction market

Langford and Male [11] identify that the market is made up of four main areas, namely building, civil engineering, repairs and maintenance and materials manufacturing and that these may be sub-divided into market sectors. For example, they state that the building market is composed of housing, industrial and commercial markets.

Lansley et al [12] suggest an alternative view, that is with the exception of housing development, contractors do not consider demand in terms of market sectors, but in terms of technologies to execute project types. They found the main features that managers stated in assessing projects were project size, project complexity and construction method. In respect of construction methods, it would seem the broad divisions that occur within the construction market, in terms of type, are for building and civil engineering work and, with respect to nature of work, for new build and alteration work.

At a national level Lansley et al [12] found that there was a clear distinction between building and civil engineering by some contractors but not by others. However, for smaller projects at a regional level there was a clear distinction for building and civil engineering work. Since civil engineering projects are less homogeneous than building projects [13], this may result in greater range of market sectors.

The foregoing discussion tends to indicate that the construction market is made up of market sectors, although there appears to be a lack of consensus on the precise boundaries of these sectors. 'In defining market boundaries it is the degree of substitutability between products or services which is the key factor ... a civil engineering firm and a speculative house builder are in the same construction industry ... but in patently different markets because a bridge is no substitute for a house and vice versa' [14].

2.3 Contract type and size

In contracting the type and nature of construction work is dictated by the make up of the contract package. This is determined by the client. For example, the contract package may consist of either or both new build work and alteration work. Some contract packages may be for construction work that is more standardised in design than other contract work. The contract package may comprise one or more building types. There is also likely to be specification differences and variability in the number of units contained in the packages. These packages are also likely to vary in content (eg. the contract package may contain either or both substructure and superstructure work).

The make up of the contract package according to type and nature will in turn influence the range and distribution of contract sizes. Contract size can be measured according to area or volume. This, however, does not take account of complexity. Hillebrandt [9] states that it should be theoretically possible to measure the degree of complexity but acknowledges that the practical difficulties are great. She suggests that the nearest readily available approximation for measuring complexity is the cost per square metre. Since cost reflects complexity to some extent, an alternative readily available measure that reflects both complexity and size of the contract package is the bid price submitted by the contractor.

3 Competition framework

3.1 Setting up the competition

Common bidding systems used by clients of the construction industry to obtain bids from contractors include open tendering, selective tendering (single and two stage) and serial/continuity tendering. The choice of bidding system used by the client will influence the ease with which the contractor can enter or exit the competition. Selective tendering systems appear more restrictive than open tendering systems as the contractor can only normally bid upon receiving an invitation directly from the client or from the client's representative, whereas for open tendering the onus is on the contractor to bid by responding to an advertisement. Related to this is the extent to which contractor prequalification procedures are formalised by the client.

Since market share reflects the outcome of the competition, both the client's choice of tendering systems and contractor selection procedures, both for competition and award of contract, will ultimately influence the distribution of market share amongst the contractors.

3.2 Hong Kong Government list of approved contractors

With the exception of contracts with special requirements, the Hong Kong Government uses a system of open tendering with approved lists of contractors as the means of obtaining bids [15]. For those exceptions single stage selective tendering is used. Lump sum contracts based on bills of approximate quantities are commonly used for civil engineering work in Hong Kong.

The Hong Kong Government lists of approved contractors are divided according to whether they are local and overseas contractors, size of contract and type of work. Local and overseas contractors are divided into List I and List II respectively. The basis of the listing is according to whether the main presence and head office of the contractor is inside or outside of Hong Kong. Following an investigation of financial status and experience, local contractors are classified as Group A (permitted to tender for contracts up to HK$6 million), Group B (permitted to tender for contracts up to HK$30 million) and Group C (no contract limit). Since overseas contractors are only permitted to tender for exceeding HK$30 million, only local contractors in Group C will be in direct competition with the overseas contractors. Contractors are further divided into the construction work categories of (1) building, (2) port works, (3) roads and drainage, (4) site formation and (5) water works.

4 The Hong Kong construction market

Hong Kong's *laissez faire* economy, together with its favourable taxation system, has attracted many foreign companies to come to Hong Kong and compete for business. Competition from foreign contractors in the construction industry is no exception. Foreign contractors are permitted to compete for building and civil engineering work in both the public and private sector. According to the 1990 Hong Kong Government list of approved contractors, they are particularly dominant in the civil engineering field (58 foreign to 47 local) and out number local contractors in the construction work categories of port works (43 foreign to 15 local), roads and drainage (45 foreign to 36 local), site formation (43 foreign to 36 local) and water works (33 foreign to 18 local). Foreign contractors are also less restricted in the type of civil engineering work they bid for. Of the 105 listed civil engineering contractors, 20 foreign and 7 local contractors are eligible to bid for work in all four work categories, 18 foreign and 6 local in three categories, 10 foreign and 11 local in two categories and 23 local and 10 foreign in one category.

Walker and Rowlinson [16] identified that the number of government approved foreign contractors grew to 58 by 1988, over four times the number in 1978. They point out that what makes Hong Kong particularly attractive is that there is no limitations on the freedom with which foreign contractors can enter the market, nor are there any advantages granted to local contractors as there are in many countries, such as Singapore. Hong Kong is also favourable for foreign investment because of its low tax rate [17] and is adjacent to the large China market. The fragmented and hierarchial nature of the construction industry in Hong Kong also makes it easier for foreign contractors to enter the market as they can easily 'hire' local expertise through subcontracting. At the other end of the scale the financial risk taken by contractors on Hong Kong's mega projects can be enormous. It is not unusual, therefore, to see the formation of joint ventures [17].

4.1 Foreign and local contractors and market share

Lee and Liu [17] identified that foreign contractors enjoy a large proportion of market share (by contract value) of public sector work in Hong Kong and considered the problems of foreign contractors have in entering the local market. They define market share as 'percentage of the total contract value an individual contractor or groups of contractors acquired in a certain defined period to that of all contractors in Hong Kong in that same period'. This paper develops Lee and Liu's work by attempting to identify the market share distribution of foreign and local contractors in civil engineering work according to type and size of contract.

5 Analysis

Data were collected from Hong Kong Government Gazettes on all government civil engineering contracts let to Group C civil engineering contractors between January 1988 to May 1993. The contracts were divided into the four construction work categories of port works, roads and drainage, site formation and water works and, using government tender price indices, updated to a common base date of May 1993. Contractors were coded as 1-30 (local) and 31-47 (overseas). Joint ventures were coded JV1 to JV9. The data sample contained HK$29.46 billion worth of work.

5.1 Contract type

Of the 189 contracts in the sample, 10 were for port work (PW), 115 for roads and drainage (RD), 38 for site formation (SF) and 26 for water works (WW) representing respective market values of HK$ 1.70 billion (5.77%), HK$ 19.18 billion (65.13%), HK$ 5.62 billion (19.06%) and HK$ 2.96 billion (10.04%). Although the Hong Kong Government classifies civil engineering work into four types, nearly two-thirds of the contracts, both in terms of numbers of awards and market value, were awarded to RD contractors. Such a contrast between RD and the other construction work categories is not, however, reflected in the numbers of listed contractors which stood at 58 (PW), 81 (RD), 65 (SF) and 51 (WW) in 1990. The ratio of number of contracts awarded / number of listed contractors stands at 0.17 for PW, 1.42 for RD, 0.58 for SF and 0.51 for WW. Given that the government operates an open tendering system, from a contractor's perspective it would seem, *prima facie*, that RD is the most attractive category within which to compete.

Such differences in the number of contract awards amongst the construction work categories are reflected in the market share distribution. The bar graphs in Figure 1 shows that the 10 PW contracts were awarded to eight different contractors, 115 RD contracts to 34 contractors and 6 joint ventures, 38 SF contracts to 17 contractors and one joint venture and 26 WW contracts to 14 contractors and 2 joint ventures. The number above each bar identifies the total number of contract awards on which the market share is based. Contract awards appear more evenly distributed amongst competitors in sectors which have fewer contract awards. This is especially so for port work where no contractor has been awarded more than two contracts.

In comparing the market share distribution of local and overseas contractors according to construction work category, the RD market shares appear to be quite evenly spread between local and overseas contractors. Although an overseas contractor and joint venture obtained the largest and second largest market share, local contractors obtained the third, fourth and fifth largest market shares. It can be seen, however, that overseas contractors have obtained the larger market shares on both SF and WW contracts. For PW it is not possible to sensibly comment on the market share distribution because of relatively few contract awards.

5.2 Contract size

89 contracts were awarded to 30 local contractors, 87 contracts to 17 overseas contractors and 13 contracts to 9 joint ventures. Local contractors, overseas contractors and joint ventures obtained market shares of 38.94%, 44.43% and 16.62% respectively. Their respective average contract sizes were HK$129 million, HK$150 million and HK$377 million. It is interesting to note that although there are more approved overseas civil engineering contractors than local contractors on government lists, more contract awards were made to a greater number of local contractors. However, because of the larger average contract size, foreign contractors ended up with a larger overall market share.

One limitation of using market share as an indicator of successful contractors is that it does not reflect the number of awards and contract size success range. For example, for RD contracts, contractor coded 31 has obtained the largest market share of 11.99 % from 13 contract awards. Yet JV2 obtained the second largest RD market share of 8.54 %, from only two contract awards.

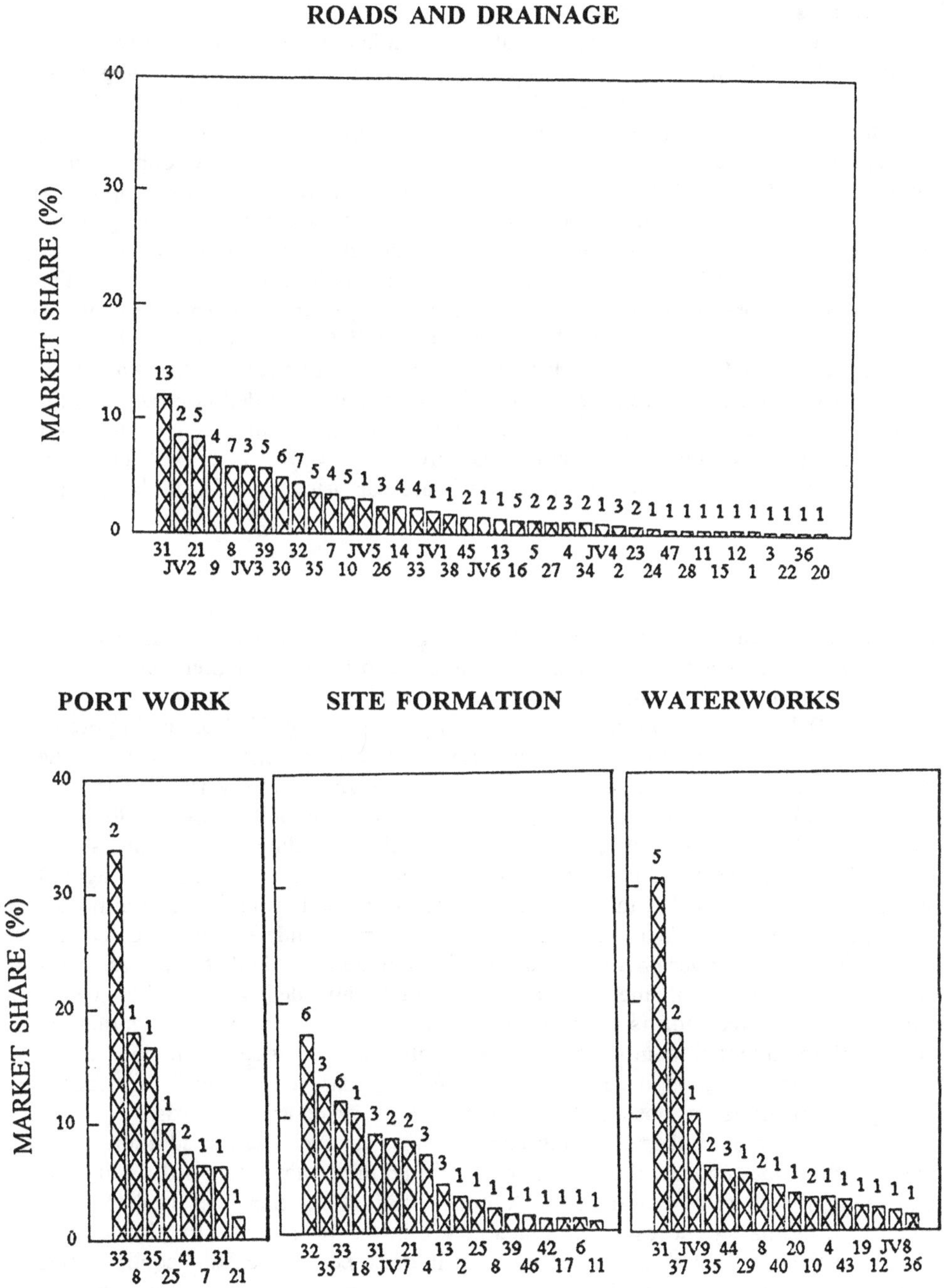

Fig. 1. Firms arranged in order of their market share

5.3 Successful contractors

Given that a dominant firm is one that attains a market share greater than 40% [6], no single firm dominates any of the four civil engineering work categories. However, clear leaders have emerged in each of the four work categories. All of these are classed as foreign contractors.

Foreign contractor coded 31 is by far the most successful Group C civil engineering contractor obtaining an overall market share of 12.84% made up of 22 contract awards. This contractor has not only obtained work in every civil engineering market sector, but has also obtained the largest market share in two of the four civil engineering market sectors. Closest rivals in terms of overall market share are contractors coded 21 (7.04% from 8 contract awards), 35 (6.24% from 11 contract awards), 32 (6.16% from 13 awards) and 8 (5.61% from 11 awards). All of these are RD contractors of which three are classed as overseas and two as local.

Each of the nine joint venture partnerships were only awarded contracts within one of the four work categories. The maximum number of contract awards given to a single joint venture partnership was three (ie. JV3). JV2 was the most successful joint venture in terms of market share capturing 5.55 % of the overall market.

6 Conclusions

Economists' commonly view markets according to two principal dimensions, that of product type and location. Definitions of the construction market indicate that it exists according to three dimensions; (1) contract size and complexity (2) contract type and (3) location. Contract size and complexity is regarded as an important additional dimension because of the wide range of contract sizes that exists in the construction market. The relative influence of these dimensions vary from market to market. In Hong Kong, for example, where many contractors operate territory wide, the construction market appears to largely exist according to two dimensions, that of (1) contract size and complexity and (2) contract type. It is generally agreed amongst commentators that the construction market is made up of sectors although there appears to be a lack of consensus on the precise boundaries of these sectors.

Hong Kong Government's four civil engineering market sectors vary enormously in terms of the number of contracts awarded and market value. However, such a variation is not reflected in the numbers of approved contractors, and therefore potential competitors, within each sector. Under such conditions, where open tendering is used, contract awards appear more evenly distributed amongst competitors in those sectors which have fewer contract awards.

The disparity of civil engineering contract sizes are such that, in addition to the number of contract awards, the size of contract is an important factor influencing the relative market share of contractors. Although market share can be used to some extent in determining successful contractors within the various construction market sectors, it seems limited in the sense that it does not reflect the contract size distribution on which the market share is based. As a result a contractor who has been more successful on many small contracts may actually end up with a smaller market share than a contractor who has been successful on just a few large contracts.

7 References

1. King, M. and Mercer, A. (1988) Recurrent competitive bidding, *European Journal of Operational Research*, Vol. 33, pp. 2-16.
2. Douglas, E.J. (1989) *Managerial economics: analysis and strategy*, Prentice-Hall International, Englewood Cliffs, New Jersey.
3. Drew, D.S. and Skitmore, R.M. (1993) Prequalification and C-competitiveness, *OMEGA International Journal of Management Science*, Vol. 21, pp. 363-75.
4. Male, S. (1991) Strategic management in construction: conceptual foundations, In Male S. and Stocks R. (eds.), *Competitive advantage in construction*, Butterworth - Heinemann Ltd., Oxford, pp. 5-44.
5. Czepiel, J.A. (1992) *Competitive marketing strategy*, Prentice-Hall International Inc., Englewood Cliffs, New Jersey.
6. Shepherd, W.G. (1990) *The economics of industrial organization*, Prentice-Hall International, Englewood Cliffs, New Jersey.
7. Skitmore, R.M. (1989) *Contract Bidding in Construction*, Longman, Harlow.
8. Flanagan, R. and Norman, G. (1982) An examination of the tendering pattern of individual building contractors, *Building Technology and Management* Vol. 28 (April), pp. 25-28.
9. Hillebrandt, P.M. (1985) *Economic theory and the construction industry*, 2nd edition, Macmillan, Basingstoke.
10. Davis, Langdon and Seah International (1994) *Asia Pacific construction costs handbook*, E & FN Spon, London.
11. Langford, D.A. and Male, S. (1991) *Strategic management in construction*, Gower, Aldershot.
12. Lansley, P. Quince, T. and Lea, E. (1979) *Flexibility and efficiency in construction management*, The final Report on a research project with financial support of the DoE. Building Industry Group, Ashbridge Management Research Unit (unpublished).
13. Raftery, J. (1994) *Risk analysis in project management*, E&FN Spon, London.
14. Newcombe, R., Langford, D. and Fellows R. (1990) *Construction Management 1 : Organisation Systems*, Mitchell, London.
15. Hong Kong Government (1980) *Rules for the administration of the Public Works Department list of approved contractors*, Hong Kong Government.
16. Walker, A. and Rowlinson, S.M. (1990) *The building of Hong Kong*, Hong Kong University Press.
17. Lee and Liu (1993) Foreign contractors in Hong Kong, *Organisation and management of construction - the way forward*, International Council of Construction Research Studies and Documentation, CIB W-65 Port of Spain, Trinidad, Vol. 3, pp. 1299-1310.

MARKETING PLAN FOR INTRODUCING A CONSTRUCTION MANAGEMENT SERVICE

G. Šarić and V. Luketa
CM-EXPERT, Zagreb, Croatia
M. Katavić
Faculty of Civil Engineering, Department of Construction Management, University of Zagreb, Zagreb, Croatia

Abstract
The aim of the marketing plan presented in this paper is to introduce a new service to the emerging class of investors in Croatia, interested in optimal execution of construction projects. In order to accomplish this challenging task CM-Expert has evolved a marketing plan in which aims and objectives are set, strategies and tactics for their achievement defined and activities needed for implementation planned. In addition a set of controls is provided for monitoring the execution of the plan keeping it always turn to set the objectives.
Keywords : Marketing plan, construction management, procurement,

Resume
Le but du plan de marketing presente dans ce travaile tend a l'introduction du nouveau service pour la classe des investisseurs recemment apparus en Croatie, interesses a l'execution optimale des projets de construction. Afin d'effectuer cette tache, CM-Expert a developpe tel plan de marketing que tous les buts et les objectifs y sont etablis, toutes les strategies et les tactiques y sont defines et que toutes les activites necessaires pour l'application sont bien projetees. En plus, le systeme des controles est envisage afin de bien observer l'execution du plan pour qu'on puisse toujour atteindre ses objectifs.

1. Introduction

The Croatian construction industry market has during the last ten years been characterised by a constant decline in demand, which recorded a serious downturn at

The Organization and Management of Construction: Shaping theory and practice (Volume One).
Edited by D.A. Langford and A. Retik. Published in 1996 by E & FN Spon. ISBN 0 419 22230 8.

the beginning of the nineties, when the war against Croatia started. Until the 1980s the construction industry accounted for between 10% and 12 % of Croatian GDP . In 1993, as a direct result of the war, the construction industry accounted for only 3,3%. From 1980 to 1990 total investments in the Croatian economy fell by approx. 50 % (from 4,5 billions US $ in 1980 to 2,3 billion US $ in 1990).The already deteriorating situation got worst when war started so that in 1992 total investments in the economy were only 1 billion US $. Throughout the observed period the share of the construction industry in overall investments was between 40 and 50%.

The economic situation in Croatia has been improving for the last two or more years, and thus the investment in the Croatian economy is slowly but surely rising as is the value of construction work, as illustrated in Table 1 [2].

Table 1. Construction industry output in Croatia (bn DM)

Investment category	1992	1993	1994	1995
Total investments in the economy	2.0	2.0	2.1	2.5
Construction industry	1.0	0.8	1.0	1.2

The construction industry market in Croatia can be divided into six segments. Their size and trends are shown in Figure 1.[2]

Figure 2 - CONSTRUCTION INDUSTRY MARKET TECHNICAL SEGMENTATION

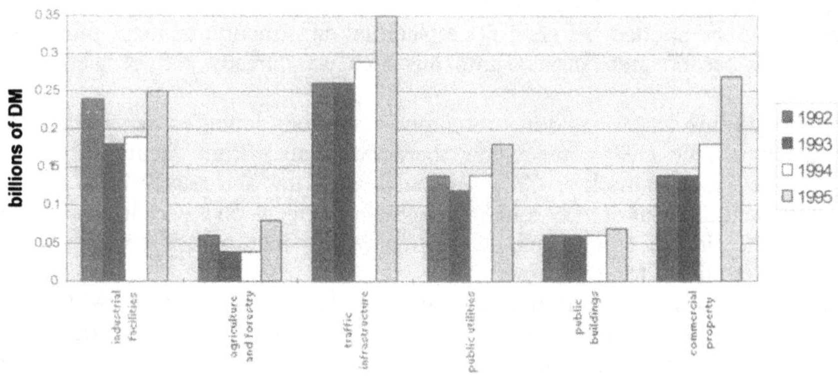

Even though the CM-Expert current service mix is spread across all market segments, the new service is in the short-term targeted only to the commercial property segment which has greater need for the new service, coming from the clients driven by profit objectives. It also has the highest growth rate. Segment size in 1995 accounts for some 270 millions DM. Main participants in the segment are private developers (housing projects, office buildings and tourism) and banks. Pension funds and real estate investment trusts are not yet fully established in Croatia for project development purposes. In the long-term they may be expected to emerge and also foreign investors who are still reluctant to invest in Croatia due to the unsettled political situation.

CM-Expert is a privately owned company with profit objectives, serving the national construction industry sector in Project management consulting services. Its generic strategy of service differentiation, accompanied by total commitment to the client's needs is the key success factor, by which the company can gain considerable competitive advantage in its service niches. This will result in increased loyalty of the company's clients and the favourable word of mouth recommendations that follow the company. One of the benefits of this approach has been the emergence of opportunities for the introduction of new services, requested directly by clients.

This developement emerged when a particular Client requested the company to help in choosing the most suitable contractor in Croatia for the delicate construction project he was developing.

His need actually triggered the development of a new service in the company's service mix. This was the service of construction contract procurement which includes elaboration of methodology, preparation of necessary documents and management of the process.

For proper development of the new service, it was necessary to make an appropriate Marketing plan. This paper contains the basics of such a plan.

2. Marketing plan for the new service

2.1. Identifying needs for the new service

The company had spotted the need for a National construction contract procurement service much earlier, and entrance into this area was already part of plans for the future.

Even though this service is quite common in western countries, where the market segment has to face great competition, there was a significant vacuum around this service in the Croatian market. Croatian construction law and related laws and codes of practice treat this matter very roughly. Actual practice is on a very low level, being always on the side of contractors, who often write conditions of contract and just submit them to clients for acceptance.

This situation was the rule in the old days of socially owned entities with no clear profit objectives. However, recent restructuring and privatisation of Croatian economy changed this situation. Some clients have became very decided in their efforts to finish the project according to their own objectives. They insist on a contract procurement procedure that gives every bidder equal transparent methodology to fit in and thus an equal chance. This way, the client will be sure that he filters out the best contractor to fulfil his objectives. Private developers and banks prevail among such objectively oriented clients and they will be targeted by this plan.

The political and economic turmoil of the nineties caused a dramatic downturn in demand for construction services, and market positions of many construction companies (contractors) deteriorated significantly. In this situation it was difficult to decide which companies were capable of executing construction works according to client's objectives of scope, quality, schedule and costs. Therefore, introduction of adequate pre-qualification and selective tendering procedures was inevitable.

CM-Expert having a thorough knowledge base, acquired in multiple international tendering procedures and construction contracts, could easily master the new service. Being aware of such an opportunity, it was important to strike first and gain an appropriate market share and competitive advantage in this service niche.

2.2. Mission, objectives and strategy

The company's aim is to provide top-quality consulting services for clients in construction management business (investors or contractors), by protecting the interests of the project and interests of the client, on a professional and independent basis.

In other words, the company must contribute in its service niches to *'implement the construction project within the client's objectives of schedule, quality and costs' [3]* and its own profit objectives.

The aim of this marketing plan is the development of a new service - National Construction Contract Procurement, for a newly emerging class of clients in Croatia that are interested in execution of construction projects within their objectives.

The planning horizon for the development and application of the new service was one year. This corresponded with the fiscal year of 1995.

In that year, the following objectives regarding introduction of the new service were planned:

- extra revenue based on the introduction of the new service of 250,000 DM;
- market leadership in the new service having a share of more than 20 %;
- augmented market share in connected consulting services due to benefits gained from the introduction of the new service (synergy effects);

If we suppose that the new service accounts for some 0.4 % of the average contract value, we come to the real segment size of interest to consultants of approx. 1.08 million DM.

This extra revenue created by introduction of the new service of 250,000 DM accounts for some 23 % of market share. It is expected that the company will become market leader in the first year, but also that it will be followed by others (about 3 consulting houses capable of entering the niche).

The strategy chosen in this case was to focus on the still unoccupied market niche in Croatia, conquer it through the development of the new service and further exploit it.

Marketing strategy centred on development of the new service in existing markets (domestic commercial property market).

Tactics for the strategy realisation will be explained through the seven P-s of marketing mix, defined by Kotler as *'the set of marketing tools that the firm uses to pursue its marketing objectives in the target market' [4]*.

2.3. Marketing mix

2.3.1. Product (service)

Specification of the service is given in the self explanatory overview in Appendix A. This shows six groups of activities, five out of which are in the strict domain of our company, covering handling of bidding, contract negotiations and procurement activities:

- pre-qualification activities
- preparation of tender documents
- pre-tender activities
- tendering activities
- preparation of software support

The first four groups represent core activities, for which adequate software support has been developed to ensure their effectiveness.

Engineering activities are elaborated by a separate organisation, but they are closely related to the company's activities, because some of them must be compiled in tender documentation (such as drawings, bills of quantities etc.)

In the pre-qualification process, the most important job has been to create forms that will ensure easy input of 'fair and true' data from candidates and provide a reliable basis for preparation of software that will evaluate candidates in the same 'fair and true' way. A key point in both activities is to agree appraisal criteria and minimum 'pass' thresholds for candidates with the Client, and to pursue what is agreed up to the end of the pre-qualification process. This way, the professional and independent approach mentioned as a the company goal is best preserved.

Regarding tender documents, two documents are of paramount importance. In the first place general conditions in which all possible contractual events between client and contractor are stipulated in a concise and unambiguous way. In order to achieve this, experience from international practice was consulted and compared with local conditions. This has greatly contributed to the competitiveness of the service. Finished general conditions were also presented to legal experts and scholars in Croatia for review and comment.

The second important document is the appendix to the bid, in which the bidder shows his capability to perform works within the Client's objectives. In this document the bidder presents its organisation, quality system, programme of the execution of work, description of preliminary work, curricula, etc. It has to be elaborated very clearly to avoid any ambiguity in input data.

The most important part of the tendering process is the comparative analysis. It is the apex of the entire process, in which the transparency of the process must be reflected. Sophisticated software support is elaborated for appraisal of bidders in all relevant elements, such as contract price, values from bills of quantities, validity period of the bid, suggested completion time, data from appendix to bid, changes in general conditions through special conditions, advance payment amount, etc. Here, the same approach shown in the pre-qualification process is applied, i.e., all criteria for the bid's evaluation and appraisal are previously agreed with the client and pursued through to the award of the contract.

Care for the variety of contract types that can be elaborated for the process is also kept in mind, because it can be a unit price contract, or a fixed price contract, or a combination of these two.

2.3.2. Price

Regarding price, the company could not apply recognised international rates for its services. In the first place, local conditions cannot be compared with international. Normal prices may be applied later, in case of successful development of the Croatian market.

The payment method is favourable for the client. The company only takes a percentage of the service charged value at the time of the pre-qualification, elaboration of tender documents and negotiations. The rest is paid during the progess of work according to the contract.

This way, the client is spared from spending large amounts of cash before the start of work. An the same time the company's loan to the client has not endangered company's liquidity.

2.3.3. Promotion

Promotion of the new service started parallel with its development. However, we must bear a number of things in mind regarding promotion. The majority of company's consulting services are characterised by a high level of credence qualities for the customer, which introduce an element of risk in his decision to engage a particular consultant. Credence qualities are '*attributes which consumers may find impossible to evaluate even after purchase and consumption*' [5].

To diminish the risk for the consumer (the Client), the company has made efforts to make the service more tangible. Completed tenders for a certain project, are attractively packaged and presented, and that way the Client can have a closer look into a new service.

This will especially be the case, if the consultant has more accompanying 'tangibilizers', such as reference letters from satisfied clients, opinions from experts on contract law or excerpts from conferences where the new service was presented and discussed by experts.

A leaflet with all necessary information concerning new service, which can be mailed directly to all interested clients is also very important in promoting the new service.

2.3.4. Place

The company's target customers for the new service are approached directly by the company's managers, with no middlemen in the process. The contact with future clients is most likely to be executed in two stages. First providing information about the new service, either by direct mail or by word of mouth. Second in detailed presentation of the new service and experience gained from former projects. Consulting services must be done at the Client's premises, but all paperwork will be carried out in Consultant's office in order to save the Client from unnecessary material expenses.

2.3.5. Process

One of the characteristics of the service is its variability. It is hard to maintain the same level of service quality on all projects, especially if the service providers are not the same. The company should aim to standardise the quality of process in order to combat this variability.

In order to do this, written procedures have been introduced and compiled in a Quality Manual, in which the whole tendering process is described step by step, with discussion of all possible situations that may occur between parties in the process. All employees are made familiar with these procedures. This was achieved through selection of employees fit for consulting in the new service and their continual in-service training. This is a part of internal marketing that is also being developed in the company.

Written procedures rely on a specific section of tender documentation, i.e., instructions to bidders, which describes the entire tendering process in detail.

Since, the process is supported and enhanced by adequate software, this has improved its effectiveness and shortened precious pre-contract time for the client. For example, any change in bills of quantities or in some other condition that took place after appraisal of the bidder, can be processed in a few minutes showing new results, such as a new contract price or a new evaluation of the bider. This has improved the client's position during negotiations with bidders.

2.3.6. Physical evidence

We have already mentioned the way tender documents must be packaged. Well presented packaging of books of tender documents in quality bindings will be persuasive not only to the client, but also to contractors, which should pay for it in order to approach bidding process.

The premises in which pre-tender meeting and tender buy-out procedure take place must be attractively furnished, with designeted sitting places for all participants, and organised secretarial services. Coffee and light beverages are available in the meeting breaks, as well as a meal after a session. The same is valid for the bids opening meeting and meetings for negotiations after bids opening. All consultant's staff should be in suits, perhaps with company ties, in order to be distinct from other participants.

An overhead projector is used to explanation the project and tendering procedure, as well as large sketching paper boards for additional clarification.

The consultant will help the client in the organisation of correspondence with bidders during the process. It embraces the design of the project's logo and letterheads, as well as writing the drafts of all letters.

2.3.7. People

People present the most important part of the marketing mix for the new service. They provide the service for the client and take pains to give them expected value for money. We have said before that a sophisticated process methodology and powerful software support greatly contribute to the differentiation of a company's service, but it is people who deliver a service and give the final touch to its quality, creating an image that will ensure that the customer's perceived value of service matches his expectations.

People are responsible for covering any misunderstanding that may arise during various relations between participants in the contract procurement process and endanger the quality of the service. Everyone involved in the process must be aware of the possibility of such misunderstandings, and trained how to overcom them. This is part of the company's culture and job of its internal marketing.

In order to have select people, careful selection of employees capable of performing the new service is carried out. They are all experienced in multiple contract procurement jobs, and they are all acquainted with project management and computer skills. Since the Croatian market lacks such professionals, training was organised within the company, where candidates were given all needed instructions in order to successfully follow the team leader during the application of the new service.

For the first company job in the new service, the task force approach was used. The team leader responsible for accelerated development and application of the service was the company's executive director. Three project managers were assigned to him, and he got all needed backup from the company.

In any new projects, team members should show commitment to the job and to the client's objectives. The team leader is responsible for public relations. He will make all necessary contacts with the client and other participants in the process, and will be spokesman of the team and the company on all formal or informal meetings.

2.3. Timetable, budget, controls

A detailed timetable of the development of the new service was made using adequate CPM software. Pre-qualification process, preparation of tender documents and pre-tender activities were planned to finish in 1994 in order to start bidding procedure in the beginning of 1995. The tendering process was planned to finish in February 1995. After successful completion of the first contract, three new contracts were planned in 1995. Two new contracts for connected consulting services were also planned in 1995. Achieving such plan, objectives mentioned in the beginning of the marketing plan will be achieved.

The detailed budget for the development of the new service amounted to 166,500 DM. Its structure is shown in Figure 3.

A set of controls was established in order to monitor the execution of the marketing plan, and to take corrective action if needed. Actually, three sets of control activities existed in this case.

The first set is connected with development dynamics. Progress of all activities, measured in time units, must be regularly recorded and compared to the original plan in order to pinpoint delays which might endanger its execution. Using CPM techniques, actual start and finish dates and remaining duration or percentage of completion of activities are entered and the plan is recalculated to show progress status. Progress information is crucial because deadlines of most activities will be contracted with the client and the company should be penalised for all delays in own default.

The second set was connected with the budget. All planned expenses were compared with actual expenses in order to find out if any deviations had occured. This is a standard part of the company's cost control procedure.

Figure 3 - THE STRUCTURE OF THE BUDGET FOR THE DEVELOPMENT OF THE NEW SERVICE

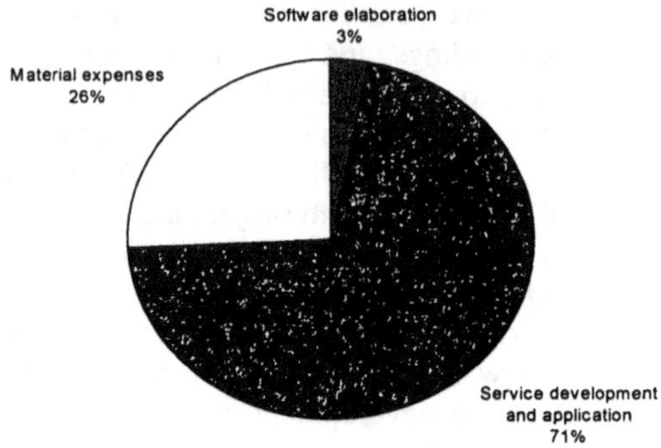

Finally, a third set of controls deals with the client's perceived value of the company's service. It may take the form of questionnaires in various phases of tendering, or it may be informal in day to day contacts with the client. However, the company's success in this area will be materialised through the client's reference letter.

3. Conclusion

If company succeeds in the execution of the marketing plan for the development of the new service, it will not only satisfy its clients, but it will also attain a market leader position in this market niche, solid extra revenue due to the new service and augmented revenues in connected consulting services.

Finally, the Croatian construction market should also benefit from its successful accomplishment, because one important segment of the project development process that is not yet fully covered in Croatia will be implemented.

4. References

1. Đukan, P. (1993) Javne investicije kao poticaj gospodarskog preporoda, *Sabor hrvatskih graditelja 1993. Proceeding,* Crikvenica, pp.15-31.
2. State Institute for Macroeconomic Analysis and Forecasting. (1995) *Internal material,* Zagreb.
3. FIDIC. (1980) *International General Rules of Agreement Between Client & Consulting Engineer for Project Management,* IGRA 1980 PM, Hague.
4. Kotler, P. (1991) *Marketing Management, Analysis, Planning, Implementation and control,* Prentice-Hall International Editions, Chapter 3, pp. 68.
5. Rushton, A.M. and Carson, D. (199) The Marketing of Services: Managing the Intangibles. *European Journal of Marketing,* Vol.23 No.2, pp. 30.

APPENDIX 1 - SPECIFICATION OF ACTIVITIES FOR THE NEW SERVICE - DYNAMIC OVERVIEW

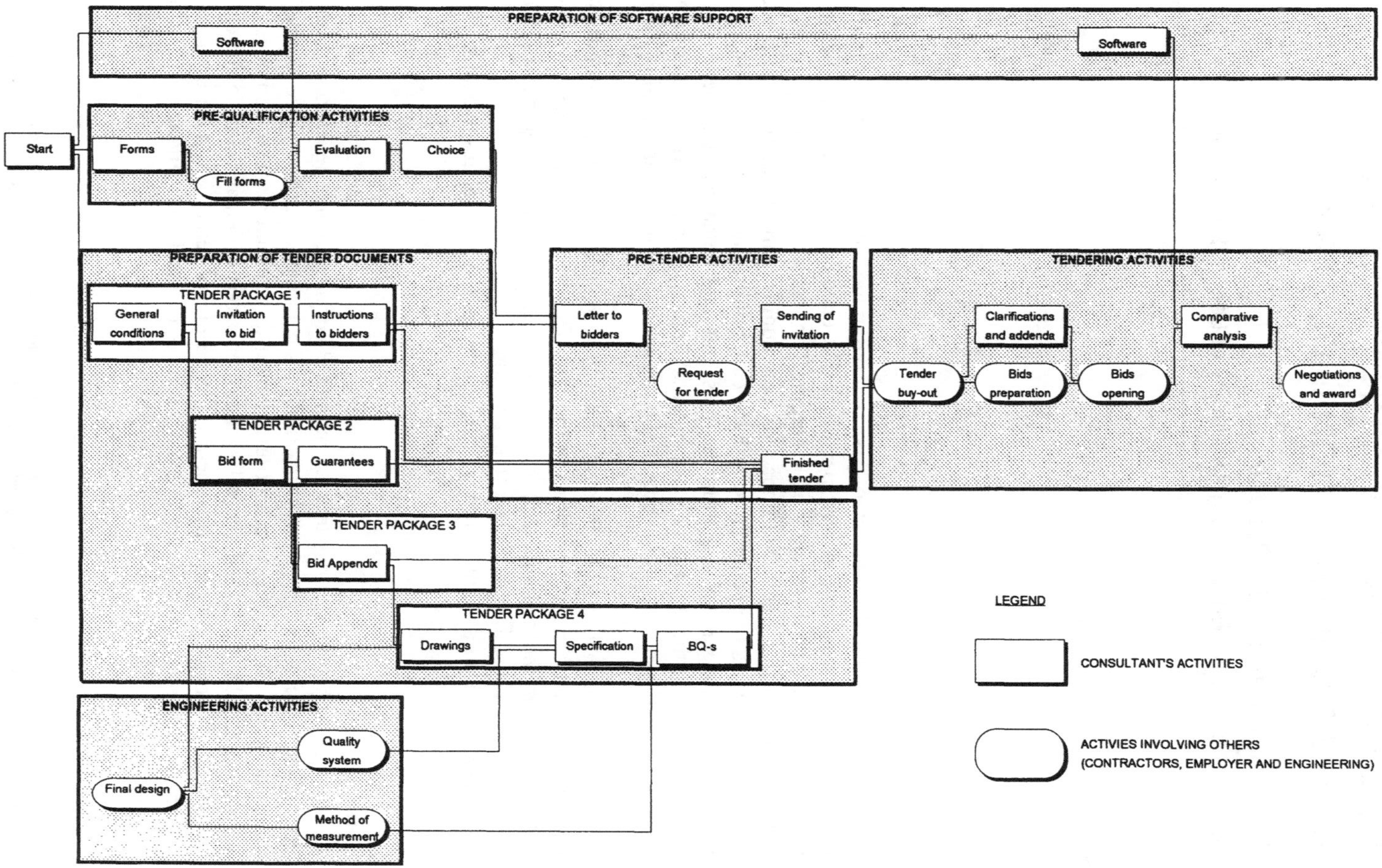

IMPLEMENTING COMMUNITY POLICIES IN THE CONSTRUCTION INDUSTRY

K. Moodley and C.N. Preece
Department of Civil Engineering, University of Leeds, Leeds, UK

Abstract
The 90's have seen the growth of social programmes and awareness of corporate responsibility by business. A new paradigm that encourages business to play a long term leadership role in society is emerging.The construction industry is a major industry which has a significant impact on communities through its physical processes and products. It has the potential to reshape communities and their environments. The industry is no longer isolated from the pressures and demands of society. There is greater scrutiny of the social and environmental impact of construction.The community and the industry can share in the potential benefits from a closer relationship.

The aim of this paper is to illustrate how an effective community policy can be developed within construction firms in the UK. This paper will provide a framework that allows for the implementation of a community policy that allows for the development of sustainable competitive advantage.
Key Words: Corporate philanthropy, community policy, social responsibility, social impact.

Sommaire
L'industrie du batiment et des travaux publics BTP joue ici un role important car c'est une industrie qui a un impact significatif sur les communes par son procede physiques et ses produits. Elle a effectivement le potentiel de restructurer les communes et leur environment. Elle n'est done plus a l'abri des pressions et demandes de la societe. Son impact sur la societe et l'environement est done nueux cerne ce qui lui permmattiait, guace a une relation plus forte avec le milieu social de beneficier d'eventuels benefices.

The Organization and Management of Construction: Shaping theory and practice (Volume One).
Edited by D.A. Langford and A. Retik. Published in 1996 by E & FN Spon. ISBN 0 419 22230 8.

Le but de cet article est de montier comment une politique de la commune pouvait etie developper par le BTP Britanique. Cet article construit donc un schema structurel que pernet la mise en placed'une politique de la commune, tout en developpant les avantages competitifs de l'industrie.

1 Introduction

Society is changing and looking at new ways of examining organisational behaviour. The socio-political lobby is becoming more influential and sophisticated. Community politics are becoming increasingly influential in shaping local decision making. Consequently greater accountability and community involvement is demanded from business by society. Business and development activity have the power to shape and alter the way society evolves through its actions and direction of resources. It has a far reaching influence over social order hence the greater scrutiny of business by society.

The construction industry is not different in its impact on society. It has the same power to shape and alter the way communities develop. Unlike many other industries it has the ability to have a more profound impact through its end products. The products of other industries are consumed and do not have a long term presence but the impact of construction can be seen all around us. Construction is a major employer and is used as a vehicle for economic and social development.

2. Corporate Philanthropy

The social compact between business and the community has a fairly established history.[1] The tradition of benevolence and social care in the UK from industry dates back to the activities of Cadbury and Rowntree which tended to be steeped in Quaker philosophy. The tradition was that business and industry had a role to play in the social welfare of its employees and the community on which the business was founded. The existence and history of social benefactors from industry has been part of industrial history.

There has also been another school of thought that believes 'the business of business is business'. This philosophy is based on the work of the economist Friedman. It follows the view that business should concentrate on what it is good at and not become involved in any social engineering. The issues and problems that society faced were not a concern of business and were to be dealt with by the state and society. These views still have strong support from many who follow a market driven philosophy

Most businesses tend to fall into these models. Donations to charity are the most clearly visible activity in modern corporate philanthropy. This is a minimalist approach that devolves the needs for 'corporate citizenship' to charities and associated organisation. In an age in which there is a growing demand for industry to do more are, these existing strategies adequate?

3. The New Philanthropic Paradigm

The latter half of the 1980's and early 1990's saw the larger part of the developed western economies go through a recessionary period. Downsizing, retrenchment and restructuring became the principal activities in the drive to reduce costs, improve efficiency and become more competitive. In an environment that was associated with cost cutting and laying off workers it was difficult to justify corporate giving. A new approach was needed that tied corporate philanthropy to strategy.

The new philanthropic paradigm encourages companies to tie their philanthropy to their strategy. The business and welfare units join forces to develop corporate strategies that will give the organisation a competitive edge.[2]

A criticism of this model may be that it serves to generate an array of programs that are self serving to the business. It is possible that this may be the case if these initiatives are purely public relations stunts without effecting any real change. This new paradigm asks firms to adopt a greater leadership role in the society it operates in by taking a long term approach to social responsibility.

4. The need for a focus - the Exxon-Valdez case

It was generally taken as accepted that society was served by three sectors; business, government, and non-profit. Each was permitted to do what it was best at and not be involved in another sector. The non-profit sector in particular described itself as independent despite funding from the private sector and government. Private sector funding tended to be about their favourite causes rather than activities associated with their line of business. This approach has widespread acceptance but the new philanthropic paradigm suggests a more directed and focused view.

The Exxon-Valdez oil spill case highlighted the need for organisations to develop a focus in their activity. The Exxon foundation was particularly well known for its educational activities and for being distant from the activities of its parent. Its activities were directed without consideration of the parent firm's activities such as building relationships with environmental groups and its consumers. Flaws in this philanthropic philosophy were not exposed until the Exxon Valdez spill in 1989. Exxon adopted a reactive strategy after the spill as it had nowhere to turn for advice on handling the crisis. This allowed Exxon to become an easy target for the anger of environmental groups. In effect Exxon had no relationships with environmental groups and insulated the company from feedback from important interest groups.[3]

5. Adopting a Stakeholder Outlook

The adoption of a more philanthropic outlook to the organisation focuses on the adoption of a total stakeholder approach to strategy. Cannon [4]focuses on the need for organisations to take cognisance of the all the stakeholders of the company. By adopting this approach greater attention can be given to external environmental influences and to people, communities and societies that are present throughout the lifecycle of a company's development and its products. A shift in emphasis means that issues outside the market also gain support thereby allowing the company to prosper in the total environment. The adoption of the stakeholder approach provides the base to justify the application of a new philanthropic paradigm.

6. Construction and the Community

The construction industry has often faced hostility over its activity. The growing concern in developed countries is the protection of the environment on which development is likely to encroach. In the UK, the major battleground between the industry and community is on roads and motorway development and how the projects degrade the environment. This confrontation pits the industry against a variety of interest groups and communities. These confrontations are by no means isolated and extend to other types of construction and development projects. Can a new focused approach that takes cognisance of the concern of people be more successful than the current approaches that often ignore concerns of society in general?

In order to function more effectively the industry needs to consider different strategies when interacting with the external environment. The construction industry is inextricable linked to the communities within which they operate through the planning processes, labour, resources, production processes and the final finished products. As communities become more sophisticated they acquire the means to disrupt, delay or ultimately stop projects from achieving completion or conversely help expedite them. It would therefore be logical acquire the support of local communities to ensure projects are completed satisfactorily.

Clients with a social remit require greater social input from their contractors and consultants. A structured approach to social policy is required in order to meet this need and provide viable deliverables. A more focused community strategy to develop closer relationships with target communities, would therefore seem ideal to help in the implementation of construction activity.

7. Community Policies

The development of a community strategy is concerned with the orientation of the company to a key external environment influences. Corporate involvement in the community are not only geographically and project specific but requires a long term outlook that requires a philosophy that is contrary to the short term approach of construction projects. This would suggest that the strategy needs to operate at two levels, a project specific level and an industry/environmental level.

The Industry/Environmental Strategy - this strategy is aimed at issues that influence the firm irrespective of the project or location. The community policy aims to support activities that are of strategic interest to the firm, as well as its norms and values. The other influential factors are general social, economic, political and environmental issues. This part of the emerging strategy is concerned with the long term approach needed to make community policy meaningful.

This part of the strategy is easily implemented because it relates more directly to the corporate ethos of the firm. The values a firm adopts will provide the springboard to community related activity. A company committed to education and training of its workforce would be able to translate these activities into a community policy. A literacy initiative, skills development programme, etc may follow as a community strategy. The community policy meets the needs of corporate goals.

Project Strategy- this part of the strategy is aimed at local and project concerns. Ideally local policies fit into the industry wide strategy but sufficient flexibility has to be built in to allow implementation at a project level to be meaningful. Local economics, local values, local politics and local infrastructure play an important part in community strategy at a project level.

Fig 1

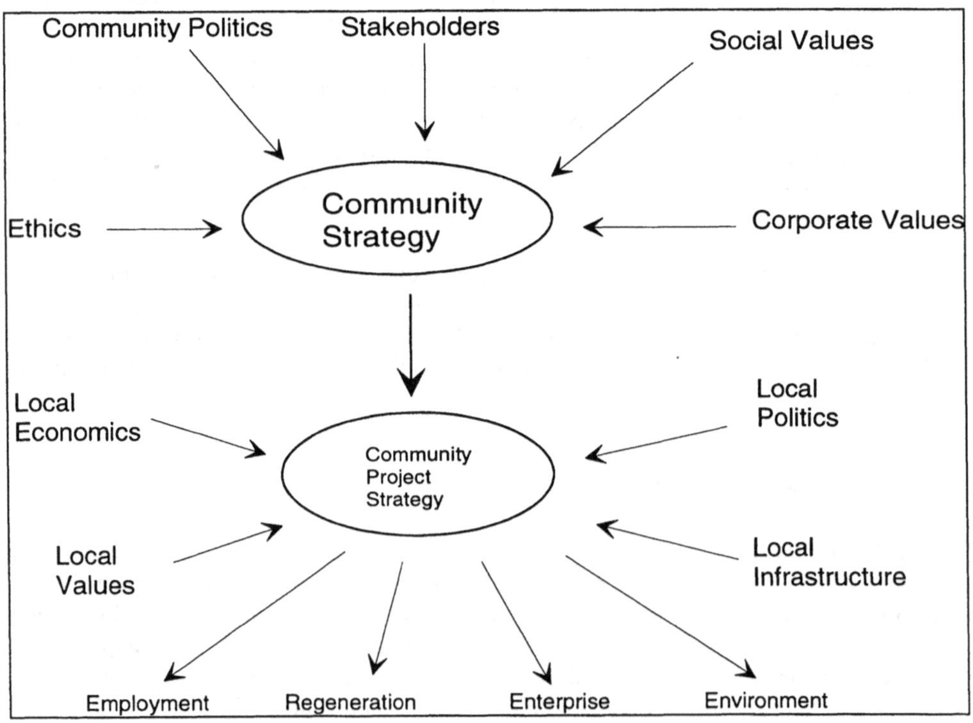

8. Implementing the Community Policy

8.1 Goal Setting

The starting point for the community policy is vested in the corporate goals and values of the company. The strategic objectives of the company should include a clear altruistic commitment. All subsequent action can then be evaluated against these aims. Clear strategic leadership is also necessary as was the case with John Laing and the Costain group. The CEO's of both organisation were instrumental in driving the need for community policy.[5][6]

8.2 Analysing the Opportunities

Once the commitment to a philanthropic policy has been established the next stage is to decide on the possible courses of action open to the firm. The activities of the firm establishes those that will most benefit from having a community policy associated with them. The current social issues and needs provide the areas from which the policy can proceed. The aim is to support a community issue that would be beneficial to the firm from more than a purely a publicity context. The activities would contribute the business interests of the firm.

8.3 Commitment of Resources

An essential part of the community strategy is the commitment of resources to the policy. This may take the form of finance, staff or equipment. The commitment of resources for what is often seen as intangible returns is often the most difficult decision to make. Inadequate resourcing and the lack of genuine commitment to the policy will lead to failure of the scheme. If the policy is tied into the corporate goals then the schemes are more likely to have adequate resources allocated to them.

8.4 Gaining internal support

The construction industry is noted for its instability. In a climate of restructuring, downsizing and redundancies, it is difficult to convince employees that a community policy is important or that the firm is concerned about social welfare. The case for the policy has to be made if there is to be commitment to its implementation. The managers and staff that are required to implement these activities need to be aware of the importance of their actions and that there is a firm commitment to the policy and that it will not be axed during the next cost-cutting exercise.

8.5 Gaining external support

Once the policy has been sold internally it has to be tested in the environment in which it will operate. Involvement in costly schemes that do not satisfy community needs are futile and often counterproductive to the aims of the policy ideals. Consultation with interest groups on the actions and activities to be undertaken are important. The practicalities of the implementation, the problems and the possible outcomes are more likely to be identified during this process. The principal danger is that there are often a wide variety of possible interest groups that may be operating to their own agendas, therefore care must be taken to ensure the programme is not hijacked. A useful dialogue will however, ensure that the policy can be refined before being introduced.

8.6 Operational Issues

The policy should be fairly clear at this stage. Its aims and mechanisms should have been identified after the internal and external consultation stages. It is unlikely within the construction industry context that the execution of these activities will be taken by people exclusively involved in community policy. It is likely that construction project managers and regional managers will more likely be the group that are involved in many of the day to day decisions on these activities. These individuals will have to be made aware of the objectives of the community policy and how they help in developing the profile of the company. The business case for the community policy must be made. A clear and easy to use management interface has to be developed between those responsible for the community policy and those involved directly in its implementation. The support system for those staff who are not familiar with the policy should be introduced with a view to assisting in decision making and action plans.

8.7 Policy Ideas

The foundations of community policy in construction have traditionally concentrated on four main areas:

The Environment
Regeneration
Enterprise and Employment
Education and Training

Construction companies are often involved in many of these activities and provide and ideal springboard to develop a fully fledged community policy. A resource and activity commitment already exist for these activities therefore a long term commitment to them could develop into a community policy.

Environment -An environmental policy is almost a prerequisite of every major construction company. This likely to be the biggest battleground between the construction industry and the public in the future. A company that is likely to be involved in activity that is environmentally sensitive should focus its activity on community schemes that will bring them into closer contact with interest groups. It reinforces the commitment of the company to the environment and provides a useful mechanism to gain feedback and advice on how to deal with future environmental problems. A pro-active response to environmental problems is a good way of setting the tone for a community.

The investment in these activities should be targeted towards meeting the corporate goals of the firms. Adams[7] identify the areas to focus on the development of environmental policy as; renewable resources, pollution, energy consumption, fossil fuels, transport and pro-environmental initiatives. These activities have to be promoted as part of an ongoing commitment to improve social conditions.

The environmental part of the scheme can be introduced at a local project level as well. Issues such as noise, dust, litter, graffiti, etc. should become part of the environmental response to any project. The environmental appraisal of construction projects' should also look at measures such as landscape rehabilitation, removal of waste, decontamination, etc. A community commitment is created to improving the environment that is closely linked to the key business objectives of the firm.

Regeneration -Construction is a key driver for regeneration particularly in urban areas. This offers another opportunity to develop a community policy and link it to the aims of the firm. The traditional views on the regeneration was to develop the infrastructure and then allow the communities to fit into the schemes. This approach was identified as being flawed and the goal of new regeneration schemes is a partnership between the local community, the constructors and the government/funding agencies.

Regeneration schemes allows construction organisations a unique opportunity to build a long term presence in the community. Opportunities exist to develop the partnership between the community and the regenerators during the planning, design and construction of the work.[8] Active involvement and a sense of ownership of the schemes can be developed. The project outcomes can be better managed.

Employer The construction industry as a major employer within communities can also be brought into the remit of community policy. In particular the extent of local labour is used allied to any additional training provided. Improved skills make people more employable and is a value added benefit to society. The extent of new jobs created and the number of local undertaking these jobs are good indicators of commitment to the community's economic regeneration.

Enterprise, general education and training need not be construction specific to help society. Community involvement means attention to areas outside construction. Companies may have to help develop literacy programmes or provide child care facilities or provide sports equipment for the local community to develop its policy. The Costain group have a scheme called "the Costain & Young people Joint venture"[9] The programme is to involve the company in the education in the community. The education programme is driven nationally by the organisation and implemented where the company has work. The scheme is centred around the work the company does and this is the basis for the learning process. This investment benefits the communities concerned as well as Costain

Another useful tool is the provision of company skills to benefit the community; for example legal aid clinics by company lawyers or promotion help by the marketing department to local business. The Laing group use this philosophy as part of its management training schemes, where trainee managers are seconded to community organisation to use and develop their skills. Effective training is provided while making a contribution to the community.

9. Conclusions

Community policy inititiatives in construction firms will have many detractors. Questions will be asked about the use of resources and the value added from the activity. Many of the same arguments can be used against public relations, charity, education and training.The benefits from harnessing these peripheral activities into a cohesive structured strategy will not only accentuate the corporate ethos of the firm, but provide better value to the firm. The foundations are already in place and all that is needed is the vision to implement these ideas.

10. References

1. Avishi B. (1994) What is Business's Social Compact. Harvard Business Review. January February. Pp 38-48

2. Smith C. (1994) The New Corporate Philanthropy. Harvard Business Review May-June. Pp.105-116

3. Smith. Op. Cit

4. Cannon T. (1992) Corporate Responsibility. Longman Group. London

5. John Laing Group Plc. Annual Reports 1986 -1994

6. Costain Group Publications. (1994) Community Involvement Policy

7. Adams R., Curruthers J., Hamill S. (1991) Changing Corporate Values. A guide to social and environmental policy and practice in Britain's top companies. Kogan Page, London.

8. Seeing is Believing (1994) Business in the Community Publications. London

9. Costain Group Publications (1994)- Costain and Young People Joint Venture.

REFOCUSING CONSTRUCTION TO MEET CUSTOMERS' REQUIREMENTS

M.F. Dulami and A.T. Baxendale
Faculty of the Built Environment,
University of the West of England, Bristol, UK
M.R. Jewell
Schal International Ltd, London, UK

Abstract
This paper investigates a Japanese design and project management system called Quality Function Deployment (QFD) in the context of its ability to solve the some of the problems that the UK construction industry is currently facing. The paper focuses on the early stages of design and the briefing process where the impact of quality improvements is greatest and provide the basis for further development to extend the use of QFD to the production stage. The study revealed the frequent occurrence of poor communication, inadequate information retrieval and the lack of positive attitude toward achieving customer satisfaction by construction consultants and their effect on customer satisfaction. The paper analyses QFD and its potential in providing a solution to these problem. QFD main advantage is its recognition of both internal and external customers and in structuring customer information so that their needs remained central to all design solutions. The evidence suggests that QFD's adoption would solve many of the construction industry's quality problems.
Keywords: Quality Function Deployment, customer satisfaction, Design Brief.

Résumé
Cet article se propose d'explorer le modèle japonais de conception et de gestion du projet, appelé Déploiement de la Fonction Qualité, et de mettre en évidence sa capacité de résoudre une partie des problèmes auxquels l'industrie de la construction en Grande Bretagne se voit confrontée. L'article se concentre particulièrement au niveau des premières phases de la conception et du processus d'information, où l'impact des améliorations de la qualité est très important et constitue la base d'un développement supplémentaire permettant d'étendre l'utilisation du Déploiement de la Fonction Qualité à la phase de production. L'étude a révélé l'apparition fréquente d'une faible communication, d'une récuperation insuffisante d'information, et un manque d'attitude positive de la part des consultants de la construction dans la satisfaction des clients. Il y a de bonnes raisons de penser que l'application du Déploiement de la Fonction Qualité pourrait résoudre un grand nombre de problèmes de qualité de l'industrie de la construction en Grande Bretagne.

1 Introduction

The construction industry is a prime example of a project based industry which has received criticism of its performance in relation to time, cost and quality. One of the main reasons for dissatisfaction with its performance and the quality of its product

The Organization and Management of Construction: Shaping theory and practice (Volume One). Edited by D.A. Langford and A. Retik. Published in 1996 by E & FN Spon. ISBN 0 419 22230 8.

may be attributed to the failure in understanding and effectively translating the needs and requirements of the clients into quality characteristics. Having said that, it is then essential that tools and techniques are developed to make this possible. *"Building Research Establishment figures show that defects cost UK clients more than £1bn a year. Research studies have shown overspecification and over-design are as much as 50%"* (1).

The economic downturn and financial uncertainty experienced during the early 1990's exposed major international construction client companies vulnerability to high levels of debt and other overhead costs, compelling companies to maintain comparatively low levels of borrowing. However, major international companies still desire to achieve the construction goals set out in their planned capital investment programmes. As a result customers are demanding from the construction industry products that perform the same functions as before but for less money. International customers *"will want to ensure that what they are procuring in terms of value for money on capital expenditure, and the way their buildings are designed (not just aesthetically) and operate in practise, is comparable with world-class standards"* (2). This view is shared by the Chief Executive of BAA, Sir John Eagan. Sir John recently expressed some doubt toward UK construction firms suitability to construct the forthcoming £1.2 billion Terminal 5 at London's Heathrow Airport, *"they can build it, but can they build it at a world-class cost, and to a world-class quality?"* (3). *"The experienced client now accounts for over 75% of project clients,"* (4) and whose influence and interest in construction has grown tremendously in recent years and continues to grow, therefore, heed must be given to their opinions.

The Japanese were employing methodologies in 1966 which focused on finding out what the customer actually wanted and produced products which gave it to them at the right price. They used Quality Function Deployment (QFD) to translate the customers requirements into designs which met the customers real needs.

2 Quality Function Deployment

The concept of QFD was first proposed by Yoji Akao but gained its first industrial application in 1972 at the Kobe Shipyard of Mitsubishi Heavy Industries in Japan. One of the main driving forces in QFD's development was the desire for manufacturing companies to bring new or improved products to the market before their competition at a lower cost and improved quality. QFD is now used by major manufacturing organisations throughout Japan and the USA, including applications in Japanese construction, but as yet there is no sign of its use in UK construction. All proponents of QFD boast of its remarkable ability to improve a companies design process and product. *"Japanese companies using QFD have achieved notable improvement in the design-development process:*

- *engineering changes have been reduced 30% to 50%.*
- *design cycle has been shortened 30% to 50%.*
- *start-up costs have been reduced 20% to 60%.*

QFD also has a positive impact on the product itself.

- *increased customer satisfaction and market share.*
- *improved designs and performance.*
- *reduced warranty claims (as much as 50%)"* (5).

Customer information gathering takes time, therefore, product definition or feasibility takes longer when using QFD, however, the design period is shortened and redesign is said to be almost eliminated.

3 What is QFD?

The Japanese meaning for deployment is *"an extension or broadening of activities"* (6) and quality function refers *" to any activity needed to assure quality is achieved"* (7). Therefore, quality function deployment means that quality is the responsibility of all those within the organisation. Stated in this fashion QFD could be confused with TQM. Akao (8) defines QFD as *"a method for developing a design quality aimed at satisfying the consumer and then translating the consumer demands into design targets and major quality assurance points to be used throughout the production stage".* Therefore, QFD can be described as the operational definition of TQM, the system behind the philosophy.

QFD forms the framework that ties all the activities of a company into one complete harmonised customer orientated package. The complete product development process is driven by what the customer genuinely wants, not what the designers think they want, nor is it driven by innovations in technology. *"If there is a low-tech, low-cost way of satisfying the customer, it should be used"* (9). However, this is not to say that QFD is a technophobic system, rather, it is the customers needs that will determine if new concepts or technologies are required.

Without the QFD framework important details may be lost in the process due to discontinuities such as personnel changes, re-design or whimsical management. Within QFD would exist a documented system, the principal documents being matrix charts. QFD practitioners have at their disposal thirty such charts, each chart is a continuation of the last, deploying the needs of the customer from design through to manufacture. The matrices translate the 'voice of the customer' into the technical language of the organisation and deploy the product characteristics that need to be controlled through the complete process. QFD does not tell the designer how to design but aids him by indicating the relative importance of each characteristic as expressed by the customer. The design team will thrash out problems and design conflicts whilst referring to the evidence contained within the matrix. Therefore allowing efforts to be concentrated on those areas that yield the greatest customer satisfaction. All departments are required to assist in forming the initial matrix 'The House of Quality' (chart A-1), see figure 1. This early involvement in the product development helps to create ownership of ideas and a commitment to the common goal. QFD does come with several health warnings. Although simple in ideology if undertaken properly QFD can be very complex, King (10) states that *"as a ritual it will not be understood".* Some companies that have implemented QFD have found that their matrices became too large, the use of pareto theory (20%/80%) is generally advised particularly for the 'House of Quality'. Newcomers to QFD are advised to select a simple, lower profile project at first to form a practical experiment.

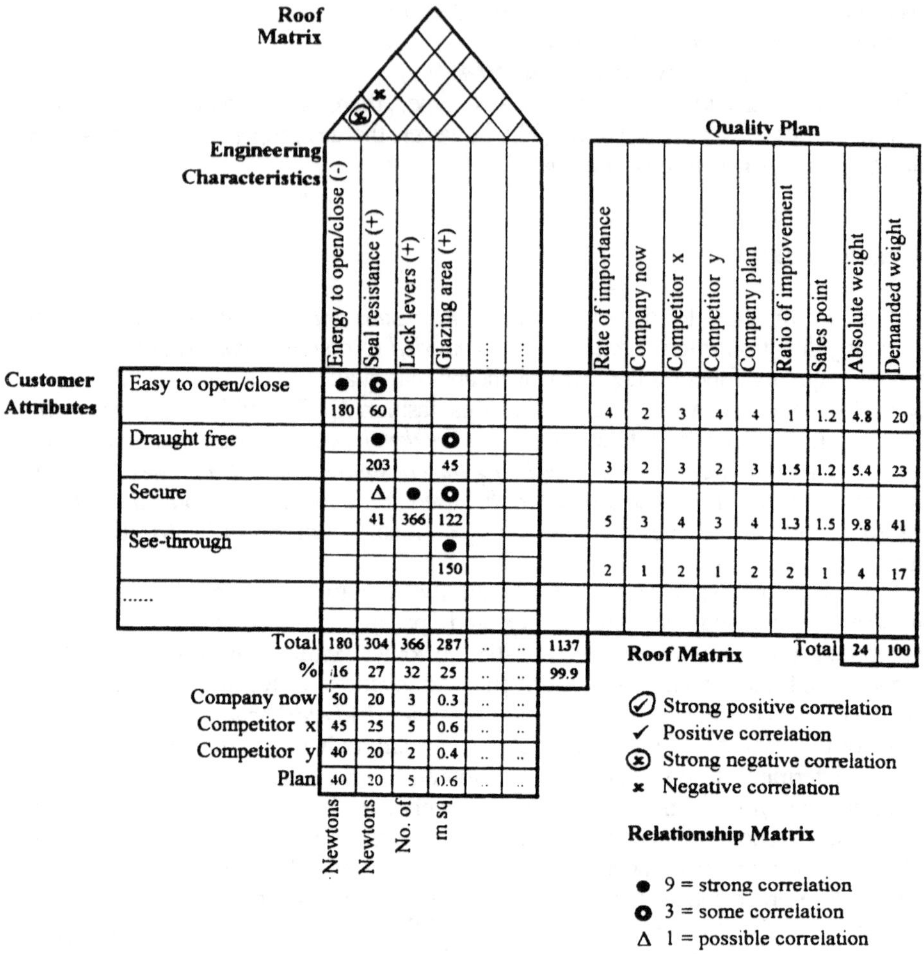

Fig, 1. The House of Quality (Matrix A-1). Example, 'Domestic External Door'

4 The Processes of QFD

'The House of Quality' is the foundation for all major design and production decisions. It maps out the customers requirements and lists how the organisation will meet those requirements. The matrix indicates the product control characteristics that must be deployed through the whole process from design through to production. The formation of the 'House of Quality' can be sub-divided into six stages and are described below.

4.1 The Voice of the Customer

Crucial to implementing QFD is the identification of who the customers (internal and external) are. Customers also include statutory regulators whose needs must be identified and met. If the customers are not identified their needs will not be catered for. Once the customers have been identified the product requirements are stated in customer terms, often referred to as 'customer attributes' (CA's). CA's can be identified using a number of techniques eg. structured surveys, interviews, market research etc. Primary requirements of basic customer wants are usually expanded into secondary and tertiary requirements in order to obtain a more definite list. The CA's should be precise, King (10) recommends using a verb plus a modifier such as 'easy to use' or a noun plus adjective such as 'high reliability'. Designers, engineers etc. tend to translate customer requirements instantaneously. Therefore, practitioners must attempt to retain the same words as used by the customer in stating the requirements, ie. 'the voice of the customer'. This of course may lead to interpretation differences but is more likely to correspond to the views of the customer than if the needs were first stated in the words of the designer. Once customer attributes have been expanded they are listed down the left hand side of the matrix (see figure 1).

4.2 Customer Preferences

This stage recognises that different requirements attain different levels of importance to the customer. Customers may want a product that has both 'high reliability' and is 'easy to use' but feel that reliability is more important. Each CA is assigned a measure of relative importance, a scale of 1-5 is often used, 5 being the of the greatest importance. Therefore, if a trade-off was necessary between 'high reliability' and 'easy to use', the product solution would be weighted toward reliability as it yields the highest satisfaction, thus the product embodies the customers preferences. Customer preferences can be obtained from surveys or from those within the organisation who have close contact with the customer. The weightings are listed in the same row as their respective CA but in a section of the 'house' called the 'Quality Plan' located on the right hand side of the matrix.

4.3 Competitive Evaluation

Market evaluations are conducted by companies to ascertain where their existing product (where applicable) sits within the market place with regard to achieving the identified CA's. This indicates areas that may need to be improved. It is also necessary to see how the competitors fare under the customers criteria, as this allows targets to be set in order to match or better the competitors performance in providing the customers with what they want. The performance in servicing each CA is ranked on a scale of 1 to 5, with 1 being very poor and 5 being very good. By examining the company's performance in relation to it's competitors and the rate of importance of a particular CA the *company plan* can be developed. Therefore, QFD maybe used as a benchmarking tool (see *the Quality Plan* section of figure 1).

4.4 Voice of the Engineer

The next stage is to describe the product in the language of the engineer. The product is sub-divided into design features which have quality characteristics, sometimes known as engineering characteristics (EC). Each CA is individually examined and the EC's that are required to satisfy them are identified. This is best explained by example; 'easy to carry' is one CA of a product. The EC's that will affect this CA are size, weight, shape, the existence and location of handles, etc.

An EC must be capable of being expressed in measurable terms since the characteristic output is to be controlled and compared to objective targets. An additional requirement is that the EC should be directly related to the customers perceptions. For example, the weight of a timber door will be *felt* by customers, therefore, it is a relevant EC. However, the structure of the door, ie. solid or hollow, is a characteristic that the customer is unlikely to directly perceive. This is because it only affects the customer by influencing the weight of the door and other EC's such as 'sound insulation' or 'resistance to fire'.

4.5 Relationship Identification

A multidisciplinary team now fills in the centre of the chart, the 'relationship matrix', indicating the correlations between CA's and EC's. Decisions are made by consensus, members rely on their knowledge and experience to assess the strength of any correlation but may also use direct customer input or data from customer experiments. Symbols are used to indicate the strength of relationships, three categories are frequently used, each being assigned a factor value, strong correlation = 9, some correlation = 3, and possible correlation = 1, those with no correlation are left blank. Some texts also distinguish between positive and negative correlations. EC's that are assigned lots of symbols can be generally regarded as important to the customer. An EC that is not assigned any symbols is a feature not required by the customer. However, if no symbols are assigned to a row of a CA then the EC list needs to be updated. Each CA's symbol factor is now multiplied by the demanded weight of that row and total put in the box with the symbol. The total of each column is listed at the bottom of the matrix and then converted into a percentage which is listed beneath. In the next row the current measurable value for each EC should be listed, eg. an EC is energy to open/close, the value is 50N. In the subsequent rows the competitor's measurable EC values are listed. The final row is reserved for the company plan or target values (measurable) for each EC. Figure 1 shows the company may need to consider increasing lock levers on their doors to match their competitor x as this characteristic has greater importance than the others.

4.6 Identifying Design Conflicts

The final stage to constructing the 'house of quality' is to complete the 'roof' at the top of the matrix. The roof highlights how EC's affect each other. Often in design conflicts occur between required features. The inclusion of one particular feature may impair the performance of others. The solution may lead to the development of a new product that satisfies all CA's or the 'roof' matrix may have identified various engineering features that need to be improved collaterally. For example, to improve the door's security you may have to improve the hinges, threshold details, and other EC's. If a desired feature impairs lots of others then the decision could be made to leave it out completely.

The important aspect of the 'roof' matrix is that it always directs designers toward balancing trade-off's in a way that best addresses the customers needs. The roof matrix uses four different symbols to depict the correlation between EC's but are not assigned a value. They four correlation categories commonly used are; strong positive; positive; strong negative; and negative. Figure 1 shows *'energy to open/close* relationship with both *'seal resistance'* and *'lock levers'*.

5 Customer Satisfaction in Construction: The practice

The techniques and procedures adopted and applied in practice to identify customers requirements and translating them into design are to be investigated. This is essential for this study for two main reasons. Firstly, to establish whether current working practice and attitudes toward producing and using a project brief could benefit from implementing QFD. Secondly, the briefing stage forms the first step into the introduction of QFD and extending it to the production stage as part of the research programme conducted by the University of the West of England.
A total of nine organisations were involved in the interviews. Their annual turnover ranged from £5m to £2bn. Projects discussed ranged in value between £100,000 to £90m. Architects, project managers and clients were interviewed. Four main areas were identified for investigation and analysis.

1. The methods of collecting customer requirements;
2. The handling of design information, internal communication;
3. The translation of customer requirements into designs;
4. Individuals/organisations attitudes toward achieving customer satisfaction.

5.1 Collecting customer requirements

Recognition of the different forms of customer was generally very low. The design team for a pharmaceutical laboratory included a project manager who conducted buildability audits, thus taking into account the needs of the constructors. However, the lead architect remarked, *"designers are let alone to do the best possible design, with little interference from the project manager".* Methods of collecting client requirements varied considerably, however, two key activities that appeared in the majority of the individual processes. The first, is clients comments on sketch designs. The second is to use minutes recorded at formal meetings.

Clients often initiated the whole process by supplying the lead consultant with a *"wish list"*. However, one of the clients, a high street bank, has conducted trials of various bank layouts and assessed their effectiveness by observing the customers behaviour. One client was an international pharmaceutical company In order to understand the client's unique technical requirements the design team employed a second lead architect who specialized in working with this particular type of client. This was the only case where the design team made it their business to learn the specific language of the client. Many of the companies relied on the fact that they had worked for the client on other projects, but none could state that they had made any effort to gain the client's perspective during their first encounter.

During detail design the two largest design and project management organisations within the sample both undertook Value Engineering (VE) studies. The VE process was seen as a method of enhancing client value by eliminating unnecessary cost whilst maintaining function. Neither company approached VE in a highly structured manner, however, both applied Value Management (VM) studies in conjunction with the VE studies. The two companies approached the VE/VM interchange in different ways. The VM study was undertaken during the first two weeks of project inception well before any design or VE exercise with one company.

The project manager of a major retail development undertook the VM study during detail design and after VE. The designers carry out VE as the project proceeds, then, once the design is complete the project manager implements a VM workshop with the client and cost consultant. The aim is to "audit" the design to achieve greater value. An example application was where the specification for the floor finish in the car park was changed from an expensive paving slab specified by the designer to a in-situ concrete finish, saving £300,000. The slabs were deemed to have no "added value" to the client.

Many of the clients had appointed a single representative to act as an "information conduit" but admitted to frequently bypassing him and talking directly to the design team. All the interviewees agreed that they could not act on an informal disclosure and took the same basic approach of tactfully raising the issue at the next meeting. Architects working for large client organisations often complained of late design changes created by clients inefficiency, *"someone decides to look at drawings sent to him months before and decides he is not happy with them"* or whimsical management, *"a new manager gets appointed and has a big enough voice to be heard in order to demand design changes."* Consultants also stated that with naive clients *"you don't get the same story two days in a row"*. This may be due to the wrong question being asked in the first instance or being asked in a manner which the client did not initially understand.

All interviewees stated that defining the clients requirements was an *"iterative process"*. Copies of all the information gathered from the client and all designs were sent to the client for approval. This is reflected by the professional choice of the two most popular information collection methods described earlier as they both involve the retrieval of information, its interpretation followed by feedback.

There was a major difference between the approach of large design organisations and smaller ones toward the length and vigour to which they would attempt to retrieve information. The larger organisations had systems in place that encouraged close client participation, whereas the smaller design practices did not. Where the small design practices were working closely with the client the closeness was always initiated by the client. The smaller practices also seemed less at ease with working too closely with the client. *"You can't afford to go through every stage with the client unless the client states at the beginning and is willing to pay for it!"* The larger organisations showed little sign of these pressures.

At the other end of the scale the design consultants on the shopping mall project took the clients to America to gain their opinion about certain requirements and to look for good ideas. One small company openly admitted to not attempting to encourage any sort of extra client disclosure thus not retrieving any information that the client didn't know would be relevant.

5.2 Handling of Design Information and Internal Communication

In order to be able to successfully relay the requirements of the customer through the design process it is essential that the handling of the information is both efficient and un-distorted. QFD illustrates customer information in a way that all of the different members of the design team know what is actually required, the decision not being left to what they think the customer wants.

Design organisation for most projects consisted of several design teams, eg. architectural, structural and mechanical/electrical. The different design teams, covered by the investigation, were often separated by over an hours travel. It appeared that the smaller design organisations had much weaker links with their fellow consultants with little synthesis. Lead consultants had the attitude that the other consultants *"know what they are doing... if they have a problem they can easily telephone or fax".*

On one project the architects, project managers, client, quantity surveyors, engineers and constructors, who are all from different organisations have been brought together and physically located in the same office building. The aim of this is to shorten the usually *"protracted communication process".*

The pharmaceutical project design team are all members of the same organisation, *"we like to think of ourselves as an integrated design team".* They attempt to encourage this in three different ways:-

1. they 'physically' force integration by seating all the different professionals that are working on the same building area along side each other (ie. architects, structural engineers, mechanical engineers, electrical engineers etc);
2. they have a 'managerial' design team leader who acts as a conduit between the client project manager and design team;
3. they have 'procedures' that ensure that the different designers meet and the work is co-ordinated.

5.3 Translation of Customer Requirements Into Design

A key feature of QFD is its ability to successfully translate the customer's requirements into real design goals and features. Professionals' differing orientation tend to manifest by producing different solutions to the same requirements thus producing a misalignment with the real needs that the client has expressed.

On the pharmaceutical project the VM study produced a graphical representation of the client and end user requirements in the form of a "value tree" which detailed the transition of lower level objectives to higher ones and weighted them. However, this technique, which is also used by QFD implementors, does not translate those requirements into design features, nor easily direct trade-offs between design elements to be made with the customer's true requirements central to the decision.

All other designers translate the clients/users requirements in their minds. They then send a copy of the drawings to the client for approval. Only the two largest design organisations attempted to convey their designs using a medium more easily appreciable to the client, ie. 3D models or 3D CAD.

5.4 Attitudes Toward Achieving Customer Satisfaction

QFD is usually implemented in conjunction with TQM. Therefore, the whole organisation has a strategy and structured approach focused toward the ultimate and common goal of satisfying the external customer. It is important to note that all design team members acknowledge 'client' satisfaction to be their goal. Beyond meetings there appeared to be no structured method or framework in place in any of the organisations that managed the progress of the whole design team and project organisation. Although the larger organisations were implementing such techniques as VM and VE they still lacked the elements that would produce a united front toward creating customer satisfaction. One lead architect commented that focusing on the common goal was *"down to individuals,.. the very bright and very dynamic individuals who have a real sense of purpose,.. they drive a project toward a common purpose... it's down to group dynamics."*

6 Discussion & Conclusion

In order to gain some insight into how QFD might be received in industry, interviewees were given a brief insight to its general concept and had an example 'House of Quality' explained to them. They were then asked to comment on its usefulness. The general consensus toward QFD was positive, the emphasis of comments were made toward the ability of the matrix to aid designers from omitting requirements from the design, an occurrence that the majority of architects admitted to. However, it was mainly the architects that expressed any doubt toward the House of Quality's effectiveness. Most of these doubts stemmed from the concern that the matrix would become very complicated.

Designers seem to rely heavily on past experience in creating design solutions. Designers use their experience wrongly and create inefficiencies in specification because they apply a previous solution to a similar situation not an identical one. QFD tailor each solution towards the wants and needs of the customer by structuring and harnessing past experiences within a multidisplinary design team. QFD requires a multi-perspective of quality which is driven by what the customer wants. This sharing within the design team ensures no domination of any one professional perceptional orientation. QFD articulates the customers needs and desires in a clear manner at the earliest possible juncture.

The two main pre-requisites for the implementation of QFD are the recognition of the internal customer, best achieved through a company-wide quality initiative such as TQM; and the collection of full, accurate and un-distorted customer requirements for each project.

7 References

1. Alden P. (1994) A Word About The Penny-wise, *Building*, No. 50, 16th December 1994, page 15.
2. Fisher N. (1993) *Construction as a Manufacturing Process?* An Inaugural Lecture, University of Reading, UK.
3. Building (1995) Better by Half, 10 February, pages 20-24.
4. HMSO (1990) *Professional Liability Study Team Report*, HMSO, London.
5. Fortuna R. (1988) Beyond Quality: Taking SPC Upstream, *Quality Progress,* Vol 21, No. 6, June, pages 23-28.
6. Quality Progress, (1988) Listening to the Voice of the Customer, Vol 21, No. 6, June 1988.
7. Sullivan L.P. (1988) Policy Management Through Quality Function Deployment, *Quality Progress*, Vol 21, No. 6, June, pages 18-20.
8. Akao Y. (1990) *QFD Integrating Customer Requirements Into Product Design,* Massachusets USA, Productivity Press.
9. Hunter M.R. and Van Landingham R.D. (1994), Listening to the Customer Using QFD, *Quality Progress,* Vol 27, April, pages 55-59.
10. King R. (1989) *Better Designs In Half The Time*, 3rd Edition, Massachusets USA, GOAL/QPC, Methuen.

MARKETING IN BUILDING SERVICES ENGINEERING CONSULTANCY FIRMS IN SOUTH AFRICA: THE WAY FORWARD

P.D. Rwelamila

Department of Building Sciences, Royal Institute of Technology, Stockholm, Sweden

T. Lethola

Department of Construction Economics and Management, University of Cape Town, Cape Town, South Africa

A. Ngowi

Department of Civil Engineering, University of Botswana, Gaborone, Botswana

Abstract

This paper examines the results of an empirical pilot investigation into the nature and extent of marketing in consulting building services practices in South Africa, using the Western Cape Region as a case study. A questionnaire survey of building services consultants was undertaken to ascertain current marketing practices and preferences in the marketing practices and preferences in the marketing strategies employed. The result suggest that marketing is, to some extent, considered a legitimate management function within these consulting firms. However, evidence exists to support the contention that the 'trappings' of marketing prevail rather than the 'substance', the latter being required if a professional practice is to have the market-led and client-driven orientations important to the implementation of the marketing concept. More specifically, the paper deals with: the role and degree of responsibility assumed by the marketing manager; the nature and extent of marketing research employed by the firm; and the marketing policies prevalent in the firm. Finally, recommendations are made for future research.

Keywords: Marketing, building services, implementation, clients, professional services.

Sommaire

Ce papier examine les resultats de l'enquete empirique pilote sur la nature et l'etendue du marketing dans les services de consultation en batiment en Afrique Du Sud. La region du Western Cape est prise comme etude de cas. Un sondage sur les consultants des services de batiments a ete entrepris pour certifier les pratiques et les preferences actuelles du marketing dans les strategies de marketing employees. Le resultat atteste qu'on attribue au marketing, a une certaine ettendue une legitime fonction de gestion dans ces firmes de consultants. Cependant des preuves existent pour soutenir que les

The Organization and Management of Construction: Shaping theory and practice (Volume One). Edited by D.A. Langford and A. Retik. Published in 1996 by E & FN Spon. ISBN 0 419 22230 8.

'signes exterieures' du marketing prevalent plutot que la 'substance', cette derniere etant requise si pour une practique professionelle, les orientations sur le marche et le client sont important pour implanter le concept du marketing. Ce papier traite plus specifiquement le role et la responsabilite assumes par le manager du marketing, la nature et l'etendue de la recherche en marketing employee par les firmes, et les politiques de marketing qui prevalent dans la firme. En fin, des recommendations sont faites pour la pratique future.
Mots clefs: Marketing, service de batiment, implantation, clients, services professionnels.

1 Introduction

Marketing is one of man's newest action disciplines, and it is also one of the world's oldest professions. From the time of simple barter through the stage of money economy to today's modern complex marketing system, exchanges have been taking place. But marketing, the study of exchange process and relationships, made its formal appearance only in the 20th century.

Although the subject of marketing has attracted increasing attention from companies, institutions and nations, marketing ideas have made singularly little penetration into the centres of influence of the construction industry [1]. According to Pierce [2], the concept of marketing is still seen as a young art, and an alien concept which does not rank high in the list of active interests of most construction professionals [3] [4].

Despite the present status of marketing in the construction industry, some observations indicate that many firms in Europe and USA are beginning to realise the need to market their services and are anxious to improve their marketing activities [5].

In South Africa many stake holders in the construction industry have shown relatively little concern for marketing. The managerial concern and commitment to the marketing concept is somewhat limited, and at best, most consulting professionals' outlook to the concept is a conservative one. There is evidence to suggest that the buoyant state of the South African construction industry contributed largely to this attitude amongst the professionals [6]. However, the image of marketing in the industry is changing [7]. Marketing is beginning to achieve a widespread acceptance as an essential part of the management practice.

2 Research paradigm

Due to the perceived incompatibility of marketing with 'professional ethics', the principles of marketing have been largely overlooked by professionals in many instances. It is apparent that most professionals do not properly understand the marketing concept and the merits that can be derived from implementing a marketing plan.

The building services engineering professional firms in South Africa, are currently under considerable pressure from a number of sources. It is necessary for these professional firms to recognise the benefits that can be derived from marketing in order to survive in the construction industry. The main objective of this study was to establish whether building services engineering firms are realising the importance of actively marketing their services. In order to achieve this it was necessary to establish if building services engineering firms are understanding, adopting and implementing marketing concepts. This being the case, it was decided that the study should follow the same format of similar studies carried out in the United Kingdom by Morgan and Morgan [8], which focused on engineering consultancy firms.

It was hypothesized that *the extent of adoption and implementation of the marketing concept by specialist building services engineering consultancy firms in South Africa is very limited.*

3 Research methodology

The data was collected via the medium of postal questionnaire. The questionnaire was divided into five main sections: *demographic data; the nature and status of the marketing information and research employed by the organisation; the marketing policies employed by the firm; and the marketing strategies adopted by the practice.*

A postal questionnaire was conducted throughout the province of the Western Cape. Questionnaires were sent to all consulting firms (31 in number), using the directory of registered practices of the South African Association of Consulting Engineers. The number of replies that were suitable for analysis amounted to 14, which resulted in a response rate of 45.2%.

4 Survey results

4.1 Demographic

The survey was found to be dominated by practices with more than 20 employees (62%), the rest reported that they were employing between 16 and 20 employees (10%), between 4 and 6 (6%) and less or equal to 3 employees (22%).

The majority of respondents (76%) reported that they have at least 5 branches nationally, 8% have 4 branches and 16% have one office. These results are depicted in Table 1.

Table 1. Number of branch offices within South Africa

Number of offices	% of firms
1	16
2	0
3	0
4	8
5	24
Between 6 and 10	52

Table 2 provides information regarding the approximate national fee income in the most recent financial year. The results indicate that 86% of the firms received a turnover of at least two million rands (L 400, 00) in the 1994/95 financial year. These results provide further evidence that the respondent firms are predominantly large in size.

Table 2. Approximate fee income in the 1994/95 financial year

Fee income	% of firms
Not exceeding R 500 000	0
Between R 500 000 and R 1m	0
Between R 1m and R 2m	14
Between R 2m and R 4m	43
Exceeding R 4m	43

It may concluded that the specialist building engineering firms in the Western Cape province (and nationally) are predominantly large in size, enjoying a high turnover and employing a comparatively large staff.

4.2 Organisation for marketing

The questionnaire first sought to establish whether or not a marketing department or a designated individual responsible for marketing exists within firms. Fifty seven per cent of the respondents reported that neither a department nor a designated individual exist, with the responsibility for various marketing tasks tending to be just one of the management roles of either one or several partners. This interpretation is supported by previous research findings in consulting engineering professions [9] [10] [11].

The majority of respondent firms (67%) were more than 5 years old, and 33% of the respondent reported that they have formed marketing departments within the previous three years.

4.3 Role of indivisual responsible for marketing

The survey sought to establish the role of the individual assigned the position of the head of marketing, *the chief marketing executive (CME)*. Table 3 records the responses relating to the title given to the head of marketing.

Table 3. CME job title

National marketing manager	% of firms
National marketing manager	10
Senior marketing manager	63
Marketing partner	27
Marketing assistant	0

In the majority (63%) of instances, the person(s) responsible for marketing comprise one or more of the partners, with no actual title relating to marketing. These results therefore indicate that marketing does not enjoy a distinct organisational identity, at least in terms of job title. Furthermore, the results appear to be in contrast with the previously reported existence of a formal marketing department in more than 40% of the respondent firms. When the reporting level of the CMEs was analysed, it was found that in the majority of cases (67%), they reported to a group or board of senior partners, 7% and 26% reported to a regional senior partner and senior partner respectively.

In order to establish the issue of CMEs reporting level further, the respondents were asked to rate the status of the CME relative to the functional heads within the firm. The majority of respondents (72%) reported that CMEs enjoy either equal or higher status to other functional heads, while 28% enjoyed a lower status. These results imply that the CMEs are widely involved in policy decision-making ranks within the firms.

To enable the marketing tasks assigned to the CME to be explored explored, a selection of responsibilities often associated with the marketing function were proposed and respondents were requested to measure the responsibility on a scale of 1-5 (*1 =full and sole responsibility; 5= no responsibility*). The results are shown in Table 4.

Table 4. CME marketing responsibilities

Area of responsibility	Full and sole (%)	Major but shared (%)	Equal (%)	Some (%)	No (%)
Advertising	0	67	33	0	0
Promotional activities	2	98	0	0	0
Public relations	0	33	67	0	0
Income forecasts	0	0	67	33	0
Marketing research	38	56	6	0	0
Marketing planning	33	67	0	0	0
Service selection & development	0	23	49	28	0
New service launches	67	33	0	0	0
Divers-ification studies	0	33	33	33	0
Marketing staff selection	57	0	0	0	43
Marketing training	67	0	0	0	33
Corporate planning	33	0	33	0	33
Fee stru-cturing	0	0	67	33	0

The CMEs were reported to enjoy the *full* and *sole* responsibility in the areas of marketing training (67%), launching of new services (67%) and marketing staff selection (57%). The CMEs were found to share *major* responsibilities mainly in the marketing communication areas, more specifically in the areas of promotional activities (98%) and advertising (67%). The shared responsibilities were also exhibited in the areas of marketing planning (67%) and marketing research (56%).

The areas at which all the CMEs were reported to share responsibilities *equally* are the fee structuring (67%), income forecasts (67%) and public relations (67%), thus giving an implication that such areas are more sensitive for policy decision-making in a firm.

When the above results were ranked based on the weighted mean score, it was found that the CMEs have the highest responsibilities in the areas of *structuring fees and forecasting income for the firms.* An average responsibility is noted in the marketing communication areas, which are: *public relations; advertising; and promotional activities.* Although there is an indication to suggest that the CMEs have excessive responsibilities, this does not imply that the potential role of marketing is being fully exploited. Instead, there is an indication to suggest that CMEs still need to improve on areas such as *service launches, diversification studies, marketing planning as well as marketing communication.*

Table 5 summarises the responses from those respondents where a formal marketing department or a designated marketing individual does not exist, in respect of the person(s) responsible for certain marketing functions.

Table 5. Person(s) responsible for marketing functions where no formal marketing organisation exists

	Person(s) responsible for marketing functions		
Marketing functions	Partners/ Directors (%)	Senior partner/ Managing/ director (%)	All employees (%)
Advertising	61	39	0
Promotional activities	93	7	0
Public relations	93	7	0
Service selection and develo-pment	33	67	0
Marketing research	60	40	0
Fee income forecasting	53	47	0
Corporate plang.	67	33	0

In the majority of instances (66%), most of the marketing functions are undertaken by the partners of the firm. The senior partners are found to be responsible mainly for service selection and development (34%). Surprisingly, all the respondents do not view the junior staff as part of the marketing team in their firms. This attitude towards the junior staff was also indicated by Hansey [12], who concluded that smaller firms tend to choose one member of the management team to be solely responsible for the overall marketing activities.

4.4 Marketing information and research

In response to whether there exists a unit responsible for market research in each consulting specialist engineering firm, an overwhelming majority of firms (86%) do not have a formal marketing department. Fourteen per cent of the respondents reported that there was an existence of a formal marketing research unit.

Focusing on data relating to the use of various types of marketing, respondents (80%) reported to be using marketing research to find out the potential for further operating in existing offices, as well as the new offices (86%). In support of this, respondents reported that they further engage in studies to determine client characteristics (72%), the bahaviour of comnpetitors (72%), as well as the fees priced in the marketing environment (86%). Interestingly, it appears that there is a relative lack of interest in research regarding the business trends (58%) and the market share analysis (43%), although these areas are of equal importance in the marketing environment [13] [14].

In order to establish the marketing culture of specialist engineering practices, firms were questioned on the extent to which they consider marketing research to be important. More specifically, respondents were asked to comment on whether marketing research emanates from the *internal* or *external services*, and the *nature* of their marketing research. The majority of respondents (91%) were of the opinion that their knowledge of existing and potential clients is such that it is sufficient to engage in research on a client-by-client basis. Furthermore, it was found that the main source of marketing research is from the informal, in-house service (72%). Few firms (32%) reported buying external marketing research. A point worth noting is that 43% of respondents claimed not to be needing any marketing research, whether in-house or external.

In conclusion, it can be argued that most of the consulting firms conduct marketing research in an unco-ordinated manner.

4.5 Marketing policies

A number of questions were devoted to establishing the marketing policies of the firms. The majority of respondents (70%) strongly feel that innovation and change come directly from the client, and therefore, the main objective is to provide a reliable, high value-added service (70%). Interestingly, these firms (86%) reported to be encouraging feedback from their clients as a yardstick in measuring the level of client satisfaction.

A significant number of firms (79%) reported that their business success rests on increased sales and long term profitability. These results indicate that many firms are rather sales-driven than market-led in theiir normal "marketing" operations. However, a general trend emerged indicating that the attitude of marketing held in the firms is such that marketing is generally an accepted business function which also considered "legitimate". These results compare well with the findings of Morgan and Morgan [8]. From the survey results, it may be concluded that the consulting specialist engineering firms are relatively client-oriented, although they still need to improve in the area of marketing communication. Many firms are more sales-led than market-led. However, marketing has been found to some extent to be a legitimate business function within the firms, although some firms still view it with some scepticism.

4.6 Marketing strategy

The concluding section of the survey was dedicated towards the marketing strategies pursued by consulting specialist engineering firms. The classification of strategies showed a slight predominance of firms endorsing a rather offensive than defensive aim in terms of the market share goals. On the aspect of routes taken to increase the market share, the the majority of respondents (86%) reported that they favour *entering the new client markets with existing services* and 72% of the respondents were *in favour of diversifying into new areas involving new services and clients*. However, many firms did not appear to be secure with developing new services for existing markets.

When respondents were asked to indicate the relative importance of certain elements within the marketing strategy from a list of probable elements, technical service excellence, professional reputation, specialised services for defined segments and utilisation of technology were rated as very important by 86%, 86%, 71% and 69% of the respondents respectively. Interestingly, the personal side of marketing (social and personal contacts of staff) was seen as of lower importance by the majority of respondents. Advertising was ranked lowest in the list of strategies elements in the list of strategic elements in the firms. This is indicative of the fact that engineering specialist firms do not view promotional tools as effective marketing strategy elements. Similar results were reported in the findings of Namo and Fellows [15].

The recent withdrawal of the statutory fees scales for consulting engineers appears to have very had very little impact on the use of fees as a marketing strategy. Only 39% of the respondents considered this aspect to be effective.

In conclusion, it has been found that consulting specialist engineers rely mainly on professional reputation, high quality of services and improved technology as their marketing strategies. A large proportion of firms do not view promotional techniques and competitive fees as effective marketing strategies.

5 Conclusions and recommendations

5.1 Conclusions

The objective of this study has been to establish the extent to which consulting specialist engineering firms have adopted and implemented the marketing concept. Based on the information and findings reported in this paper, conclusions could be drawn:

1. *Organisation for marketing*: the existence of marketing organisation within firms is limited. Many firms neither have marketing department nor an individual designated for marketing responsibilities. Even where marketing departments exist within the firms, the consulting firms still struggle to implement marketing functions effectively.
2. *The marketing responsibilities*: marketing is not seen as an activity applicable to

all employees of the firm. The senior staff, especially partners, carry out most of the marketing responsibilities, whilst the junior staff is almost not involved in most of the activities.

3. *Marketing information and research*: the consulting specialist engineering firm seldom conduct marketing research to gather information about their marketing environments. For those that do engage in research, this is conducted in an unco-ordinated manner and the firms rely mainly on in-house information for research.
4. *Marketing policies*: the consulting specialist engineering firms are distinctly client-oriented in approaching their business. The sign of marketing direction is manifested by the engineers' attempt to discover what the clients really want and orient their firms to satisfy those wants. The client oriented approach to business is close to the current marketing thinking in satisfying the clients' requirements profitably.
5. *Marketing strategy*: the relatively recent withdrawal of the statutory fee scales for consulting specialist engineers has had little impact on the use of fees competitively as a marketing strategy. Instead, consulting engineers frequently choose to rely on their professional reputation, the use of improved technology and by offering high quality services to market their services.
6. The hypothesis is proved that: *the extent of adoption and implementation of the marketing concept by specialist building services engineering consultancy firms in South Africa is very limited.*

5.2 Recommendations

Based on the findings and conclusions of this paper, the following recommendations are made:

1. Further research is needed nationally to establish the role of marketing communications within the specialists consulting engineering services. The research should also encompass client perceptions of such communications.
2. Further research should be taken to establish the cultural barriers of the implementation of marketing concept within the consulting specialist engineering profession. This work could be extended to include a comparison with similar barriers in other professional services.

6 References

1. Jepson,W.B. and Nicholson,M.P. (1972) Marketing and Building Management, Medical and Technical Publishers, Manchester.
2. Pierce,P. (1992) Construction Marketing: A professional approach, Thomas Telford, London.
3. Morgan,R.E. and Morgan,N.A. (1991) An appraisal of the marketing development in engineering consultancy firms. Construction Management and Economics, Vol.9,pp. 355-368.
4. Rutland,P. (1987) Going to the market. Chartered Quantity Surveyor,March,p.15.

5. Arditi,D. and Davis, L. (1988) Marketing of construction services. Journal of Management in Engineering, Vol.4(4), pp. 297-313.
6. Phillips, R.S.B. (1990) Marketing and the quantity surveyor. Quantum, Vol.29, pp.17-21.
7. Van der Merwe, S. (1976) Some research findings on South African top executives and the marketing concept. Marketing Research and Media, Vol.1(2), pp.21-28.
8. Morgan, R.E. and Morgan, N.A. (1991) An appraisal of the marketing development in engineering consultancy firms. Construction Management and Economics, Vol.9, pp.355-368.
9. Tessler, A. (1983) Success in Invisibles - Report on the marketing of professional services overseas. ITI Research, London.
10. Hensey, M. (1986) Strategic planning: Its purpose, value and methods. Journal of Management in Engineering, Vol.2(1), pp.213-221.
11. Port, J. (1991) Still room for improvement. Chartered Quantity Surveyor, March, pp.17-18.
12. Hensey, M. (1986) Strategic planning: Its purpose, value, and methods. Journal of Management in Engineering, Vol.2(1), pp.213-222.
13. Brunjes, B. (1995) Marketing series - part4: Tenders that talk! The Civil Engineering Contractor, Vol.29(6), pp.22-25.
14. Hansen, K.L. and Tatun, (1989) Technology and strategic management in construction. Journal of Management in Engineering, Vol.5(1), pp.67-82.
15. Namo, F. and Fellows, R.F. (1993) The role of advertising in marketing civil/structural engineering consultancy firms. Construction Management and Economics, Vol.11, pp.431-441.

COMPETENCE-BASED CONTRACTING BUSINESS

P. Huovinen
Institute of Construction Economics and Management,
Helsinki University of Technology, Espoo, Finland

Abstract
The core competence perspective is here applied to the construction sector. Construction is defined as the generic contracting business which is further differentiated into five distinctive businesses by investment sector. A single contracting business is then redefined as a vision, namely a competence-based system consisting of four interlinked elements: (1) investors and their investment needs, (2) solution contractors and their investment solutions, (3) solution integrators and their core solutions, and (4) competence developers and their core competencies. The new kinds of "Solution Contractors" act as the leaders of leveraging networks. The business processes will involve complete investment solving, core solutions commercialization, core competencies productization and future investors/investment needs anticipation. The primary benefits are radical investment profitability and productivity improvements over the life-cycles. The most important precondition (barrier) is the renewal of the current price-driven and capacity-driven procurement methods.
Keywords: Core competencies, construction, contracting business, competition

Sommaire
La perspective des noyax de compétence est ici applicueé au secteur de BTP. Le secteur de BTP est défini comme activité commerciale d entrepreneurs, se divisant en cinque activités commerciales différentes en fonction du secteur d investissement servi. La naissance d une activité commerciale basée sur la compétence nécessite le développement et une vraie intégration des quatres éléments qui suivent: (1) investisseurs et besoins d investissement, (2) entrepreneurs et solutions d investissement, (3) intégrateurs et noyaux de solutions, et pour servir comme base aux précédants, (4) developeurs et leurs noyaux de compétence. Les avantages d une activité d entreprises basée sur les noyaux de compétences sont une amélioration radicale de la rentabilité et de la productivité.

The Organization and Management of Construction: Shaping theory and practice (Volume One).
Edited by D.A. Langford and A. Retik. Published in 1996 by E & FN Spon. ISBN 0 419 22230 8.

1 Introduction

It is here argued that **leading contractors play a role as the primary investment solvers or facilitators**, who, by definition, undertake contracts to provide all the technologies, designs, production lines, materials etc. needed to execute capital investments. Thus, construction or investment markets are treated as **contracting markets**.

In her pioneering theory [of growth], Penrose [1] states that [a contractor] is seen as **a collection of productive resources** [i.e. competencies] for the purpose of producing and selling investment solutions (end products). A contractor aims at growth in order to survive, and this growth is ... based on continuous use and development of resources. His competitiveness increases or decreases due to the ability to combine resources [in a unique way] to compete successfully in changing market conditions. More recently, Prahalad and Hamel [2] stated that [the contractor's] competitiveness derives from an ability to build, at lower cost and more speedily than competitors, **the core competencies** that spawn unanticipated investment solutions (end products).

The aim is to apply these two concepts to construction by answering the following questions:

1. What are the most common weaknesses of the construction sector, world-wide?
2. Can construction be defined as generic contracting business(es), as a prequisite before applying the core competence perspective?
3. How competence-driven are large international contractors, already? Can generic strategic competitor groups be detected among these contractors?
4. What is competence-based contracting business? Can this new competence-based approach fix the construction sector's current weaknesses?
5. What are the preconditions for the adoption of the competence-based approach?

While being in the first stages of the research process, my primary argumentation is understandably **partly hypothetical and visionary**, and not yet based on empirical evidence on possible industry transformations already taking place on a global basis.

2 From construction into a set of distinctive contracting businesses

2.1 Features of traditional construction

It is argued that the following **weaknesses** apply to most of the national construction, or capital investment markets, world-wide:

- **Target quality, costs, and throughput time** of the new investments are seldom attained. The procurement methods result in quality variations, cost overruns, and delays. The actual costs will be known often many years after the handovers due to claims.
- **Price-driven and capacity-driven procurement methods** are widely used preventing design expertise, production, and life-cycle knowhow from cumulating at single organizations. Traditionally, the investor hands out first a design task. A procurement task is handed out to the same designer, who subcontracts the erection.
- **National technology and design rules, and formalized practices** prevent the best proven technologies and investment solutions already widely used in one country from penetrating neighbouring countries.
- **"Murky" practices** are common in national and local markets. Besides, the non-ethical dealings are often detected in relation to international multibillion projects.

- **The construction sector is regarded as an "economic regulator"** by national governments trying to adjust to fluctuations. Public investments and subsidies are increased or decreased based on reigning governments' opportunistic conduct. The ad hoc, and often consecutively turnaround policies prevent the construction industry from adopting the long-term perspective.

For example, **WS Atkins** states in their sector report to the European Commission that for the most part ... construction will remain a geographically concentrated industry, with national and local characteristics. The level of demand fluctuates unpredictably. Survival depends on minimizing fixed capital assets and being extremely flexible. There has been **'no European construction industry'** - firms engage in ... **cut-throat competition** ... and take a very short-term approach [3].

2.2 Generic contracting business

The generic scope of capital investments includes feasibility studies, engineering, design, supply of a production line, construction of facilities and infrastructure, start-up as well as after-investment services [4]. The investment needs, i.e. project and task contents, vary markedly within this generic scope. Accordingly, construction is defined as **the generic contracting business** having the following, partly emerging features:

- Domestic and foreign contractors are solving and realizing investors' needs by participating in their complex investment processes in targeted locations, across the globe.
- Competing networks managed by the leading contractors offer, and one or several of them are chosen by the investor to carry out his particular investment.
- Each investor applies a specific procurement method (in the UK or EU context), or a contracting mode (in the US context) to achieve the objectives in terms of functionality, buildability, quality, money, time etc. This method determines the roles of contractors to become involved, as well as the nature of competition, by project.
- Dominant characteristics are project specific, and gradually more long-term, partnerships and networking among all the parties to be involved on a project. As a rule, leading partners possess core technology that investors prefer most.
- Typical operations include technology transfer, systems selling, packaged deliveries, alliances, subcontracting, as well as management and services contracts.

2.3 Differentiation of the contracting business by investment sector

The capital investment sectors of economies form **a sustainable basis** for a contractor to define its mission, the investment types to be executed, as well as to choose its competencies, targeted markets and investors. It is argued that that the generic contracting business can be differentiated across investment sectors into **a set of distinctive businesses** as follows (Fig. 1):

1. Natural resources contracting, i.e. execution of investments in exploration, use and processing of minerals, oil and gas, etc. including related facilities and infrastructure
2. Power utilities contracting, i.e. execution of investments in energy production based on using coal, water, oil, gas, wind etc., incl. related facilities and infrastructure
3. Industrial plants contracting, i.e. execution of investments in manufacturing and process plants incl. related facilities and infrastructure
4. Telecommunications networks contracting, i.e. execution of investments in mobile

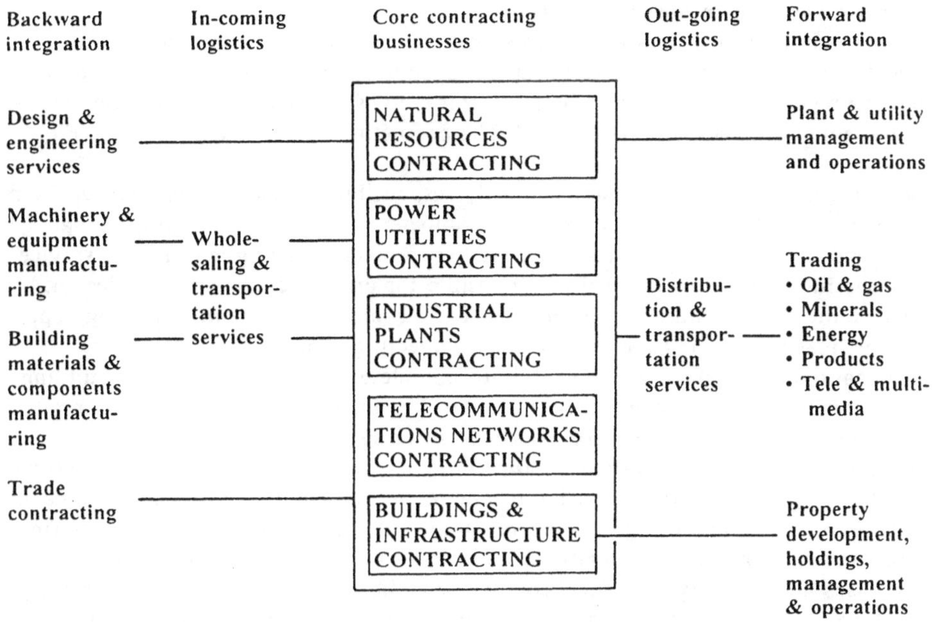

Fig. 1. From construction into five distinctive contracting businesses with options to integrate backward and forward.

and fixed networks incl. related facilities and infrastructure

5. Buildings and infrastructure contracting, i.e. execution of remaining investments in housing, leisure, commercial and public buildings as well as roads, railways, airports, harbors, waterways and other infrastructure.

Many **large, diversified contractors** are currently stretching and deploying their competencies across two or more or all five contracting businesses. In addition, large contractors are typically integrated backward and/or forward [5]. In general, Japan has gone further [than the US or EU industry] in integrating design and off-site manufacture and creating a fully [backward] integrated construction industry [6].

At this stage before applying the competence-based approach to contracting business, it is hypothesized:

H1. Each of the contracting businesses, defined by investment sector and needs, is so different from the other ones that long-term success requires sector-specific technologies, competencies and their combinations as well as customized solutions:

(a) This is more often true when the investors consider technology and effective investment solutions over life-cycle as the primary criteria for making investment decisions and choosing contractors.

(b) This is less often, or not at all, true when the investors consider initial investment costs (price) as the primary criteria for decision-making.

2.4 Generic strategic competitor groups

Large international contractors are grouped into the five strategic competitor groups [7]. Contractors belonging to the same group have similar competitive behavior based on similar competencies. In addition, their offerings, roles and contractual responsibilities cover the same, broad or narrow scope [8]. The competitor groups are as follows (Fig. 2):

1. **General (or turnkey] contractors**: typically responsible for turnkey delivery at fixed price and by set execution period. Their competencies cover capital project management and investor services, feasibility studies, project financing and trading services.
2. **Engineering contractors**: typically responsible for process engineering & design, plus additional CM and training services. Their competencies lie in technology applications, feasibility studies and advanced engineering capacity.
3. **Plant contractors**: typically responsible for complete plant deliveries (excl. buildings) Their competencies lie primarily in machinery design and manufacturing, and production line delivery, in many cases also in plant or special process design.
4. **Design-construct (or design-build) contractors**: typically responsible for total delivery of a building or infrastructure without process investment contents. Their competencies lie in project and design management and often in components manufacturing.
5. **Construction contractors**: typically responsible for building or civil engineering works, purchases and site services. Their competencies lie in project and purchase management, works techniques and equipment, and often components manufacturing. They represent the simple, price-driven end of the contracting business spectrum.

A single contractor very seldom owns all the technologies, competencies and resources for a turnkey project, but general, engineering and plant contractors are able to bundle these competencies through partnerships and networking. It is hypothesized:

> H2. In the advanced end of the contracting business spectrum, the internationally renowned general (or turnkey) contractors, engineering contractors and plant contractors have, in fact, designed and realized competence-driven strategies for years. This requires that there have emerged advanced investors as well.

This is supported, for example, by WS Atkins [9] stating that "Japan has the world's largest... contractors with excellent research and training organisations, providing integrated design and CM services. The US industry has the largest heavy engineering specialist contractors which design processing plants... and undertake the R&D needed.".

Towards the simple end of the spectrum, it is hypothesized:

> H3 (a) In the simple end of the spectrum, most of the buildings and infrastructure are still first designed by designers, and then carried out by trade contractors who all have access to the same resources and competencies, in regions with similar geographical conditions. There are more, and more severe weaknesses, i.e. quality defects, costs overruns etc. (b) Exceptions are a few contractors with design-driven strategies, as well as many specialty contractors with patented techniques and equipment.

This is supported a.o. by the Design-Build Institute of Amerixa (DBIA) stating that "the number of the US ... **design-builders** is relatively small. They still think of themselves as [construction] contractors ... who offer a variety of delivery methods" [10].

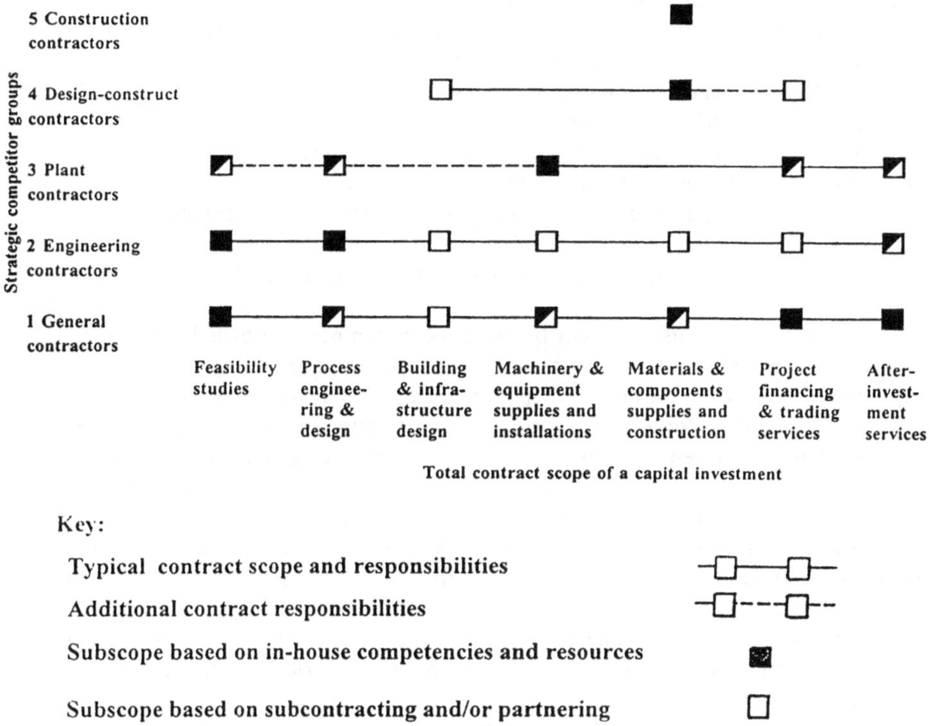

Fig. 2. Generic competitor groups related to contracting businesses.

3 Competence-based approach applied to construction

3.1 Definition of the competence-based contracting business

So far, construction is differentiated into five distinctive contracting businesses, and contractors are divided into five competitor groups. What relevance will the core competence approach add to this contracting business approach? According to the concept of Prahalad and Hamel [11], management's [primary task is] to consolidate contractor-wide technologies and skills into competencies that empower to adapt quickly to changing opportunities. A contractor, like a tree, grows from its roots. The core solutions are the physical embodiments of one or more core competencies, which engender business units, whose fruits are investment solutions (end products).

According to the "followers" of Prahalad and Hamel, the competitive locus is a contest for acquisition [internally or through outside means] of new competencies and leveraging these across contracting businesses and investment solutions [12].

When the competence-based concept of Prahalad and Hamel is applied to any of these five contracting businesses, the supply side is divided into three elements of core competencies, core solutions (core products) and investment solutions (end products). The investment needs (and investment projects) is the fourth element [13]. Each of these elements will have **a primary stakeholder** who is interested in developing and leveraging the specific deliverable within the total "contracting chain". Besides the investors, the stakeholders are contractors as well as collaborating designers and suppliers.

The competence-based contracting business, as the system, consists of (Fig. 3):

- Investors and their investment needs
- New kinds of "Solution Contractors" and their investment solutions
- New kinds of "Solution Integrators" and their core solutions
- New kinds of "Competence Developers" and their core competencies.

Currently, the emphasis is on the "end product" competition. It is envisioned:

H4 Each contracting business will become more competence-based, although with varying degrees. This will take place nationally, internationally (such as EU-wide) or globally, depending on how widely integrated "the system" in question will become. **The driving force** is the four enabling linkages, or business processes:

(a) **Complete investment solving**, which requires replacing current procurement processes and competition rules with the new ones that encourage execution of investments by competence-based solutions (investors <-> solution contractors)

(b) **Core solutions commercialization**, through applying and reusing core solutions, networking subcontractors and creating delivery processes (solution contractors <-> solution integrators)

(c) **Core competencies productization**, such as integrating and modularizing core competencies into core solutions, and anticipating generic applications (solution integrators <-> competence developers)

(d) **Future investors/investment needs anticipation**, leading to symbiotic building of new core competencies (competence developers <-> investors).

H5 More and more investors will prefer "Solution Contractors" who will offer complete solutions, based on networking (incl. core solution integrators and competence developers), and re-engineering delivery processes. This means reorganizations of knowhow and resources among current stakeholders **to form national, international and global networks** that possess technology, designs, life-cycle expertise, building systems, prefabrication, modularization, automation and miniaturization technologies as well as project management and logistics expertise.

H6 All or most of the members of these networks solving investment needs may belong to the same corporation, or all or most of the members are independent firms.

Leading contractors will adopt and foster **competence-based management** to integrate all the investment factors. This is supported a.o. by WS Atkins stating that the future improvements in productivity and value-for-money will come from **restructuring and managing the whole supply chain** from raw materials to site erection [14].

For example, Brown & Root has a profession ... "that new ... [communications] techniques will [enable the creation of new alliances]. In these alliances, the organizations are totally integrated, saving costs up to 30 % ... [15]. Similarly, John Brown E&C, of the UK, "is forming alliances with clients and seeing more ... lump-sum turnkey work. **Technology** is a key way to distance a firm from competitors". It has a target of doubling its portfolio of technologies over the next four years [16].

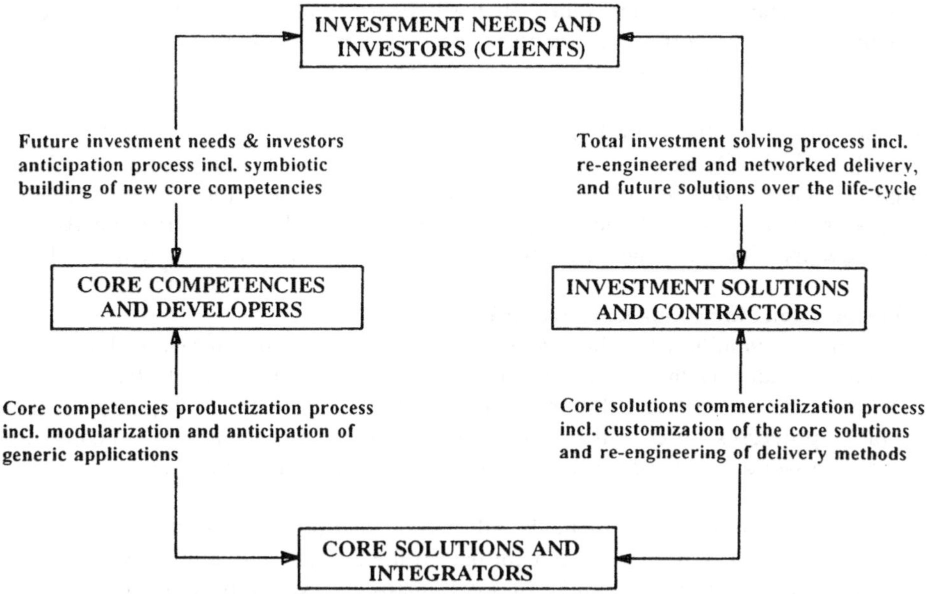

Fig. 3. Definition of competence-based contracting business as a system consisting of four elements and their enabling linkages.

3.2 Benefits of the competence-based contracting business

A new approach is only justified when this will fix the current weaknesses of construction. **The primary benefits** of introducing the competence-based contracting business are envisioned as follows:

- **Radical investment profitability and productivity improvements over the life-cycles**: enabling investors and end users to define and control the realized value for money. For example, in the EU, US and Japanese scale of about the ECU 500 bill. investment markets, a 10 % savings in the initial investment costs would be worth ECU 50 bill.
- **New investment solution-driven competition**: instead of the price only, all the primary attributes of the investment will be included in a set of investor's decision-making criteria, incl. life-cycle costs. New solution-driven codes of competition, and procurement methods will be adopted to replace the current price-driven competition.
- **New radical investment solutions creation:** to attain at least 20-30 % more productive solutions, higher quality, faster execution, lower construction and life-cycle costs.
- **New procurement and performance specification systems:** to allow competition among new kinds of investment solutions which integrate all the elements over the life-cycle incl. divestment aspects. A new performance specification system will offer the basis for quality and productivity measurement, to enable a solution-driven competition.
- **Competence building and leveraging**: managerial cognition and organizational learning will become the driving force of change, and determine the search for competencies. Accumulations of competencies and patterns in uses will determine the future behavior of [strategic competitor groups] as short-run conduct in competence leveraging, as well as long-run conduct in competence building. [17]

- **Sustainable platform for integrating all the industry developments**: to enable the effective use of supporting technologies (such as computer-integrated-contracting systems), as well as concepts and tools (such as TQM) on a large scale.

A superior competence in building competencies through continuous learning ... [ensures] sustainability of competitive advantages better than unique resources [18]. For example, Bechtel Corp. is rolling out **seven standard designs** for gas and coal-fired powerplants [which will] cut delivery time and cost around the world dramatically while assuring quality. Each package ... is tailored ... to meet an owner's requirements. [19]

3.3 Preconditions for competence-based breakthrough

The primary precondition for the emerging competence-based contracting business(es) lies in raising radically the investors' objectives which would ultimately lead to radical improvements in contractors' offerings. This means the adoption of new procurement and cooperation processes which will enable solution-driven competition between competing, integrated networks managed by Solution Contractors. It is hypothesized:

> H7 Unless **the procurement methods** will be renewed by all the parties involved in a contracting business, or market, a major part of the development efforts by single companies - such as re-engineering, lean, total quality (TQM), activity-based (ABM) and time-based management (TBM) - will turn out be ineffective. This is so because in a particular project, today's inflexible methods prevent from combining those quality, time, costs and services improvements that any party has alone achieved.

WS Atkins states that innovation is most likely to come from [those EU investors] demanding better service, not from initiatives by the industry itself. A new framework [to be developed jointly] is suggested to permit investors to choose from **a range of procurement processes** to suit their own circumstances and capabilities - within the context of a system of guarantees, insurance and legal liabilities [20].

The other preconditions lie a.o. in establishing networks, and creating competence developers and solution integrators. It is hypothesized:

> H8 The core competence concept best suits those leading contractors who already possess core technologies and investment solutions, have multiple businesses, as well as operate on large domestic and/or versatile international target markets. They are best equipped **to renew** competitive practices, in cooperation with advanced investors.

WS Atkins states that the EU construction sector ... needs **to remedy structural problems**, such as the separation of design from construction and the uncertain role of the manufacturer in the design process. [The re-engineering] means putting research, ... investment and computer control where it is most effective: in industrial manufacture..., distribution and site process management. The contractors should help suppliers to develop better products ... , to adapt site processes to these, ... and to combine the flexibility of ... specialist and local firms with ... world-beating contractors [21].

On the other hand, WS Atkins recommends the EU becoming the style leader ... world-wide, building on the reputation ... of **Europe's designers**. This would build also on ... engineers, managers and high quality product manufacturers, as opposed to cheap labour, materials and energy, low cost finance and strong government intervention.

4 Concluding remarks

Why should the core competence approach be preferred? The author argues that this concept will, more likely than many others, enable (a) investors to raise radically their investment effectiveness, (b) contractors (as solution contractors) and other parties (as core competence developers and core solution integrators) to ensure profitability and sustainable competitiveness, as well as (c) the contracting [it is today's construction] industry to become one of the most respected sectors in the society.

It remains to be seen, which of the leading EU, US and Japanese contractors will take the lead, and in which of emerging competence-based contracting businesses.

5 References

1. Penrose, E.T. (1959) *The theory of the growth of the firm.* Basil Blackwell, pp. 24-5, 65-6.
2. Prahalad, C.K. and Hamel, G. (1990) *The core competence of the corporation.* Harvard Business Review, May-June, pp. 79-91.
3. WS Atkins International Ltd (1994) *Strategies for the European Construction Sector,* Construction Europe & Office for Official Publications of the EC. p. 14, 22.
4. Huovinen, P. (1984) *The adjustment of construction industries to international investment markets.* Research paper. European International Business Association (EIBA), Proceedings. Dec 17-18, 1984, Oslo. Vol. III, 14 p.
5. Huovinen, P. (1990) *What business directions and concepts an international construction company should consider in the 1990s.* A speech. The "Conditions of Success" Symposium. Oct 4-5, 1990. 12 p.; reviewed in Hawk, D. (1991) *An Agenda for the Future.* Report Two. Stockholm, Sweden, pp. 33-43.
6. WS Atkins International Ltd (1994), p. 81.
7. Heikkilä, J., Huovinen, P. and Silventoinen, J. (1987) *A method for competition analysis in international contracting.* HUT. Construction Economics and Management. Report 82. Espoo, Finland, pp. 29-31. (in Finnish)
8. Compare with Porter, M.E. (1980) *Competitive strategy.* Free Press, pp. 161-3.
9. WS Atkins International Ltd (1994), p. 80, 82 and 84.
10. Engineering-News-Record (1995) *Industry embraces design-build.* May 22, p. 74.
11. Prahalad, C.K. and Hamel, G. (1990) *The core competence of the corporation.* Harvard Business Review, May-June, pp. 79-91.
12. Sanchez, R., Heene, A. and Thomas, H. (1995) *Towards the theory and practice of competence-based competition.* John Wiley. pp. 1-2.
13. The original system is designed by Heikkilä J., Huovinen P. and Warras T. (1995) *Renewal of paralized building construction processes - Finland as the initiator in the EU?* Espoo, Finland. 5 July 1995. A policy paper (in Finnish, draft). 9 p.
14. WS Atkins International Ltd (1994), p. 103.
15. Engineering-News-Record (1995) *Going global, contracting ... .* Aug 28, p. 30
16. Ibid, p. 36
17. Sanchez, R., Heene, A. and Thomas, H. (1995), p. 33.
18. Ibid
19. Engineering-News-Record (1995) *Delivery being streamlined.* July 24, p. 16.
20. WS Atkins International Ltd (1994), p. 6, 14.
21. Ibid, p. 15, 103, 108.

2 The firm and its competitive partners

ADVERSARIES OR PARTNERS? DEVELOPING BEST PRACTICE FOR CONSTRUCTION INDUSTRY RELATIONSHIPS

A.J. Hinks and S. Allen
Research Centre for the Built and Human Environment,
University of Salford, Salford, UK
R.D. Cooper
Research Institute for Designm, Manufacturing and Marketing,
University of Salford, Salford, UK

Abstract
The Construction Industry has traditionally operated through competition, driven by the client's need to achieve the lowest-cost bid. As a result, relationships have tended to be adversarial, with the parties concerned resorting to contractual claims, which lengthen timescales and drive up costs. The Latham Report identified this sort of adversarial relationship, together with industry fragmentation, as the greatest barriers to improving quality and productivity. If Latham's recommendations for improving the Construction Industry are to be achieved, then it is necessary to develop better relations between clients, contractors, and suppliers.
Key Words: Communications, Relationships, Partnering.

Sommaire
L'Industrie du batîment fonctionne traditionellement dans l'environnement de la compétition poussé des besoins du client pour obtenir le meilleur prix. En conséquence, les rélations ont été hostiles, tous les interessés font des réclamations de contrat, çela rallonge les delais convenus en faisant monter les prix. Le Report de Latham indiqua que ces rélations hostiles considerées conjointement avec la fragmentation de l'industrie sont les plus grandes barrières á l'amélioration de la qualité et la productivité. Si l'on va pouvoir réaliser les récommendations de Latham pour améliorer l'industrie, il faut qu'on améliore les rélations entre le client, l'entrepreneur et le fournisseur.
Mots Justes: La Communications, Les Rélations, L'Association.

The Organization and Management of Construction: Shaping theory and practice (Volume One).
Edited by D.A. Langford and A. Retik. Published in 1996 by E & FN Spon. ISBN 0 419 22230 8.

1.0 Introduction

This paper will work from the premise that *co-operation* in business and commerce, including construction, is an essential ingredient for success. Through co-operation both risks and benefits can be shared, allowing businesses to develop and improve product / service quality. If construction is to be considered a team activity [1], then co-operation and effective communication are assumed to be essential for the production of high quality structures, on-time and within budget. With the contemporary internationalisation of competition and construction markets, and the supremacy of a client-oriented culture in other industrial and commercial sectors of UK society, there is a profound need to address the Construction Industry's stereotypical shortcomings over communication and conflict.

A generally poor economic climate [2], allied to the gathering momentum for reform of the industry's traditional ways of doing business, seems to have resulted in an industry-wide perception of a 'last opportunity' for change. These pressures have created a mood for change at a high level within construction companies. Against this backdrop of change, the need for new working philosophies and techniques, such as those already adopted and refined in the 'sister' industries of manufacturing and process engineering, is clearly recognised at the policy-making level of the Construction Industry. Catalysed by Latham's contemporary focus, the industry response has now turned its focus onto relationships.

This paper discusses the conflict and commercial practices within the Construction Industry, the types of relationships which exist, and the scope and potential for more co-operation. To date , co-operation has not been a universal premise for business within the industry. The paper also reviews *partnering*, as the term is understood in the construction and manufacturing industries, and the possibility of mutual lessons is discussed. Also, industry requirements for adopting such an approach successfully is considered, as is how the degree of success of such arrangements can be evaluated.

1.1 Conflict in Construction

Historically, the process of construction has been associated with conflict. Traditional procurement systems and the contractual and legal framework by which participants are bound together are often criticised as being confrontational and adversarial [3]. This is in combination with competitive tendering which seeks to drive costs down initially. Subsequently this action leads to disputes and conflict, which revolve around financial self-interest, between the various stakeholders (e.g. client, design-team, consultants, main contractor, sub-contractors, suppliers) throughout the construction process [4]. A consequence is that final costs of the project usually exceed the initial agreed price, and the conduct of the process is characterised by confrontation.

Numerous enquiries spanning 50 years have echoed the recurrent causal themes of poor communication compounded by fragmentation within the industry [1,5,6,7,8]. Thirty years ago Higgin and Jessop [8] identified *'stubborn difficulties of communication between various groups and elements'* involved in construction, affecting productivity and quality. They went on to state that if the relationships between stakeholders could be improved, then the issue of communications would take care of itself.

In the 1990's this is still a central problem for the UK Construction Industry, as highlighted by the Latham Report and the industry-wide recognition of its significance. Partnering is being proposed as a possible solution to this problem of communications and conflict by both industry leaders and business consultants [9,10,11], and for many is regarded as a component of the solution to the Construction Industry's problems with relationships. For many it is *the* central component. This concept of working together with common goals and objectives has seemingly worked successfully in other industries, such as the automotive industry [12,13]. So what is partnering, a radically new business philosophy or merely a technique which has the characteristics of good business practice? How does it differ across industries?

2.0 Partnering in Construction

Evidence indicates that the concept of partnering, of common goals and shared objectives between firms, has been around a long time in the construction industry with many developed relationships. One such example is Bovis, who have developed a culture and tradition of non-adversarial relationships with particular clients since the 1930's. Recently attempts have been made to label such arrangements as 'partnering', a new phrase for existing situations and relationships. In practice, the term partnering covers a broad spectrum of relationships and each is flavoured by the particular stakeholders within the arrangement. Further, the speed and urgency of 'partnering' developments is in itself contradictory to the philosophy of earned trust and developed relationships. The underpinning element of all relationships is people, and a central requirement of a more long-term industry is the attitude of holding onto its people to allow long-term relationships to develop.

Such a partnering approach to construction activity has been the subject of much recent research [9,10,11,12]. Whilst contemporary reports clearly advocate the use of partnering, published material to date does not take an objective and rigorous approach to qualifying and quantifying the benefits of partnering, merely operating as a vehicle to reiterate unquestioningly the positive views of consultants and salesmen. Indeed, the criteria for measurement are not agreed and the industry appears not to fully comprehend what it is claiming. Whilst it is reasonable to assume that co-operation is a good thing, unless any benefits can be objectively quantified and presented convincingly to the industry, then the necessary cultural changes will simply not occur or endure. This is a barrier to change in an industry oriented towards measurements of hard criteria (such as time, cost and possibly quality), an industry which will not easily recognise and adapt to soft and intangible criteria (such as stakeholder expectations and perceptions).

In construction, partnering currently has many definitions which makes analysis difficult. One form of partnering is that of an *attitud*e (or philosophy) which itself summarises 'good practice' in terms of dispute / conflict avoidance and minimization (as opposed to dispute resolution). There is also 'partnering' which could be termed a building procurement *technique*. This tries to operationalise the good practice, to bring about cultural change and thus create a more cohesive 'team'. However, one thing is certain, partnering is a complex process and will involve greater difficulty than current advocates would suggest, but benefits are by no means inaccessible.

The Construction Industry must consider whether the current widespread enthusiasm for partnering signifies a change of philosophy or merely the introduction of another procurement technique. There is the risk that the naive optimism designed to encourage uptake of a practice which is not yet fully understood, will create unrealistic expectations which could retrench attitudes if partnering appears to fail. Can an industry epitomised by the centrality of its adversarial relationships and managerial procedures change *en masse* to embrace all that a partnering philosophy implies and requires? Is partnering an end in itself or the means to an end? Is partnering a philosophy of process and relationship, or the outcome of a novel technique of process and relationship?

2.1 Styles of Partnering in Construction

Understanding such issues needs to be prefaced by a definition of partnering and its context within construction. Recent work [10,11] falls into the category of regarding partnering as a central component for alleviating construction's problems, almost the 'universal panacea'. Writing more with the conviction of consultants than with the objectivity of academics, the above work is based on the premise that partnering is a good thing and can be implemented through a *formal approach* including such procedures as preselection of contractors, teamworking workshops, the identification and promotion of 'champions', a role for a faciliator and a form of contract. A plausible case is presented. In this sense partnering provides a mechanistic and prescriptive solution for improved management of the building process. Again the work does not present adequate analysis of the measurement of costs and benefits attributed to the partnering approach. What else does partnering mean?

> *'Partnering involves contractors and clients linking up to work with each other on a long term basis to slash project costs and timescales' (Consultant view)*

If partnering is understood to be the creation of closer relationships between client and contractor with the joint exploitation of sub-contractors and suppliers in the traditional spirit of adversarial cost minimisation, then this is partnering short of the philosophical. Alternatively, if partnering is a 'quickwin' marketing tool then it may operate at a rhetorical level only: *'dream teams producing quality buildings' (Consultant view).* This may be sufficient to create a project 'feelgood' factor and so achieve mutual benefits. Furthermore, partnering can mean the development of relationships built on long term, earned trust and the sharing of knowledge, technology and even personnel and finance. This is the embodiment of a shared philosophy, the result of an attitude of continuous improvement and reflection:- *'Partnering is a very positive way forward for the industry and something that should be encouraged as an alternative to the adversarial nature of many current contracts' (Construction company director).* Such an understanding of partnering is closest to that of the manufacturing industry, but more of this later.

So, conceptual differences in the meaning of partnering clearly exist within the construction industry. Can a common set of parameters be identified and used to assess these diverse forms of relationship. Moreover, is this desirable and over what period of time should these be measured? Further, does partnering emerge from an existing relationship, or can it be used as a philosophy or technique to design the relationship? Can partnering, as a technique, address the causes of relational conflict evident in the industry?, or is nothing short of a philosophical phase change needed? Without a clear definition of the expectations and outcomes of *all* the parties involved, partnering in construction could be destined to induce further widespread dissatisfaction and dissillusionment amongst stakeholders.

Perhaps partnering, seen in this ambiguity, prior to rigorous analysis and subject to inflated expectations, is and can be all these things. Research must address this ambiguity, confirming good practice in partnering and considering measurement of success or failure of a partnering approach and the design, selection and appraisal of relationships based upon appropriate criteria. So, partnering is a particular approach to business, an attitude and a necessary part of a co-operative industry. This needs to be understood, appreciated and acted upon before it is possible to set down guidelines and before partnering can be used as a procurement strategy in its own right. Through examination of partnering in a different context, in this case the manufacturing industry, insights for the development and implementation of partnering attitudes may be gained.

3.0 Partnering in Manufacturing

As previously stated, 'partnering' and co-operative working is a well established feature of other industries, as market pressures in the early 1970's created the need to innovate and to collaborate in order to compete [13]. Firms in the manufacturing and particularly automotive industries [13,14] have adopted and developed long-term business relationships successfully and now have more mature and sophisticated partnering arrangements. In the last 20 years, Manufacturing has identified that the two central factors determining the competitiveness of manufacturing firms are *internal efficiency* and the *management of external relationships with customers and suppliers* [13]. Companies have turned to partnering, using preferred suppliers and developing customer and supplier relationships. A recent study [15] demonstrated, for instance, that 63% of UK engineering companies planned to enter into partnership agreements with preferred suppliers in the next three years.

Research work on partnering in manufacturing has now advanced to the stage of developing and refining customer-supplier relationship models. Examples of this include Lamming's four phase model and his more recent developments, beyond partnership, of a lean supply model [13]. The 'four-phase' model identifies the phases chronologically as traditional (pre-1975), stress (1972-85), resolved (post-1982) and partnership or 'Japanese' (post-1990). The degree of *integration* and *collaboration* between firms being the crucial differentiating factors between these phases.

This development of long-term relationships in order to understand customers needs, improve quality, and reduce design and production timescales has been used extensively in the automobile industry. Here first and second tier suppliers have been encouraged to work closely with the automobile manufacturer to solve problems and develop innovative design solutions, with consequent production and manufacturing benefits. Furthermore, it has been broadly recognised by the Manufacturing Industry that design and development is a team activity. In order to ensure that the needs of the end-user and all stakeholders in the manufacturing process are addressed, and that the process maintains a learning culture, all these parties must be represented throughout the process. Hence it is accepted that successful product development requires a joint venture between the manufacturer, client, and all key suppliers.

Hamel et al [16] identified that successful collaborators are those which see joint ventures as learning exercises. Contractor and Lorange [17] conclude that there are a number of synergistic benefits for the supplier and the supplied with partnering: such as *risk reduction,* since the supplier is more willing to invest capital and staff in the shared mode; *economies of scale;* and *technology exchanges.* In addition, *increased stability,* allowing business systems to be customised for a known set of transactions; *better control,* as attention and resources are focused on few sharply defined goals; and *more direct investment,* as goals and objectives are shared, are also possible.

4.0 Learning Lessons

Some initial work has already been conducted into the appropriateness of applying manufacturing based concepts in construction. Relational contracting [18], transaction cost analysis [19], supply chain management and lean productivity [20] are all areas which seem to offer scope for wider application in construction with possible resultant improvements in quality productivity. The basis of these concepts being the development of good relationships built through collaborative working attitudes. Further, considering the Japanese concept of 'Kaizen', which requires continuous improvement and the adoption of Total Quality Management, and of the 'Keiretsu', which are horizontally connected and vertically integrated production networks, from the Japanese model of manufacturing development [21], will also provide the possibility of insights. Analysis of these methods of working may offer means to identify, and then to facilitate the cultural changes required within the Construction Industry.

Transferring concepts from one industry to another is complex and fraught with difficulty. The manufacturing and automotive industries provides potential scope for improvement but caution is required. For example, is manufacturing dealing with the same inherent product complexity, and is it faced with the particular attitude and cultural nuances which exist within the Construction Industry? Also, whilst it is widely accepted that these industries have been at the forefront of developing partnering relations, they are now considering what lies beyond partnering. Indeed, it is reported they are considering the 'network' structure of the construction industry! It seems that both industries may learn from each other.

4.1 Analysing Partnering Relations

The analysis of relationships is another complex area, yet it is necessary if best-practice guidelines are to be developed. A useful starting point would be to define the parties to the relationship. Who are the stakeholders? Where is the balance of power? What are the culture and management styles? This can then be followed by the development of approaches to changing them in order to improve the industries productivity. The aim would be to change the philosophy to one of collaborative rather than adversarial relations, whether it is called 'partnering' or not.

It can be dangerous to give too much credence to generalisations, but, they can form the basis of useful analysis and subsequent action. So, in construction it is widely accepted that the commercial activity of companies within the industry tends to be adversarial, driven by the need to achieve the lowest prices through competitive tendering. A direct result of this is the prevalence of project disputes and a general air of conflict and even animosity between various stakeholders. Such relations could be said to constitute a *spectrum of relationships*, running between adversarial and collaborative relations. This is directly linked to a 'procurement strategy' spectrum which is defined in terms of contractual obligations. A basis for the development of these spectra may be Sako's spectrum of relationships from *arms length contract relation* to *obligational contract relation* (ACR-OCR) [22]. Using this model, any construction project should be able to be located through the identification and measurement of particular job specific attributes and criteria. Partnering, as a technique and measured according to stakeholder specific criteria, would be able to be located on this spectrum.

Also from the field of manufacturing, Lamming has identified 9 factors used to analyse customer-vendor relationships [13], and has carried out work to measure the success of relationships and is developing a model for inter-organisational relationship assessment. Such models could provide ideas for identifying parameters, and developing measures and ideal outcomes applicable to the construction industry. Analysing relationships and measuring corporate cultures and management styles will provide insights into the potential for partnering arrangements and also the possibility of identifying 'good' partners in advance of contractual obligations within the Construction Industry.

5.0 Action Summary

When analysing relationships in the construction industry, Systemic Stakeholder Analysis, a 'soft-systems' tool for use in messy problem situations, could be used to help rethink who is involved, who should be involved and potentially facilitate a redefinition of the 'power-balance'. The construction industry must go beyond 'technique' to a new 'philosophy' or 'culture' if the benefits from a manufacturing interpretation of partnering are to be achieved. The task of this research, an EPSRC funded project, will be to identify costs and benefits of partnering. To do this there is a need to develop measurement criteria for the Construction Industry to monitor its performance and ensure fitness for purpose. To meet these requirements the research will identify measurement criteria (tangibles / intangibles), based upon the critical *Cost, Time, Quality* factors, and the critical client specific factors or 'hot-buttons' which may be more important. This research will identify what is needed for construction partnering by examining developed relations in manufacturing and drawing parallels. This process could be the beginning to addressing the transformation of partnership arrangements as a whole in construction.

It is necessary to advise caution in any advocacy of partnering and to be wary that expectations are not falsly raised before delivering disapointment. Radical cultural change requires time for it to be successful. Sustainable benefits will not be achieved overnight. A long-term view is essential to allow relationships to develop and the Construction Industry and its clients will need to become more reflective when assessing relationships and the criteria which are used.

The Construction Industry must understand the committments required of the various stakeholders to fully appreciate the concept of partnering and to adopt this as successfully as manufacturing. A sense of realism needs to be brought to bear in the current partnering debate; the industry's pragmatism may be its greatest asset here in setting achievable targets which will avoid disapointment and further harm to an already suffering industry.

The contracting sector of the Construction Industry has taken a central role in the maintenance of adversarial approaches, and is accordingly in a prime position to change them. The Latham Report identified the adversarial nature of the industry as a major failing, and focused on the need to reduce conflict and improve efficiency. The report suggested that the solution to the chronic problem of an industry entrenched in an adversarial philosophy lay in teamwork and partnering. Indeed, Latham recommended that *'advice should be issued on partnering arrangements'*, with clients the key as already recognised in manufacturing. The task of developing a partnering culture in construction does have something to build upon as developed historical relations exist, some of which were established over 60 years ago. These partnering experiences need to be considered objectively and good practice identiifed to help the industry and its companies become stronger in light of a changing business environment and increased competition. The industry must truly want to adapt and change its culture if the potential benefits are to be realised. Current liaison with academia is a start, and this will aid the process of reflection and help to identify good practice and meaningful measurement criteria.

NOTE: The authors would be pleased to hear from other researchers or industrialists with specific interests in collaborative working practices and the development of best-practice guides to partnering in the Construction Industry.

6.0 References

1. Latham, M. (1994) *Constructing the Team: Joint review of Procurement and Contractual arrangements in the UK Construction Industry.* London, HMSO.
2. Department of the Environment. (1995) *State of the Construction Industry* February 1995, Issue 3.
3. Fenn, P. and Gameson, R. (1992) *Construction Conflict Management and Resolution,* London, E & FN Spon.
4. Loosemore,M. (1994) Problem Behaviour. *Construction Management and Economics,* Vol. 12, No.6, pp.511-520.
5. Simon (1944) *Report of the Committee on the Placing and Management of Building Contracts,* HMSO, London.
6. Emerson, H. (1962) *Survey of Problems Before the Construction Industry (The Emerson Report),* HMSO, London.
7. Banwell (1964) *Committee on the Placing and Management of Building and Civil Engineering Work,* HMSO, London.
8. Higgin, G. and Jessop, N. (1965) *Communications in the Building Industry,* Tavistock Publications

9. Bennett, J. and Jayes, S. (1995) *Trusting the Team; The best practice guide to partnering in construction,* The University of Reading, Centre for Strategic Studies in Construction.
10. Baden-Hellard, R. (1995) *Project Partnering: Principles and Practice* Thomas Telford Publications, London.
11. Lorraine, R.K. (1994) Project Specific Partnering. *Engineering, Construction and Architectural Management,* Vol.1, No. 1. pp. 5-16.
12. National Economic Development Council. (1991) *Partnering: Contracting without Conflict.* NEDC, London, June 1991.
13. Lamming, R. (1993) *Beyond Partnership: Strategies for Innovation and Lean Supply,* Prentice-Hall.
14. Cooper, R.D. & Leverick, F. (1995) *Supplier Role in Product Development: A study of the automotive industry,* Final Report, University of Michigan's Japan Technology Management.
15. Benchmark Research. (1993) *The 1993 Manufacturing Attitudes Survey,* Basingstoke, UK: Computervision Ltd, January.
16. Hammel, G., Doz, Y. and Prahalahad, C. (1989) Collaborate with Your Competitors and Win. *Harvard Business Review,* Jan-Feb.
17. Contractor, F.J. and Lorange, P (1988) *Co-operative Strategies in International Business: Joint Ventures and Technology Partnerships between Firms,* Lexington Books, Lexington, MA.
18. McDermott, P. and Alsagoff, S.A. (1994) Relational Contracting: A prognosis for the UK construction Industry? in *CIBW92, Procurement Systems Symposium, East Meets West,* Hong Kong, December 1994.
19. Alsagoff, S.A. (1995) Construction Transaction Cost Conflicts: Analysis of Dispute Triggers in Construction Contracts, in *TG15: Construction Conflict: Management and Dispute Resolution,* Japan, October 1995.
20. Koskela, L. ((1992) *Application of the New Production Philosophy to Construction.* CIFE Technical Report No.72, Centre for integrated Facility Engineering, Stanford University, Stanford, California.
21. Miyashita, K. and Russell, D.W. (1994) *Keiretsu: Inside the Hidden Japanese Conglomerates,* McGraw-Hill, Inc., NY.
22. Sako, M. (1992) *Prices, Quality and Trust: Inter-firm relations in Britain and Japan* Cambridge University Press: Cambridge.

IMPROVING THE MATERIALS SUPPLY SYSTEM IN SMALL SIZED BUILDING FIRMS

C.T. Formoso
NORIE – Federal University of Rio Grande do Sul, Porto Alegre, Brazil
V.H. Revelo
Fundacion Ecuatoriana de Ingenieria, Quito, Ecuador

Abstract
This paper presents a research study aimed at developing a method for improving the materials supply system in small sized building firms using TQM principles. The study involved three companies from the Brazilian building industry, which worked cooperatively through several stages of TQM implementation. The proposed method is based on simple well known quality techniques for problem identification, analysis and solving, such as flowchart, brainstorming, checklist, Pareto diagram, etc. The paper points out some of the difficulties in applying such techniques and principles in small sized building firms and discusses some typical problems found in materials supply systems of such companies in Brazil.
Keywords: Quality management, materials management, materials supply system

Resumé
Cet article présente une recherche qui a pour but le développement d'une méthode pour améliorer le système de fourniture de matériaux dans les petites entreprises de bâtiment, en utilisant les principes de gestion de la qualité totale. L'étude a entouré trois entreprises de l'industrie du bâtiment brèsilienne, qui ont travaillé conjointement au travers les différents stages de l'implantation de la gestion de la qualité totale. La méthode proposée s'est basée sur des techniques de qualité bien connues pour l'identification de probléme, analyses et résolution, telles que diagramme de flux, "brainstorming", "checklist", diagramme de Pareto, etc. L'article met en evidénce les difficultés de l'application de telles techniques et principes dans les petites entreprises, et met en question quelques problémes typiques rencontrés dans les systémes de fourniture de matériaux de ces entreprise au Brèsil.
Mots clés: Gestion de la qualité, gestion de matériaux, système de founiture de matèriaux

The Organization and Management of Construction: Shaping theory and practice (Volume One).
Edited by D.A. Langford and A. Retik. Published in 1996 by E & FN Spon. ISBN 0 419 22230 8.

1 Introduction

Materials account for a very large percentage of building costs. In recent years, the proportion of building materials in terms of cost has increased more than labour. In Brazil, the cost of materials can amount to as much as 60 to 65% of the project cost [1]. In addition, the lack of materials is one of the most common causes of delays in construction [2].

Despite the high percentage of materials in the total cost, the materials supply process is often neglected. While the manufacturing industry spends 1% of its costs in the supply system management, the construction industry spends only 0.15% [3].

Some studies in the field of materials management have concluded that effective materials management systems will result in 6 to 8% improvement in labour productivity, improved cash flow, reduced bulk materials surplus, reduced materials management human resources, improved vendor performance, reduced requirement for physical warehouse facilities, quantity purchasing discounts, minimized cost impact of change orders, and fewer project delays [2] [3].

The process of supplying materials to building sites is fairly complex, even for a moderate construction project. It involves many steps, carried out in several different departments, such as design, purchasing, production, accounting, etc. The number of materials is very large, and there is a wide variety of suppliers, ranging from small sized low technology firms to modern capital intensive industries.

The efficiency of materials supply is also strongly related to the strategic posture of the firm, because it is dependent on the kind of relationship that is established with each supplier.

This paper describes the main results of a research project aimed at developing a method for improving the materials supply system of small sized building firms using TQM principles and techniques. The study involved the development of improvement projects in three different companies which were working cooperatively in quality management programmes.

2 Materials supply process

Leenders & Blenkhorn [4] classify the contribution of the materials supply system towards achieving the objectives and aims of the organization in three stages:

- Negative stage: materials supply is organized in a bureaucratic way, rather than strategic. Many important decisions in the organization are taken without considering the materials supply system. Each element in the net works independently, without taking into account the global impact of their decisions.
- Neutral stage: supply operations are managed by a specific department or sector, which works towards eliminating some of the inefficiencies in the supply process. However, materials supply still has a limited role in the establishment of the company strategy.
- Positive stage: the materials supply system is thoroughly integrated in the organization, participating in the definition of important strategic guidelines. The staff involved in materials supply have to identify important strategic opportunities

through their work. Besides, the other sectors are encouraged to support the effort to improve the materials supply system.

Materials supply should be considered a process, rather than a department, which involves a complex net of internal clients and suppliers. Decisions are made at distinct managerial levels. The supply net can be seen in three levels [5]:

- Internal supply net: includes all departments of the firm involved in the supply process, such as design, planning and costing, purchasing, accounting, production, etc.
- Immediate supply net: includes direct suppliers and clients.
- Total supply net: includes all suppliers involved, such as suppliers of suppliers.

It is necessary to understand the needs and limitations of other intervenients in the supply net, in order to improve the performance of each individual operation and turn the whole process profitable. If there are inefficiencies in any part of the net, this may affect the performance of other participants.

It may be difficult to consider the total supply net in decision making since most of the net is out of the firm's control. However, such a view brings several advantages to materials supply management: (a) places the materials supply operations in a competitive context; (b) helps identifying key partners; (c) changes the focus from short term opportunism to long term profitability; (d) avoids sub-optimal solutions; and (e) enables the organization to perform major changes in its strategy [5].

This study is limited to improving the performance of the materials supply net by intervening in the internal supply net, because of the limited time and resources for carrying out the research. There is another work being developed at the Federal University of Rio Grande do Sul which aims at improving the relationship between small sized construction companies and their suppliers.

3 Methods and techniques for process improvement

One of the main difficulties for improving a process is the fact that decision making tends to be *ad hoc*, intuitive, not based in data and facts. Besides, errors tend to be attributed to particular circumstances, such as the performance of individual people or departments, rather than to the efficiency of the whole system [6], because of the lack of a systemic understanding of the organization.

Shared decision making is one of the key principles of TQM. It ranges widely, from joint consultation to full delegation [6], and is usually based on the work of quality teams.

There are some simple well known techniques which can make the work of quality teams more effective and organized. The most common ones are: flowchart, brainstorming, checklist, Pareto diagram, cause-effect diagram, histogram, and control chart. These techniques inherently require participation and involvement by all members [7]. They also encourage the use of data and facts in decision making and the development of a systemic understanding of the process by all the people involved, including the notion of internal client-supplier relationships.

Ishikawa [8] points out that those seven quality techniques are effective in 95% of the problems. However, although previous work has shown that learning each of the techniques is relatively easy, there are major difficulties in carrying out an improvement project, which are mostly related to the lack of involvement of the quality team members and the need for a consistent method, i.e. the logical sequence in which such techniques are applied.

4 Materials supply improvement project

The method for improving the materials supply system was developed in three different small sized building companies which were carrying out quality programs in a cooperative way within a group of six firms. All of them had less than 100 employees, used traditional building technologies, and were mostly involved in speculative house building.

Cooperative TQM implementation has already been practiced in other countries, such as Norway, and has been widely adopted in Brazil for implementing TQM in small sized companies. It brings a number of advantages, as pointed out by Hansen & Sjoholt [9]: (a) companies learn a good deal from each other; (b) each firm tends to engage themselves thoroughly in the project, because of the pressure from the other ones to report on the internal progress at each meeting; (c) they get to know each other better, a factor that can be useful in other contexts; and (d) the cost of consultancy work becomes cheaper.

Each of the three companies had a quality committee, formed by four to eight people, who met regularly - every one or two weeks. The chairperson of the committee was the company quality manager. The members of the quality committee were representatives of the different departments or sectors of the firm. The work of the committee was supervised by a researcher from the University. Usually once a month there was a meeting gathering directors and the quality managers of the six companies and the research team to present the work carried out so far and to plan the next stages.

Differently from the concept of quality control circles, the quality committees had a more strategic role in the quality program. The main objective of the committee was to plan and control the quality program, and also to carry out interdepartmental improvement projects.

Initially, it was suggested that the directors of the company should not participate in the committee, since the aim was to delegate some of the higher management responsibilities to that body. However, the participation of the directors in the committee was of key importance in some of the companies, since in a small sized firm even the owners get directly involved in materials supply, production, and other operational processes.

One of the initial actions of the quality program was to give a 16 hour course on quality concepts and tools to the participants of the committee. After the course, the decision was made to develop a quality improvement project which would be common to some of the companies in the group, in order to practice the application of concepts and techniques.

An improvement project concerned with the materials supply process was chosen by three of the companies for the following reasons:

- There was a need to consolidate the quality committee, and to train their members on the application of quality concepts and techniques. As the supply system involves several departments, it was possible to engage most members of each committee in the project.
- It was desirable to develop in each member of the committee a broader view of the firm and a clear understanding of the interrelationships between different departments, as well as a notion of internal supplier-client relationships. Developing an improvement project in the materials supply system was an opportunity to understand the inter-dependency between departments.
- All companies had defined as a priority to deal firstly with problems mostly related to the main office, and only later to work on issues related to site management.
- Deficiencies in the supply system had been pointed out as one of the three most important problems in those firms.

5 Application of the method

5.1 Problem identification

The objective of the first stage of the method was to identify the most important deviations in the process and their causes. It was divided into three main steps: (a) definition of the actual process flow and identification of internal client/suppliers relationships; (b) brainstorming; and (c) prioritization of problems.

A flowchart was used to represent formally the materials supply process. Only one of the companies had a material supply flowchart before the project started. However, that flowchart did not represent the real process, since it had not been produced with a strong commitment to model reality, and based on the views of all the people involved.

Some problems were actually been solved while the flowchart was being built. For instance, there were problems related to the lack of responsibility definition which had to be negotiated during the meetings of the quality committee. Also some unnecessary zigzags were identified and eliminated. That had not happened before since the process had never been formally registered or discussed in a systematic way.

The initial improvements carried out at that stage were mostly related to documentation formatting, introduction of new documents, minor changes in the process flow, and decentralization of responsibilities.

The resulting flowcharts were relatively complex, because of the complex nature of materials supply. A typical materials supply flowchart can have more than 50 steps, which makes it difficult for some of the members of the committees to understand the process thoroughly.

The production of the flowchart was one of the most time consuming stages of the improvement project. This was because of the initial difficulties in starting the group work in the quality committee. People were not used to making decisions in groups, and several of them found it difficult to apply the flowchart technique to such a complex process.

Brainstorming was applied in identifying the main causes of deviation in the materials supply process, based on the perceptions of all the people involved. A comprehensive list of possible deviations and their causes was produced for each company.

Compared to the other problem identification, analysis and solution techniques, brainstorming was the easiest one to be applied. It provided an opportunity for exchanging views and experiences and allowed the participants to learn more about each other's role in the process, increasing the consciousness of internal supplier-client relationships. Very often there were conflicts between different participants of the team, usually caused by disagreements on the way the process works or should be working. However, such conflicts had a positive influence in the project, since they encouraged people to have a constructive attitude towards solving the problem [6]. As the participants defended their point of view, they tended to assume more responsibility in the process.

Once all possible deviations were listed, they were classified according to the department which they were related to. These were: production, design, finance, accountancy, central depot, site depot. An Ishikawa (or cause-effect) diagram was built for visualizing the relationships between deviations and their causes.

The perceptions of the participants was again considered for ranking the deviations, in terms of seriousness and frequency. Twenty six of them were chosen as the most important ones for further investigation.

5.2 Data collection

5.2.1 Check list

A number of check lists were made to register the frequency of different deviations in the departments involved. Each department was responsible for collecting data related to the sub-processes in which they played the role of internal client. For example, the purchasing department were in charge of counting the number of urgent material orders.

The check list was applied for a period of two to three weeks. It allowed the comparison of the quality committee perception with reality.

The application of the check list was not straightforward. Although that technique seems to be very easy to understand in theory, it was necessary to carefully train the people responsible for collecting data. In addition, a pilot study had to be carried out to avoid ambiguity in the definition of each deviation.

5.2.2 Time monitoring

A time sheet was used for monitoring both the processing time and the cycle time for each task in the process. This data collection technique was applied during a one month period in each company.

Processing time is the time effectively spent to perform a task, without considering the waiting time. It can be used for estimating the cost of labour in the whole process.

The cycle time is measured by observing the exact moments when the task starts and finishes. It includes both the processing time and the waiting time. The difference between the processing time and the cycle time for each task is a measure of the process efficiency, since it allows the detection of excessive waiting time.

Time monitoring was one of the most time consuming stages of the materials supply improvement project in terms of preparation and training, since it was necessary to instruct all people involved both in defining very clearly when each task starts and finishes and on how to estimate the processing time. One of the difficulties was the fact that the process was not properly consolidated, and there was some confusion on the way some documents should be processed or approved.

Although the resulting data were approximate, they provided a level of magnitude on the way time was spent, which was helpful for identifying some bottlenecks in the process.

5.3 Data analysis

5.3.1 Pareto Diagram

The deviations were ranked by the frequency using Pareto diagrams. A general Pareto was built including all deviations. In addition, some Pareto diagrams were produced considering the deviations related to each department.

5.3.2 Cycle time and processing time

Table 1 shows the resulting cycle time and processing time for each materials supply task in one of the companies involved in the study. The long waiting time in the tasks "checking stocks" and "transport and delivery" indicates these are the stages in which there is potential for improvements in that company.

Table 2 presents the average cycle time and processing time for the three companies. Companies A and C have similar cycle time and processing time, while these figures in Company B are considerable higher. The proportion between cycle time and processing time in all companies were very similar. The former is approximately three times the later.

Table 1. Average cycle time and processing time for each task in one of the case studies

Task	Cycle time (h)	Cycle time (%)	Processing time (h)	Processing time (%)
Materials order	0.14	0.52	0.14	2.54
Checking stocks	7.47	27.98	0.66	12.45
Price survey	4.03	15.09	0.93	17.55
Purchase approval	0.44	1.65	0.43	8.11
Acquisition	0.23	0.86	0.18	3.40
Transport and delivery	12.77	47.83	1.66	31.32
Receiving materials from depot	0.53	1.99	0.50	9.43
Receiving materials from suppliers	0.56	2.09	0.53	10.00
Payment	0.53	1.99	0.27	5.10
Total	26.70	100.00	5.30	100.00

Table 2. Average total cycle time and processing time in the three case studies

Company	Cycle time	Processing time
A	26h29min	14h52min
B	119h21min	32h52min
C	26h42min	15h18min

The total cost of the process was estimated by considering the cost of labour, based on the processing time and other costs, such as telephone, fax, and stationery. It does not include capital costs. Table 3 presents the percentage of the total cost in relation to the monthly amount of purchases made in each of the companies. In all three companies the materials supply process was mostly done manually.

It can be observed that all companies have very similar costs. However, the percentage in relation to the total cost of the purchased materials varies considerably, from 4.41% to 14.58%. It must be pointed out that, although the relative cost of the materials supply process is a measure of its efficiency, its optimization should not be seen as a major objective of the company, since this could jeopardize the aim of global efficiency.

Table 3. Percentage of the total cost of the process in relation to the monthly amount of purchases

Company	Number of sites	(a) Materials supply process cost (US$)	(b) Monthly amount of purchases (US$)	a/b (%)
A	4	4,343.94	98,450.00	4.41
B	8*	4,192.00	48,023.63	8.73
C	3	4,447.00	28,538.00	15.58

* Refurbishment work

5.3.3 Prioritization

Based on the analysis of Pareto diagrams and the average cycle time and processing time for each task, a prioritization of the problems detected was made. Such prioritization was important to establish a corrective action plan (see item 5.4). The results of this stage were very similar for the three companies. The main problems detected were:

- Problems related to design, such as delays, incompleteness, lack of details, and inconsistencies between different designs (e.g. structural and services);
- Lack of planning and organization of transportation and delivery of materials;
- Materials ordered on short notice or verbally;
- Incomplete or inconsistent materials specification;
- Lack of estimation of the amount of materials needed;
- Delays in price surveys (or suppliers bidding), and in ordering materials;
- Delays in checking stocks.

5.4 Corrective action plan

Through discussions involving all participants of the quality committee, a corrective action plan was made. A clear definition of aims and responsibilities was achieved by describing each action in terms of 5W1H (What? When? Where? Why? Who? How?), which is another well known quality management technique. Because there was a concern related to cost-benefit relationships, another "H" (How Much?) was added to the plan.

The improvements chosen for the plan aimed at blocking completely the causes of deviations. For that reason, the plans included both long and short term improvements. They were mostly related to supply planning, design phase management, qualification of suppliers and designers, and changes in the process flow.

5.5 Documentation of the new process

The implementation of the plan was performed by each company. The materials supply process was then monitored in order to check whether there were secondary deviations as a result of the changes made.

Once the improvements in the process were consolidated, the new flowchart was made. A simplified 7 x 7 flowchart was produced in order to disseminate the new process for everybody in the company. This flowchart divides the process into seven major sub-processes, each of them being further divided into seven tasks.

6 Conclusions

6.1 Work carried out by quality committees

The role performed by the quality committees varied to a great extent, in terms of frequency and continuity of meetings, form of participation, and scope and emphasis of the improvement project. It was largely dependent on the commitment of high management to the quality program, the previous degree of shared decision making, and the sort of problems which were considered priorities.

In all companies, there were difficulties in the initial work of the quality committees, because of the lack of experience in holding meetings efficiently, since it was the first formal mechanism for shared decision making established in any of the companies. Some of the participants did not have a clear idea about the objectives and aims of the committee.

It was necessary to establish a number of rules for running the meetings, in order to give everybody the chance of speaking out, avoid deviation, and to create a positive attitude towards the aims of the committee work. Rules had to be established for time management, documentation of decisions, participation, etc.

Training had a key importance in the performance of the committees work. Although quality concepts and techniques seem easy to understand (several books on TQM claim that), their application in complex processes, such as materials supply management tend to be rather difficult and time consuming. The study suggests that training should contain both formal classes and the application of concepts and techniques in real problems - learning while working.

The committee played an important role in terms of improving the integration of different sectors of the company. It worked as a link for the distinct departments as well as a forum to discuss problems and negotiate changes in inter-departmental processes. It was also used for disseminating principles, aims and actions of the quality program throughout the organization.

One of the difficulties in carrying out the work by the committees was the rate of turn over among the members of the committees, which was mostly caused by the fact that the firms were in a process of rearrangement of people due to the quality program.

6.2 Application of the method and techniques

Working with data and facts proved to be very useful for the improvement project. Many of the problems identified in the supply process were different from what people expected in the initial stages (based on the perception of the group). For instance, the Pareto diagram indicated that the most frequent problems were not related to the performance of suppliers but to internal inefficiencies of the organization, often related to previous stages, such as inconsistent design and lack of planning.

In general all techniques were helpful in terms of making the meetings more objective. One of the main advantages was that all techniques produced some kind of documented output which gave some sort of concrete aim to the meetings.

Very clearly, people involved found some of the techniques more useful compared to others. The cause-effect diagram was not considered very useful for the kind of process studied, and the flowchart was considered too complex, even by managers and engineers, not helping in the visualization of the process.

As far as data collection is concerned, check lists and time sheets should be carefully planned and thoroughly tested in pilot studies, before they are applied.

Some problems detected were relatively easy to solve. In some cases, the fact that communication had been improved in the company through the establishment of a quality committee was enough to create the conditions for identifying, analyzing and solving problems effectively.

In general, the development of the improvement project helped the three companies to consolidate the quality program and the work of the quality committee.

7 References

1. Picchi, F.A. (1993) *Sistemas da qualidade: uso em empresas de construção*. São Paulo, EPUSP. Doctorate Thesis
2. Construction Research Institute (1987). *Project materials management handbook*. Austin, Texas, USA.
3. Marsh, J.W. (1985) Materials management: practical application in the construction industry. *Cost Engineering*, ASCE, 27(8), pp. 18-28
4. Leenders, M.R. and Blenkhorn, D.L. (1991) *Marketing reverso: um novo conceito no relacionamento comprador/vendedor*. São Paulo, McGraw-Hill.
5. Slack, N. (1993) *Vantagem competitiva em manufatura: atingindo competivividade nas operações industriais*. São Paulo, Atlas.
6. Hedenstad, K. and Meyer, B.O. (1993) *Establishing a quality system: pitfalls and psychological problems*. Oslo, Norwegian Building Research Institute.
7. Rosenfeld, Y. et al. (1992) Using quality circles to raise productivity and quality of work life. *Journal of Construction Engineering and Management*, ASCE, 118(1), pp. 17-33
8. Ishikawa, K. (1982) *Guide to quality control*. Tokyo, Asian Productivity Organization.
9. Hansen, R. & Sjoholt, O. (1989) *Quality management: a challenge for the construction industry*. Oslo, Norwegian Building Research Institute.

BENEFITS OF LONG-TERM RELATIONSHIPS: UK CONTRACTORS' EXPERIENCE

A.M. Haksever and H.S. Kim
Department of Construction Management and Engineering, University of Reading, Reading, UK
G. Pickering
Faculty of Construction Management, Chongqing Jianzhu University, Republic of China

Summary
This paper presents the results of a study into the benefits of long term relationships (LTR) between contractors and their subcontractors in the UK construction industry. The research was prompted by the awareness that although the practice of LTR is common in Japan, it is not generally considered so in the UK. Consequently, many potential benefits of such relationships are not widely taken advantage of. Initially, the Japanese experience is examined and then used as a basis for a questionnaire survey of UK contractors. In UK business environment, commercial pressures tend to reduce the weight placed on a successful working relationship. This is confirmed by demonstrating commercial factors are considered more important than relationship-related factors in the selection process. However, the gap between these two factors does not appear to be significant. Furthermore, the results show that contractors who are in a mode of LTR with their subcontractors have mainly experienced indirect benefits such as effective communication, less conflict and risk. Far fewer claim to have experienced lower costs or shorter times. Some implications of the results are discussed and suggestions for further LTR research are made.
Keywords: Contractors, Japan, long term relationship, subcontractors, UK

Sommaire
Cet article présente les résultats d'une étude concernant les bienfaits d'une longue relation (LTR) entre entrepreneurs et sous-entrepreneurs dans le domaine de l'industrie de la construction au Royaume-Uni. Cette recherche vient du fait que, bien que la pratique de LTR soit répandue au Japon, elle ne l'est pas en Angleterre. Par conséquant, plusieurs bienfaits potentiels de ce genre de relations ne sont pas très bien exploités. Initiallement, l'expérience japonaise est examinée puis utilisée comme base pour une enquête sur les entrepreneurs anglais. Dans l'environnement des affaires

The Organization and Management of Construction: Shaping theory and practice (Volume One).
Edited by D.A. Langford and A. Retik. Published in 1996 by E & FN Spon. ISBN 0 419 22230 8.

anglaises, les pressions commerciales ont tendance à reduire le poids placé sur une relation couronnée de succés. Ceci est prouvé en les démontrant que les facteurs commerciaux sont considérés de facon plus importante que facteurs relationnels dans la sélection du processus. Cependant le fossé entre ces deux facteurs ne semble pas considérable. En outre les résultats demontrent que les entrepreneurs qui sont dans le mode de LTR avec leurs sous-entrepreneurs, ont experimenté des bienfaits, comme par exemple une communication effective, et moins de et conflicts et risques. Quelques implications de ces résultats sont développés et discutés ainsique d'autres suggestions concernant les recherches de LTR.

1 Introduction

The UK construction industry in the mid-1990s is very different from the industry in the 1980s. The industry has been experiencing an exceptional change mainly due to political and economical pressures. One of the major changes is that an increased amount of construction activity is now being subcontracted. It is reported that over 90% of construction work is subcontracted on the majority of contracts. This suggests that the contractor is at greater risk if the sub-contractors fail, as more risk is passed to the latter [1].

This changing situation makes the role and performance of the subcontractor critical to project success, yet subcontracting does not have a glamorous image in the UK [2, 3]. This is due to many reasons such as large numbers of bankruptcies, high levels of disputes and claims, and poor performance. The uneasy and short term nature of the relationships between contractors and subcontractors could partly account for this negative image. A great amount of friction, dissatisfaction and mismatching of goals were reported by various authors in their investigations into the relationships between contractors and subcontractors in the UK [1]. Each party's viewpoints have been reported and contrasted. It is interesting to quote a statement which could reflect contractors' point of view regarding the working relationships with their subcontractors:

> "Whilst we try to maintain a good working relationship and ensure repeat business with approved specialists, the competitive tendering system means we must see price as the overriding feature and frequently this means an inability to offer repeat business."[1]

The statement above could be a fair indication of the reason why relationships between contractors and subcontractors are short-lived in the UK. As a consequence, an opportunity to accumulate skills and knowledge gained from working with the same subcontractor is lost. To overcome this limitation, Gray and Flanagan[1] call for a new working relationship where the contractor can maintain all his sub-contractors at a high level of efficiency and thus have the ability to complete the work. It is suggested that this can be achieved by working with the same subcontractor consistently over a long period of time, that is, on a long term relationship (LTR) basis [4, 5]. This suggestion is based on a close investigation of the Japanese construction industry where LTR plays a major role in the achievement of efficiency and business success [6].

It is suggested that the great success of the Japanese building industry depends on long term relationships. Japanese contractors overcome many of the problems discussed above by working together with tightly knit families of subcontractors for decades. These long term relationships are based on trust and a sense of brotherhood. Contractors set tough standards but help their subcontractors to achieve them. They pay a fair price for work and feel responsible for ensuring that their subcontractors are profit making and have opportunities to grow. In return, subcontractors strive to deliver the agreed works to the specified standards. They are willing to try new methods and feel responsible for proposing ways of improving quality, productivity and so on [4]. While the practice of LTR is very common in Japan and its benefits are widely acknowledged in practice and literature, it is not generally considered so in the UK. As a consequence, an opportunity which LTR could offer seems missing in the UK construction industry [7].

The Japanese experience could be a good example to support the argument that LTR should be exercised in the UK as well. However, one may argue that, due to different business and corporate environment in the two countries, there is a slim chance that LTR could be used in the UK. As Table 1 shows, relationship-related business environments are quite different in the UK and Japan.

Table 1. Relationship-related business environment in UK and Japan

	UK	Japan
General Environment	• Individualism	• Group orientation
	• Business relations based upon competition	• Business relations based upon trust and human relations
Corporate Level	• Short-term view of the business	• Long-term view of the business
	• Business contact based upon competition on price and quality	• Human relations based business contact forging long term relationship

Source: Adapted from Capital & Countries Report: Japanese Construction Industry (1987)

It may seem that the table above supports the claim that LTR is merely something Japanese and difficult to be implemented in UK environment. This might be a reason that makes UK practitioners hesitate to accept that LTR is workable. Furthermore, there is little empirical evidence as to whether or not the use of LTR in the UK generates the benefits proposed in the literature.

The discussions above pose two important questions in studying LTR issues: (1) To what extent are relationship-related factors considered important above others for invitation to bid and in final selection when sub-contracting? (2) Have the benefits of LTR been experienced by UK contractors or are those benefits merely speculated in the literature? If experienced, which benefits are most common?

The objective of the paper is to investigate these two issues in order to demonstrate the current status of LTR practice in the UK construction industry. It is hypothesised that benefits of LTR are widely experienced by the contractors in such relationship but relationship-related factors are overtaken by commercial factors at the point of subcontracting. To fulfil the research objective, a questionnaire survey was conducted with randomly selected contractors in the UK.

Following this introduction, the paper next presents the research methodology employed for this investigation. Then, findings regarding the importance of relationship-related factors in the selection process and contractors' experienced benefits of LTR are discussed. Finally, some suggestions for further LTR research are made.

2 The survey

A questionnaire survey of 53 contractors in the UK was carried out, each firm receiving 5 questionnaires. Altogether, 32 completed questionnaires were returned representing a 12.1% response rate. Only top level management personnel were asked to complete the questionnaire. Job positions of the respondents vary and include titles such as chief executive officer, procurement director, contract administrator and so on. Common characteristics of the respondents are: first, they have a vast amount of experience in the industry with an average of 23 years; second, they are in a position which could reflect overall company policy. These two criteria justify the reliability of the data source for a LTR study in that they are the right people to be approached to seek information on the experienced benefits of LTR and assess the importance of the selection factors for both invitation to bid and final selection.

Subcontractors chosen by the respondents for the questionnaire vary in trade. More than half of the respondents have had a relationship with the subcontractor for more than 10 years, see Table 2.

Table 2. Longevity of relationship of the respondents with the subcontractors

Longevity of relationship with the subcontractor (years)	less than 5	5 - 9	10 - 19	over 20
Number of the respondents	3	13	11	5

Although it is difficult to draw a norm by which LTR could be distinguished from non-LTR, it is reasonable to propose that the minimum of over 5 years of relationship is sufficient to allow an opportunity to have repetitive business and thus to experience its benefits. All the respondents with LTR are either satisfied or very satisfied with their chosen subcontractors.

3 Selection factors for subcontractors

The selection process is important because the nature of the contractor-subcontractor relationship is project-based. As discussed earlier, commercial factors are generally seen as overriding features in the selection process. However, it is important to analyse where relationship-related factors are placed in the process in order to investigate how the value of long term relationships is perceived by the contractors. This analysis enables us to see the relative importance of relationship-related factors over others for invitation to bid and final selection.

A total of 12 selection factors were identified from various literature sources and presented to the respondents. A three-level scale of not important, important and very important was used to indicate the level of importance. In order to rank the relative importance of the selection factors for invitation to bid and final selection, each answer is given a weight of; not important (1), important (2) and very important (3). A score of each factor is calculated by the following formula: $\sum$ n x w where n is the total number of respondents giving a particular answer and w is the weight of that answer. For this analysis, only 30 respondents' answers are usable. This means each selection factor has got a chance of scoring a minimum of 30 (n x 1) and maximum of 90 (n x 3). The ranks and scores of the selection factors are presented in Table 3.

Table 3. The ranking of importance of selection factors for sub-contractors

Rank	For invitation to bid	For final selection
1	Experience of similar projects (74)	Previous project performance (78)
2	Previous project performance (72)	Current work load (76)
3	**Previous disputes (71)**	Lowest bid (75)
4/5	**Previous experience with the sub (66)**	**Previous working relationship (74)**
	Previous working relationship (66)	**Previous disputes (74)**
6	Technical/managerial capability (63)	Experience of similar projects (73)
7	Current work load (61)	**Previous experience with the sub (71)**
8	Reputation (58)	Technical/managerial capability (69)
9	Bank reference (56)	Bank reference (67)
10	BS 5750: quality systems (50)	Reputation (65)
11	Location of the sub-contractor (49)	BS 5750: quality systems (53)
12	N/A	Location of the sub-contractor (51)

As Table 3 shows, the top three factors considered important for invitation to bid are: experience of similar projects, followed by previous project performance and previous disputes with a close range of scores. The top three places in final selection are previous project performance, current work load and lowest bid, also with a close range. Overall, commercial factors are considered more important in both stages. On the other hand, relationship-related factors which are highlighted are ranked middle-top for invitation to bid and middle for final selection. This result confirms that commercial factors are considered more important in the selection process reflecting the short-term and commercial-driven attitude of the industry. It also confirms earlier studies in that the industry's attitudes has not changed at least as far as contractors are concerned.

However, it must be noted that the gap between commercial and relationship-related factors in terms of their relative importance is not very significant. An implication of this is that if it is demonstrated that benefits of LTR are experienced by the UK contractors, this could support an argument that relationship-related factors should be given higher priority. In this regard, the next section examines whether or not respondents experience the benefits of LTR when it is used.

4 Experienced benefits of LTR with the sub-contractors

This section aims to find out whether or not contractors in a mode of LTR with their sub-contractors have experienced benefits and which ones are experienced most. To fulfil this objective, 15 major potential benefits of LTR from the literature have been identified and presented. These benefits are categorised into 6 groups which are cost/price, time, quality, risk, conflict and communication. The respondents were given the freedom to choose a subcontractor they wish to consider, as long as the chosen subcontractor fits the criteria which are that they know the subcontractor very well and they have been working together for a long time. They were then asked to indicate to what extent they experienced the benefits of LTR with their subcontractors. A five point agreement scale of strongly disagree, disagree, undecided (neither disagree nor agree), agree and strongly agree was used to indicate the level of experienced benefits. However, in the analysis of the data, strongly disagree and disagree options are combined into one group due to a small sample size. The same is also done for strongly agree and agree at the other end of the scale.

The result shows that contractors who have worked with their sub-contractors in a mode of long term relationship have experienced most of the benefits of LTR. The summary of the analysis is shown in Table 4. The most experienced benefit is better co-operation. It is followed by better team spirit, more confidence of success and effective communication. 8 out of 15 benefits have been experienced by more than 70% of the contractors.

Table 4. Experienced benefits of long-term relationship

Category	Benefit	Disagree	Undecided	Agree
Time	fewer time overruns *	4 (12.9%)	5 (16.1%)	22 (71.0%)
	shorter project completion time	4 (12.5%)	13 (40.6%)	15 (46.9%)
Cost/Price	lower bidding price	6 (18.8%)	13 (40.6%)	13 (40.6%)
	fewer cost overruns *	4 (12.9%)	8 (25.8%)	19 (61.3%)
	less supervision cost	8 (25.0%)	13 (40.6%)	11 (34.4%)
Quality	better quality	1 (3.1%)	11 (34.4%)	20 (62.5%)
	fewer defects	2 (6.3%)	9 (28.1%)	21 (65.6%)
Risk	more willingness to share risk	4 (12.5%)	6 (18.8%)	22 (68.8%)
	more confidence of success	0 (0.0%)	3 (9.4%)	29 (90.6%)
	less project risk	4 (12.5%)	4 (12.5%)	24 (75.0%)
Conflict	fewer disputes	1 (3.1%)	6 (18.8%)	25 (78.1%)
	fewer claims and litigation	2 (6.3%)	2 (6.3%)	28 (87.5%)
	better co-operation	0 (0.0%)	0 (0.0%)	32 (100.0%)
	better team spirit	0 (0.0%)	2 (6.3%)	30 (93.8%)
Communication	more effective communication	0 (0.0%)	4 (12.5%)	28 (87.5%)

* Sample size is 31 due to an incomplete answer.

It is interesting to note that most of the benefits which achieved over 70% agreement fall into the risk, conflict and communication categories. These categories could be characterised as indirect benefits while time, cost/price and quality as direct benefits. The contractors have experienced more indirect benefits than direct ones. In addition, a close examination of the experienced direct benefits of LTR reveals that the least experienced are in the cost/price group, followed by time and quality.

This aspect of the result can be a possible explanation for the less common practice of LTR among contractors in the UK. Contractors who have not been in a LTR mode may be reluctant to use it due to the characteristics of indirect benefits which are qualitative in nature and difficult to observe in an explicit manner. It may be a common sense that these indirect benefits should eventually contribute to direct ones by creating better working environment and reducing risks. However, contractors who do not use LTR may need to be convinced about this contribution in the first place, although it is very difficult to prove it in the short term. Nevertheless, it should be emphasised that although most of the contractors that participated in the questionnaire have not achieved as many direct benefits as indirect ones, they all are satisfied or even very satisfied with their sub-contractors and maintain a relationship for a long period of time. This provides strong evidence that LTR is not merely something Japanese and should be beneficial to UK contractors if they are in such a relationship.

The important implications of the findings above are two-fold. First, researchers in LTR studies should not concentrate only on such direct benefits when attempting to measure and promote the benefits of LTR. This may lead to a narrow understanding of the benefits of LTR. Alternatively, further research should look into the qualitative nature of LTR benefits in terms of risk reduced, effective communication and better working environment. Second, in order to promote LTR in the construction community, there is a need for a technique which can be used to demonstrate in a more explicit manner the contribution of indirect benefits to overall project performance.

5 Conclusion

The paper demonstrated that the benefits of LTR mentioned in the theory and achieved in Japan have been experienced by most of the UK contractors who have worked with their subcontractors in a LTR mode. Most experienced benefits are indirect ones such as less risk, less conflict and more effective communication. The direct benefits such as cost/price, time and quality are not experienced as much. Furthermore, it is confirmed that relationship-related factors are not considered as important as commercial factors at the point of subcontracting. These findings explain why LTR is not commonly practised between contractors and subcontractors in the UK construction industry characterised by short-term and commercial-driven attitudes.

The result of this paper suggests that future studies should give emphasis to the qualitative aspects of the LTR benefits and take a global view. Also, more research is needed into inter-organisational relationships in construction upon which further LTR research can be based. Furthermore, more studies are needed into the subject to shape up LTR theory in the context of the construction industry and their outcomes should constantly be disseminated to practitioners to promote the implementation of LTR.

Today's business environment requires more co-operation and collaboration between organisations to improve business performance and thus to achieve value for money for their clients. LTR is a pro-active response to it.

6 References

1. Gray, C. and Flanagan, R. (1989) *The Changing Role of Specialist and Trade Contractors*, Chartered Institute of Building, Ascot, UK.

2. CSSC. (1989) *Investing in Building 2001*, Centre for Strategic Studies in Construction, The University of Reading, Reading, UK.
3. CSSC. (1988) *Building Britain 2001*, Centre for Strategic Studies in Construction, The University of Reading, Reading, UK.
4. Bennett, J. (1991) *International Construction Project Management*, Butterworth.
5. Bennett, J. and Jayes, S. (1995) *Trusting the Team*, Centre for Strategic Studies in Construction, The University of Reading, UK.
6. Bennett, J., Flanagan, R. and Norman, G. (1987) *Capital and Countries Report: Japanese Construction Industry*, Centre for Strategic Studies in Construction, Reading, UK.
7. Haksever, A.M., Kim, H.S. and Pickering, G. (1995) Collaborative long-term relationships in UK construction industry: A lost opportunity?, *Proceedings of the 11th ARCOM Conference*, University of York, York, UK.

Acknowledgement

The authors wish to thank Professor John Bennett, Director of Centre for Strategic Studies in Construction for the insight and guidance he provided during this research project.

ORGANIZATIONAL EFFECTIVENESS OF THE CONSTRUCTION FIRM: A FRAMEWORK FOR ASSESSMENT

V.K. Handa and A. Adas

Department of Civil Engineering, University of Waterloo, Waterloo, Canada

Abstract

Assessment of organizational effectiveness in the context of the construction firm is typically based on project performance measures where firm's effectiveness is judged by its ability to achieve specific project goals that relate to costs or time of completion. These are crude indicators at best and could be misleading in certain instances. An assessment framework that incorporates the study of important organizational attributes and their relationships to overall firm's performance would give the construction firm a better tool to monitor its organizational effectiveness.

First, a brief review is given of the various theoretical approaches used by researchers in the study of effectiveness and its assessment. Secondly, the *competing values* approach is discussed as a suitable approach. The approach consolidates most criteria of effectiveness into four ideal organizational configurations or types with dominant values that could be pursued by a firm in its quest for effectiveness. These configurations are utilized to define variables from four general categories of organizational characteristics relevant for examining effectiveness: structural context; person-oriented processes; strategic means; and organizational flexibility, rules and regulations. These variables are hypothesized to influence effectiveness of the construction firm. The hypothesized relationships forms the basis on which an assessment methodology can be developed and validated. Further research steps are outlined as how to develop an assessment methodology based on the proposed framework.

Keywords: Competing values approach, effectiveness, assessment, construction firm

Abstract: L'évaluation de l'efficacité organisationnelle dans le contexte d'une entreprise de construction est généralement basée sur la mesure des performances lors de projets, où l'efficacité de l'entreprise est jugée sur son aptitude à atteindre des objectifs particuliers du projet relatifs au coût ou à la durée de son accomplissement. Ces données sont brutes et peuvent mener à des conclusions erronées. Un cadre d'évaluation incluant les proprietés d'organisation importantes ainsi que leur impact sur les performances de l'entreprise devrait donner à ces dernières un outil plus approprié pour visualiser leur efficacité organisationnelle.

En premier lieu, une revue générale des differentes approches théoriques utilisées par les chercheurs dans l'étude de l'efficacité et son évaluation est faite. En second lieu, l'approche des valeurs concurrentes est discutée comme étant une approche appropriée. Cette approche regroupe les critères d'efficacité en quatre configurations ou types d'organisations idéales avec des valeurs dominantes sur lesquelles l'entreprise peut se baser dans sa quête d'efficacité. Ces configurations sont utilisées pour la definition de variables à partir de quatre catégories générales de caracteristiques, capitales dans l'examen de l'efficacité: le contexte structurel, les processus relatifs au personnel humain, les moyens stratégiques, les règles, la réglementation et la flexibilité d'organisation. Ces variables sont hypothétiques afin d'influencer l'efficacité de l'entreprise de construction. Les relations hypothétiques forment la base sur laquelle une méthodologie d'évaluation peut être développer et valider. D'autres étapes de recherche sont soulignées, entre autres, la manière de développer une méthodologie d'évaluation basée sur la cadre proposé dans la présente étude.

1 Introduction

Principles and tools of emerging management strategies such as total quality management (TQM), strategic management and planning, and conformance to ISO

The Organization and Management of Construction: Shaping theory and practice (Volume One). Edited by D.A. Langford and A. Retik. Published in 1996 by E & FN Spon. ISBN 0 419 22230 8.

9000 standards, have stirred structural and strategic change in many organizations, mainly because they are being advocated to improve organizational effectiveness. Researchers have offered a variety of models for examining organizational effectiveness, yet there is little consensus as to what constitutes a valid set of effectiveness criteria. In order to properly evaluate the impact of adopting these principles, management in construction firms must address two fundamental questions related to organizational effectiveness: what theoretical approach or model is most appropriate to define and understand it? and how suitable is that approach as a framework on which a practical methodology can be developed to assess effectiveness?

The study of organizational effectiveness represents a complex part of organization theory due to the various definition and approaches that can be utilized to model it Many researchers have offered a variety of models for examining effectiveness, yet there is little consensus as what constitutes a valid set of effectiveness criteria [1]. In the literature many criteria and techniques were used in developing measurement. The first is traditional and is done by applying one of the common approaches such as the goal model as the case in Pennings and Goodman's framework [2]. Other techniques such as Hitt's [3] also utilize a single approach however they rely on the more elaborate models such as the resource model or strategic adaptation model.

However, most commonly used methodologies are based on the goal approach. They measure the level of achievement of a specific goal or goals. Three commonly used indicators in construction ask if work was completed on time, within budget, and if it met certain quality standards. Other criteria such as the level of achievement of certain goals in term of profits and desired levels of productivity are commonly used by construction firms. Other indicators measure levels of achieved profits and/or levels of workers' productivity and compare these levels to specific levels that were established as goals. Profit indicators by themselves are crude and shortsighted. Clearly an organization can make a profit without being effective. Ineffective organizations make profits by cutting corners and improper practices; such organizations do not stay in business long. These indicators do not give enough information about organizational effectiveness of the firm. The use of inappropriate measures is one of the reasons that leads to low performance. Improper assessment of effectiveness by firms lead to inaccurate conclusions which in turn result in sub-standards performance. Although most managers use some indicators (mostly financial), these do not capture all of the salient elements of effectiveness and can not be relied upon as predictors of effectiveness. Development of better methods to measure effectiveness should be a high priority in order to achieve and maintain improved performance.

2 Effectiveness

2.1 An Overview

There are various views of organizations advanced by different theorists that compete for the attention of researchers, each has its strength and weakness. Theorists of the classical school, developed universal principles or models that would apply in all situations and treated organizations as closed systems. For example, Weber's theory of bureaucracy that identified characteristics of bureaucratic authority in the organization, how formalized rules and regulation are used to govern and control the firm through hierarchically structured positions in order to achieve effectiveness. One of the principal drawbacks of classical theorists is that they tend to treat all organizations as machine like closed systems and that organizational control and hence effectiveness can be achieved by division of work, establishing lines of authority, and discipline. Influence of external environment is not recognized.

The classical view of organization was later challenged by some theorists who advanced the social nature of organizations. These theorists view organizations as made up of both tasks and people and represent a human counterpoint to the classical machine view of the organization. They proposed that management's roles in achieving effectiveness were to facilitate communication and to stimulate subordinates to high level of effort. An example here is the human-relation school with its participative decision making principles which emerged from McGregor's Theory X- Theory Y which promotes the creation of responsible jobs for workers and developing good group or team relations.

Afterwards, an open-system perspective for examining the functioning of the firm and its effectiveness emerged. Systems approaches emphasize the organization as a system and attempt to assess the effectiveness of the system in terms of its inputs, transformation, and outputs. Examples here include the internal process model and the strategic adaptation model. In the internal process model effectiveness is reflected in the efficiency of the processes inside the organizational systems. Theoretically, looking at the internal processes of the system, especially when the firm has little control over its environments seems to be more revealing of the effectiveness of the system than any other approach. However the approach has a narrow perspective of the functioning of the organization. It has no focus on the external interactions of the firm. Strategic adaptation model suggests that effective firms monitor their external environment constantly, receive feed back regularly, and take corrective actions to achieve their goals in the short term and ensure survival in the long term. This model recognizes the open-system nature of organizations, and their susceptibility to external forces. A major limitation of this model is that it pays little attention to what goes on inside the organization.

Another perspective advanced that in order for firms to achieve organizational effectiveness, they should study their conditions or environment in which they function. This perspective is known as contingency theory. It states that effective design of organizational structures' parameters (such as job specialization, unit size, centralization, etc.) are contingent or influenced by the various characteristics of its environment such as complexity and stability; the age and size of the organization; its technical system (technology) used for production; and its power system, for example, who controls the organization. Many contingency theorists have investigated environment-structure relationship and defined many types of environments and organizational structures that best fit for them.

Yet, another perspective draws attention to the various stockholders or constituencies in the internal and external environment around the organization. Stakeholders are groups or individuals affected by the organization performance who seek to influence the organization to satisfy their goals. Stakeholders could include any number of the groups that include owners of the organization, government and regulating agencies, local community organizations, customers, competitors, workers, special interest groups, environmentalists, suppliers, and the media. This perspective is based on the political view of the organization and it integrates the criterion of effectiveness for each group or constituency within or outside the organization that has a stake in the firm's performance. An example here is the strategic constituencies model. A major limitation of this model is assignment of proper weights to constituents to indicate the relative importance of satisfying their goals [4].

According to contingency theory firms must change their internal attributes-structures, strategies, and processes-to cope with changes in the environment. It is argued that organizations opt for internal consistency or stable configurations as long as possible because of reasons that include: (a) environmental change can sometimes prove to be temporary and therefore it is sensible to delay reaction to it; (b) internal changes are costly and therefore it will be resisted especially when a successful integration of structural and process attributes have been achieved; and finally successful organizations are never sure of the attributes that lie at the roots of their success and thus would avoid tampering with their tried and successful configuration [5]. Usually adaptation is avoided until a major threat is perceived because change must eventually come. In the face of worsening environmental fit, firms opt for totally new organizational configurations changing all their attributes drastically rather than piecemeal attributes change.

In the literature, a number of configurations were uncovered such as "mechanistic" structures in firms dealing with stable environments and "organic" structures in firms found in dynamic environments. The prospector, the analyzer, the defender, and the reactor forms of organization resulted from classifying the organizations according to their strategies, structures, and managerial styles [6]. The effective structuring of organizations into five configurations that include: the simple structure where force of direction that the various activities of an organization take to achieve a common goal, and results in the entrepreneurial form when this force dominates an organization; the machine bureaucracy where the force for efficiency becomes dominant, which attempts to ensure a viable ratio of benefits gained to costs incurred, the professional bureaucracy where the force of proficiency is dominant, which makes organization carry out tasks with high knowledge and skill; the divisionalized form where the force for concentration helps concentrate efforts on serving particular markets; and the adhocracy form that develops in response to an overriding need to innovate a new product [7].

Organizational effectiveness study according to configurations is justified on grounds of attempting to understand commonalties across organizations. Organizations are driven toward configurations in order to achieve consistency in their characteristics and rather than trying to do well on everything, effective organizations concentrate on configurations and try to bring their elements into line with these configuration [8]. A configurational approach that models organizational effectiveness criteria pursued by organizations has been proposed and is called the competing values approach [9]. It identifies ideal configurations or types based on dominant values of structural context, focus, and strategic means and ends. An organization can pursue the values of these ideal configurations and depending how close it is to these values determines its effectiveness.

3 The Competing Values Approach

The competing values approach is based on the premise that there is no one criterion for evaluating effectiveness. It organizes, consolidates, and integrates multiple criteria in the domains of effectiveness into three sets of incompatible dimensions. These are flexibility versus control, internal versus external focus, and means versus ends. The first set contrasts two dimensions of an organization's structure: flexibility values innovations, adaptation, and change while control favors stability, order, and predictability. The second set deals with whether focus and emphasis should be placed internally on the well-being and development of the *people* in the organization or externally on the well-being of the *organization* itself. The third set relates to organizational means versus ends; the former stressing internal processes and the long term, the latter emphasizing final outcomes and the short term.

These three sets are depicted into four organizational models or configurations shown in Figure 1. The models are the open system model, the human-relations model, the rational goal model, and the internal process model. In the figure there are axes of contrasting values that define the four models. Each model represents a particular set of values and has a polar opposite with contrasting emphasis. The vertical axis pertains to organization structural context and it contrasts stability and control with flexibility. The horizontal axis pertains to focus, which is whether dominant values are internal or external to the organization. The two inner axes pertain to organizational means and ends for each model and they contrast the processes or means (e.g. goal setting) to organizational outcomes and the outcomes or ends (e.g. productivity) themselves.

In brief, each model has characteristics that differs from the other and that which influence the level of effectiveness in the firm differently. The rational goal model emphasizes control and organizational focus as dominant effectiveness values; planning and goal setting are means, and productivity and efficiency are ends. The open system model emphasizes flexibility and an organizational focus as dominant effectiveness values; readiness and flexibility are means, growth and external support are ends. Dominant effectiveness values for the internal process model are control and internal focus, stressing communication processes as means and control as ends. The human-relations model emphasizes flexibility and internal focus, with cohesion and morale as means and skilled workers as ends.

3.1 Related Methodology

Maloney and Federle [10] developed a method for assessing the effectiveness of construction and engineering organizations based on assessment of configurations of organizational culture. The method is based on typologies or configurations of cultures identified by the four models of the competing values approach. Assessment was achieved by classification of the organizational culture according to perceptions of management and workers along the properties of four types of cultures identified in the competing values approach. The respondents' ratings are compared with each other to check for consistency. Particular attention is placed by the method upon the comparison of the ratings of the manager and that of his subordinates because if there is significant differences between the perceptions of the manager and his subordinates, the potential for major problems exists.

By delineating variables from the competing values four models, Ostroff & Schmitt [11] studied configurations of organizational effectiveness and efficiency. Their findings indicate that effective and efficient organizations are influenced not only by strength of culture, but also by other variables such as participation in decision making, goal emphasis, attitude toward change, and level of structural contextual emphasis.

Although the study did not target construction organizations, their findings suggested the inclusion of other organizational characteristics or variables along with culture in the study of effectiveness seems to be appropriate.

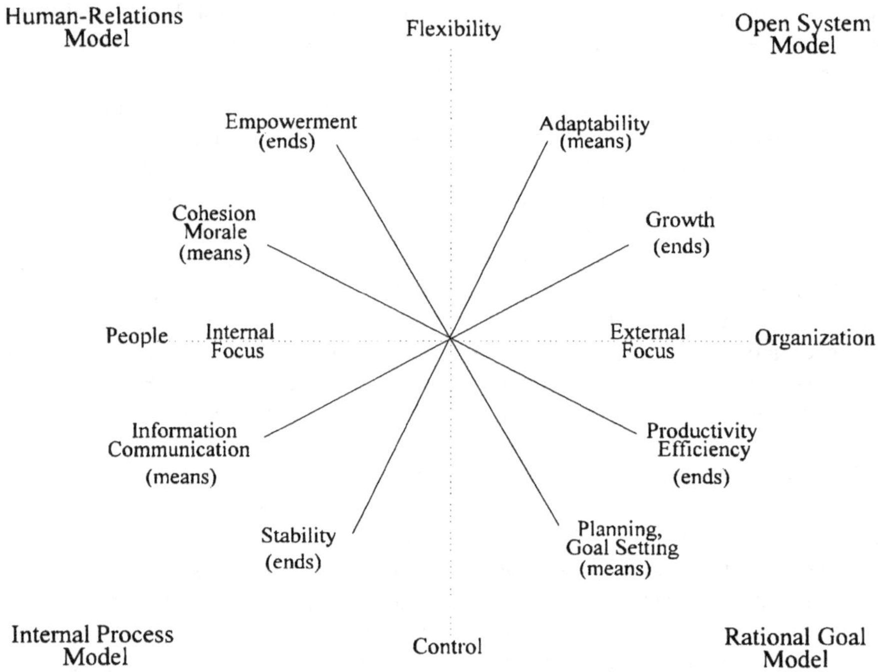

Figure 1 Four ideal models of organization in the competing values approach and their effectiveness criteria

4 Proposed Framework

4.1 Logic

A configurational-based approach towards organizational analysis such as the competing values can be used to develop a methodology to assess effectiveness of construction firm. The approach can serve as a conceptual model to categorize organizational characteristics of construction firms along its four ideal configurations. The reasons being, first its four models or configurations, emphasize characteristics that represent the integration of most effectiveness criteria already used by researchers and managers. Second, it is been validated for construction organizations [10].

The use of multiple criteria in assessment can be represented by the hybridization of values in the four models of the approach. The hybridization results in certain tradeoffs between levels of conflicting or paradoxical values based on the specific environmental situation faced by the firm. For example stressing a moderate level of competitiveness and external focus by a firm does not exclude it from placing some emphasis on the development of its workers, and adopting strategies to enhance morale and cohesion among them. The inclusions of paradoxical or conflicting criteria in assessments of organizational effectiveness in order to achieve a better assessment has been argued because firms could operate in multiple domains and may perform well only in a limited number [12].

This multidimensional view of effectiveness implies that there are different patterns of relationships between organizational effectiveness and its determinants. It is suggested that in order to develop accurate measurement at the organizational level, variables/measures must be combined into some overall model that indicates performance in the multiple domains of effectiveness. From this perspective, viewing effectiveness as the result of hybridization of certain levels of values of the four models as discussed above represents a more realistic view of the nature of the organization, and how it organize itself to achieve effectiveness. A firm could pursue any of the four models of organization in the competing values approach.

3.2. Methodology

Three main steps are suggested in order to develop valid quantitative models based on configurational inquiry [13]. First organizational configurations in an identified approach must be conceptualized and modeled as ideal types where effectiveness is highest because the fit among the contextual, structural, and strategic factors is at a maximum in these configurations. Second, organizational characteristics/variables that are relevant for examining effectiveness of firm and which represents the different effectiveness domains in the ideal types must be identified. Third, those variables must be integrated into an overall multivariate profile or model to empirically determine the significant variables and their levels against a valid indicator of effectiveness.

Firstly and as discussed previously, the competing values approach forms the basis of the proposed framework. This is because the four models in the approach can be modeled as ideal types or configurations with identified effectiveness criteria. Furthermore only, the hybrid of the four configurations is considered here because it is a more realistic modeling of how construction firms organize to achieve effectiveness.

Secondly, the result of hybridization of the criteria in the four models led to identifying four categories in which variables can be grouped and they are as follows: structural context; organizational flexibility, rules and regulations; organizational focus; and strategy (means vs. Ends). Analysis of construction firm's characteristics along the dimensions of effectiveness in these four categories helps to identify the important variables that represent the backbone of the proposed framework. In the structural context category, a number of variables can be used to indicate the effectiveness of the structural component of the construction firm in dealing with its external and internal environment. The level of subcontracting used in majority of construction projects has been shown to influence the complexity of structure in the construction firm and reflects the firm's attitude toward risk sharing and enhancing its costs effectiveness [14]. A construction firm that pursues strategic flexibility must have a structure that is better suited for contractual arrangements that provide flexibility such as subcontracting. A construction firm, in its quest for flexibility of services offered, may integrate vertically upstream by offering financing and A/E design and material supply services, or downstream by offering maintenance services [15]. Therefore the level of integration in services offered by the construction firm can be considered as another variable that reflects the structural complexity of the organization in its attempt to control the quality and range of its construction product. Other variables that can be identified in this category include:

- level of multi-project handling which reflects the ability of the construction firm's structure to handle simultaneous work at different locations to increase its volume of business and hence its profits.
- level of partnering, joint venturing, and strategic alliances which reflects how successful the organization's structure is in accommodating the added structural complexity resulting from such relationships. Typically construction firms enter into such relationships to access technology, share risks, secure financing, enter new markets, improve competitive position, and meet project requirements [16]
- level of organizational coordination which reflects how effective the organizational structure is in coordinating its internal and external relationship.
- level of information flow which indicates the degree of openness and quality of communication in the organizational structure.

In the second category of flexibility, rules and regulations, a number of variables can be used to indicate the effectiveness of the organizational structure in mediating between the flexibility-control points of the flexibility dimension. These include the extent of using regulation in work procedures, instruction, and evaluation by the firm, the level of adherence to rules and regulation by management and workers of the firm, the level of control the firm attempts to exert over its processes, and most important, the firm's attitude toward change. This variable would reflect management and workers' attitudes in encouraging, adopting, and accepting organizational changes in order to adapt the firm to new challenges in market

The third category of variables is based on the dimension of focus which represents whether the firm is internally or externally focused according to the competing values approach. Internally focused construction firms that value control in the assessment of effectiveness stress adequate communication and information management through a strong culture of hierarchy with clear rules and regulations for performance. Externally focused construction firms which value flexibility in the assessment of effectiveness emphasize adaptability though a strong culture that promotes readiness and innovations. Externally focused firms which value control in the assessment of effectiveness emphasize planning, goal setting, and a strong culture that promotes productivity and accomplishment [10]. Two variables which are deemed important in determining the focus of the construction firm and its effectiveness are:

- strength of organizational culture which reflects the internal climate in the firm and its strengths or weaknesses which are directly tied to organizational effectiveness of the construction firm [10].

- level of workers' participation in decision-making which underlies firm's attempt to empower and motivate its workers. Internally focused construction firms that value flexibility in the assessment of effectiveness are more sensitive to their workers. They achieve this by allowing a higher degree of participation in decision-making through a strong culture of team work. Therefore the level of workers' participation in decision-making could be considered.

The fourth category represents strategic means used by the construction firm to achieve effectiveness. Two important variables can be identified in this category and they are: the level of strategic planning, and the level of goal-setting. The former underscores the firm's attempt to adopt effective strategies that adapt and fit its environment. The latter reflects level of importance that the firm puts on setting goals such as increasing profits levels, increasing cost effectiveness, growth into other construction sectors, improving level of process quality, improving client satisfaction, and increasing workers' empowerment.

The variables identified in the four categories are not inclusive of all variables that can be used in the proposed framework. However, they represent the most significant variables that can be identified by this research and which can be included in the proposed framework. Further research is needed to investigate and build a model of the relationships between these variables and a valid measure of effectiveness. In order to develop this model, a field study must be carried out to collect pertinent data from a sample of construction firms. In this study, the identified variables must be operationalized and measured by accurate measurement scales. Questionnaires can be used in soliciting the desired data and they must be checked. in order to ensure their reliability in measuring the underlying attributes of the variables. Statistical techniques could be used to build and select a model that relate these variables with a valid measure of effectiveness such as overall performance in the three domains that are commonly used in construction as indicators of effectiveness. Namely, time of execution and whether work completed by the firm is performed within scheduled time or not, cost of work completed and whether it is within budget or not, and quality of completed work and whether it is according to specs or not. The resulting model would be a very valuable tool to management of construction firms in predicting their level of organizational effectiveness.

4 Summary

A framework for assessment of organizational effectiveness of the construction firm is proposed. The competing values approach towards analysis of effectiveness criteria is used to group criteria proposed for measurement in four categories of organizational characteristics. A number of variables are identified in each category. These variables possibly represent the most significant variables that are deemed relevant for examining effectiveness in the context of the construction firm. Future research steps are outlined to utilize the proposed framework to develop an assessment model that relate the identified variables with a proposed measure of effectiveness.

5 References

1. Lewin. Y. A., and Minton. W. J. (1986) Determining Organizational Effectiveness: Another Look and Agenda For Research. *Management Science*. Vol. 32, No. 5, pp. 514-537
2. Pennings, J. M. and Goodman, P. S. (1977). *New Perspective On Organizational Effectiveness*. San Francisco, California, Jossey-Bass, Inc.
3. Hitt, A. M. (1988) The Measuring of Organizational Effectiveness: Multiple Domains and Constituencies. *MIR* Vol. (28), pp. 28-39
4. Tsui. A. S. (1990) A Multiple Constituency Model of Effectiveness: An empirical Examination of the human resource subunit model: *Administrative Science Quarterly*, (35), pp. 458-483
5. Miller, D. and Friesen, H. P. (1984) *Organizations: A Quantum View*. Prentice-Hall.
6. Miles, R.E., and Snow, C. C. (1978) *Organizational Strategy,Structure and Process*. New York: McGraw-Hill
7. Mintzberg, H. (1979) *The Structuring of Organizations*. Englewood Cliffs, NJ: Prentice-Hall
8. Meyer, A. D., Tsui, A. S, and Hinings, C. R. (1993) Configurational Approach to Organizational Analysis. *Academy of Management Journal*, 36, 6, pp. 1175-1195
9. Quinn, R. E. and Rohrbaugh, J. (1983) A Spatial Model of Effectiveness: Criteria Towards a Competing Values Approach to Organizational Analysis. *Journal of Management Science*, 29, pp. 363-377.
10. Maloney, W. F., and Federle, M. O. (1993) Practical Models For Organizational Assessment. *ASCE Journal of Management in Engineering*. 9(1), pp. 64-81
11. Ostroff, C. and Schmitt, N. (1993) Configurations of Organizational Effectiveness and Efficiency. *Academy of Management Journal*, 36, 6: pp. 1345-1361
12. Cameron, K. S. (1986) A study of Organizational Effectiveness and Its Predictors *Management Science*, 32, pp. 87-112
13. Dotty, D. H., Glick, W. H., and Huber, G. P. (1993) Fit, Equafinality, and Organizational Effectiveness: A Test of Two Organizational Theories. *Academy of Management Journal*, Vol. 36, No. 6, pp. 1196-1250
14. Eccles, G. R. (1981) The quasifirm in the Construction industry. *Journal of Economic Behavior and Organization.* (2), pp. 335-357
15. Krippaehne. C. R., McCullouch. G. B., and Vanegas. A. J. (1992) Vertical Business Integration Strategies For Construction. *ASCE Journal of Management in Engineering*, Vol. 8, No. 2, pp. 153-166
16. Badger. W. W., and Mulligan. E. D. (1995) Rationale and Benefits associated with international alliances. *ASCE Journal of construction engineering and management*. Vol. 121, (1), pp. 100-111

ARCHITECTS' NEED STRENGTHS AND SATISFACTION LEVELS, AND ARCHITECTURAL ORGANISATIONS' CLIMATES

P.J. Rutland
Faculty of Architecture and Design, UNITEC Institute of Technology, Auckland, New Zealand

Abstract

This paper describes research undertaken within private architectural practices. It is part of a longitudinal study to determine changing patterns of need strengths, satisfaction levels, and climate typologies in professional architectural firms.

The paper explains how a stratified convenience sampling method was used to gather data on need strength and satisfaction, and of organisational climate using eight sections, namely: Authority, Standardisation and Formalisation (Routine), External Environment, Ambient Personality System, Nature of Work, Interpersonal relations, Creativity and Innovation, Professional Presence.

The climate results are analysed and discussed, both from an overall viewpoint, and by significant difference analysis between Director/Senior Partner and Architectural Staff. The need strengths and satisfaction results are discussed on an individual basis and considered in the light of accepted need motivation models and other studies (e.g. MacKinnon's) of architects and other construction professionals.

Some correlation analysis is performed between the need satisfaction and climate scores in order to examine possible conflicts between need spectrums and identified climate types, since climate has been shown to significantly affect organizational performance.

This research is being conducted on a longitudinal basis to determine changes in the measures so as to present possible reasons and then prognoses of trends and their likely impact on construction professionals.

The Organization and Management of Construction: Shaping theory and practice (Volume One).
Edited by D.A. Langford and A. Retik. Published in 1996 by E & FN Spon. ISBN 0 419 22230 8.

Résumé

Le présent exposé décrit les recherches conduites dans des bureaux d'architectes privés. Il fait partie d'une étude de longue durée, menée afin de déterminer l'évolution schématique de la force des besoins, des niveaux de satisfaction personnelle et de la typologie des ambiances dans les entreprises d'architecture professionnelle.

Cet exposé explique l'utilisation d'une méthode d'échantillonnage structuré par paliers, en vue de recueillir des données sur la force des besoins et les satisfactions personnelles, ainsi que sur l'ambiance des organisations, ces données étant répertoriées dans huit sections, à savoir : Autorité, Normalisation et Formalisation (Routine), Environnement extérieur, Système d'ambiance personnelle, Nature du travail, Relations interpersonnelles, Créativité et Innovation, Présence professionnelle.

Les résultats obtenus sur l'ambiance sont analysés et examinés, tant collectivement que par rapport à l'analyse différentielle significative entre le directeur/associé principal et le personnel des bureaux d'architectes. La force des besoins et les niveaux de satisfaction personnelle sont traités individuellement et examinés sous l'angle des modèles d'analyse standard pour la motivation des besoins, ainsi que d'autres études (p. ex. celle de MacKinnon) portant sur les architectes et sur d'autres professionnels de la construction immobilière.

Quelques analyses ont été réalisées pour étudier la corrélation entre la satisfaction des besoins et les résultats tirés de l'ambiance des bureaux, et ce, afin d'examiner les éventuels conflits entre l'éventail des besoins et les types d'ambiance repérés, puisqu'il est démontré que l'ambiance affecte sensiblement les performances d'une organisation.

Ces recherches à long terme sont menées pour évaluer les modifications des mesures, afin d'en dégager les raisons éventuelles puis de formuler des pronostics sur l'évolution des phénomènes et l'effet probable qu'ils auront sur les professionnels de la construction immobilière.

Keywords

Architects, architectural practice, management, motivation, organisational climate.

1 Introduction

Status, prestige and role play an important part in the working lives of professionals generally and in no less manner with architects. Architects' claim 'to be different' from other professions and there is some truth in this. Carr-Saunders remarked on aspects of this difference in his contribution to the discussion on W. Allen's paper to the RIBA on 8 March 1960, when he said "one saw that other professions were using a technique founded on the application of some natural or social science: the architects were doing more and introducing aesthetics ... They (Carr-Saunders and Weston) find that the proceedings of most professions were conducted in a somewhat humdrum way; but when they looked at the history of the architects' profession and read accounts of their meetings, they seemed to detect a certain emotional atmosphere". [1].

This is no less true today and throughout the emergence of the architectural profession the architect appears as someone highly involved with intellectual and emotional activities. It is important therefore to identify and recognise the characteristics of professional architects since they constitute such a critical part of the current construction team, to know if these characteristics have changed over time and if they have what effect will this have on the architectural firm, the profession and the construction industry.

In searching through the literature, there appears at every twist and turn, messages that define characteristics of the professional architect. Some are explicit even though subjectively derived, as in the case of Carr-Saunders, "in temperament, architects are touchy, liable to take offence when they conceive their dignity to have been affronted and therefore anxious for the moral recognition as the part of the community, ... and for the more material recognition implied by a higher (relative) standard of remuneration. At the same time they are solicitous for their art ... [2].

MacKinnon's study [3] identifies the varied skills needed by an architect as being "... juggler-like ability to combine, reconcile and exercise the diverse skills of a business person, lawyer, artist, engineer and advertising person, to say nothing of author-journalist psychiatrist, educator and psychologist." MacKinnon's study was concerned mainly with creative architects (architect I), and he identifies amongst them characteristics of dominance, freedom from conventional restraints and inhibitions, rather low sociability and a readiness to admit views that are unusual and unconventional.

It is important that the needs driving architects in the quest for excellence in their work and careers are understood so that they may be satisfied within the context of a climate formed in an organisational setting. Such characteristics ascribed to architects as above are not easy to assimilate in an organisational system that meets both the creative and business demands of society.

This study looks at the needs of architects in private architectural practices in Australasia, assesses their strengths and levels of satisfaction. At the same time, the climate of the organisations in which these architects work is measured and some cross matching analysis is performed.

This study is to be conducted as a regular longitudinal piece of research to see if the needs and levels of satisfaction change over time and if the climate typologies change also. The study reported here has established the base instruments and analysis for this longitudinal study.

Changes in these factors over time can be related to changes in the political, economic or social scenes and can be used to prognosticate changes necessary in architectural practice operations into the future.

2 The Research

Twenty-six private architectural practices were surveyed, with ninety seven architects and forty one partners forming the final individual sample. This constituted an 8% sample of the total population and was size stratified with 90% having more than 10 employees and 56% having more than 20.

Convenience sampling was used, as previous experience had shown the author that random sampling of the architectural profession produced an extremely poor response rate.

The data was gathered by questionnaire distributed to each firm and returned directly in a stamped addressed envelope, so that no censorship could take place.

The need strength questionnaire used a set of 52 statements subdivided into 13 categories which were scored on a 5 point Likert scale. Strong influences on the choice of the 13 categories came from the work of Hall & Nougaim, Alderfer, Maslow and Murray. Table 1 shows the need categories and question numbers:

Table 1. Need strength categories

	Category of need	Question numbers	Related needs from other sources
1	Achievement	2,26,38,40	Similar to self-actualisation on Maslow scale
2.	Dominance	3,14,39,41	Self-esteem on Maslow scale
3.	Social welfare	4,15,27,42	Altruistic desire
4.	Self-expression	1,25, 20,43	Component of self-actualisation
5.	Creativity and Challenge	5,16,29,44	Component of self-actualisation
6.	Affection and interpersonal relations (internal)	6,17,30,45	Social need (internally met)
7.	Recognition and approbation (external)	7,18,31,46	Part of self-esteem
8.	Affection & interpersonal relations (professionally)	8,19,32,47	Social need (career-line item as in Wood/Lebold (1970) safety
9.	Economic security	9,20,33,48	
10.	Dependence	10,21,34,49	Safety
11.	Independence	11,22,35,50	Factor in self-actualisation
12.	Socio-economic	12,23,36,51	Security as 'autonomy' in Payne's sense
13.	Recognition and approbation (professionally)	13,24,37,52	Author of Lebold's 'career line' type items.

The need satisfaction section was based on a measure of deficiency of need fulfillment, as described by Porter (1961). The 13 need strengths were reduced to 7 by combining some of them, since deficiency was felt harder to establish than strength of need. Fourteen statements about a job were produced similar to that shown below:

e.g. Item 7	It enables me to do work that is highly regarded by people in my profession	(a) How much does it do this now?	1,2,3,4,5,6,7
		(b) How much do you think it should do this?	1,2,3,4,5,6,7

The difference (b)-(a) gave the deficiency in need fulfillment. If (b) was lower than (a), the negative sign was ignored and the difference was counted as a deficiency (or dissatisfaction) since the respondent considers that need is being given too much attention than is necessary.

The seven needs used for deficiency measurement are shown in Table 2.

Table 2. Needs strengths and need satisfaction items.

Need	Title	Needs from need strength section	Statement numbers for need satisfaction
A	Socio-economic status	12	1,9
B	Security	9	4,12
C	Social (internal)	6,3	2,11
D	Esteem and Power	7,2	10,14
E	Self-Actualisation	1,4,5	5,8
F	Personal direction	10,11	3,6
G	Professional attachment	8,13	7,13
	Total	13	14

A climate assessment questionnaire was developed based on the work of Payne and Pheysey, Tagiuri, Litwin and Stringer and Campbell et al. The final instrument contained eight sections, which were subdivided into twenty-six subdivision and eighty-nine statements were scored on a binary true/false scale. The data from the climate instrument was subjected to cluster analysis which reduced the scale length for the climate measure from 86 to 41 statements and Table 3 shows the outcome.

Table 3. Climate dimensions and interpretations

	Dimension	Interpretation
1.	Authority	Measures (1) Questioning authority (2) Egalitarianism (3) Management concern for staff Thus a high score means authority is not over exerted. Thus firm is non-authoritarian.
2.	Standardisation and Formalisation	Measures (1) Administrative efficiency (2) Conventionality Thus a high score means high efficiency and routine.
3.	Ambient Personality	Measures : open-mindedness

System	Thus a high score means a very open-minded firm.
4. Nature of Work	Measures (1) Future orientation (2) Practical orientation (3) Scientific and Technical orientation (4) Intellectual orientation (5) Job challenge (6) Task orientation A high score means the firm has a good motivational content of nature of work, whereas a low score indicates most of the above are missing and hence under Herzberg's theory, the firm will not produce a good motivational climate.
5. Interpersonal Relations	Measures (1) Altruism } both measure consideration for (2) Sociability } others and a feeling of togetherness. Thus a high score means an "all pull-together" type of firm.
6. Innovation and Creativity	Measures (1) Variety in physical environment (2) Efforts to stimulate creativity (3) Readiness to innovate A high score means firm has a creative climate.
7. External Environment	Measures (1) Relation to Community (2) Relation to other components in the Industry A high score means the firm is good at extending beyond its own boundaries.
8. Professional Presence	A high score means the firm has a climate that is very much pro the Institute and Professional activities on a broad basis.

3 The Results

The following tables show the results of the need strength survey (Table 4), the average need satisfaction data (Table 5) and the climate analysis (Table 6). Table 7 shows a Pearsons product moment correlation results for Climate versus Need Satisfaction.

Table 4 Need strength N=97

Need	Mean score	Mean rank	Overall rank	Split-half reliability of rank across all raters (after S-B correction)
Achievement	15.13	4.75	3	.977
Dominance	14.78	4.73	2	.983
Social Welfare	14.26	5.43	6	.965
Self-Expression	15.67	3.77	1	.982
Creativity and Challenge	14.46	5.31	4	.968
Affection and Inter-personal	14.05	6.05	7	.975

Relations (Internal)				
Recognition and Approbation (Internal)	13.65	5.39	5	.982
Affection and Interpersonal Relations (Prof)	11.05	9.76	10	.981
Economic Security	10.79	10.06	12	.984
Dependence	12.51	7.84	9	.975
Independence	13.25	6.80	8	.973
Socio-Economic	11.59	9.10	11	.918
Recognition and Approbation (Professionally)	10.37	10.61	13	.987

Table 5 and Table 6 - please see Appendix

Table 7 Climate correlated with need satisfaction N=26

Need Satisfaction Climate	A	B	C	D	E	F	G
Authority	□-.74	□-.59	□-.83	•.42	.30	□.48	•.37
Standardisation Formalisation	-.08	-.13	-.19	.25	-.16	.27	-.12
Ambient Personality System	.32	0	.17	•.38	.30	•.57	•.41
Nature of Work	□.46	□-.52	•.41	.19	□.56	□.55	□.60
Interpersonal Relations	.19	-.14	.05	.25	•.37	•□.46	•.41
Innovation and Creativity	□-.89	□-.88	.004	-.08	.17	.19	•.41
External Environment	.01	-.22	.01	.29	•.39	□.61	□.50
Professional Presence	.28	-.16	□-.78	.04	.28	.19	.28

□ = significant at 0.01 level

• = significant at 0.05 level

4 Discussion

Not unexpectedly, the five most predominant needs for the architect are shown to be:

- Self-expression — high
- Dominance
- Achievement
- Creativity/Challenge
- Recognition and Approbation internally — low

The placement of Social Welfare at sixth rank (Table 4) indicates a desire of the architect to do more than merely satisfy economic and commercial funding in their work.

Recognition and Approbation professionally, in other words the need to have a study and career line in the profession of architecture ranks last of the thirteen scores.

This does not support the contention that prestige is a strong modivator for entry into a profession, as Move and Kohn discovered in their study of dentistry [4].

The combination of the 13 needs into the 7 item need satisfaction scale, produced:

- Personal Direction (F)
- Self-Actualisation (E)
- Esteem and Power (D)

as the top three needs, with F & D being significantly satisfied and Professional Attachment is again last on the hierarchy.

Table 8 shows that three of the lowest level needs are the lowest satisfied (C,A,B) whilst the top three needs (F,E,D) are the highest satisfied (with exception of G).

Table 8 Mean need strength in relation to mean need satisfaction.

Average satisfaction level	5.8	5.4	5.	5.3	5.	5.2	5.5
	4	5	7	5	3	4	4
	F	E	D	C	A	B	G
	Highest						Lowest
	⇒Descending order of need strength						

The climate results show firms to be non-authorisation, yet fairly high in routine with open-minded attitudes. The highest score came from the 'all-pull-together' type of climate, which is interesting in that architects in general are individualistic in nature.

The lowest score was on the external environment item, indicating the firms were not good at extending beyond their own boundaries. It is interesting in to see that this climate dimension has a 0.39 correlation factor (Significant @ 0.05 level) with the satisfaction of self-actualisation needs (Table 8) which indicates that architects do not find satisfaction of their second highest need through this dimension.

The next two lowest items were Innovation and Creativity and Nature of Work. When put together as almost equal low scorers, the picture presented is salutary as this indicates a low motivational and creative climate for work. This relatively low score on Innovative/Creative Climate is of concern, particularly in the light of Frederiksen's findings [7] that innovative climates encourage greater productivity from people who are high on the personal direction and independence need levels. This should also be considered in relation to the 7 rank (out of 13) of the need strength measure Affection and Interpersonal relation (internal) (See Table 4).

The concept of organisational climate is critical in determining the motivation of members [5] and it is generally considered to affect the behaviour of members of the organisation [6].

When the measure of climate by partners/directors is looked at separately from that by the architectural staff, then on every dimension of climate, the partners saw their firms climate as being better than did their staff in terms of the type of climate that has been identified as good for self-actualising motivatated people.

5 Continuing Research

This study was undertaken as a means of establishing appropriate instruments and analysis systems to continue with a longitudinal study. It also provided a set of base data on which to undertake a comparative study from repeated studies over time.

A further study is to be undertaken in 1996/97 to establish changes from this study with preferred reasons and explanations.

Some significant areas for consideration are:

- The low needs of architects in the Social Welfare dimension. Will this continue or has it changed? Does the pressure of commercial existence overpower this dimension in the minds of architects or is there some other reason if no change is evidenced?
- External professional recognition was not a high need in the study and current indicators are that this has not changed. The professional associations have been seriously reviewing their roles in the light of falling membership.
- The highest level needs were most satisfied and the lower level needs less well satisfied. Does this reflect a disinterest on behalf of architects in the lower level needs or will these become more important in the late 90's?
- The low score on the external environment dimension of climate will surely be much higher - but will it be significant or be merely a move to generate an expected profile for architects?
- The low motivational/creative climate in organisations could well be a factor in some of the poor quality work produced in the last decade or so. Has this factor improved and, if so, do the architects recognise it and respond to it?
- The significant gap between the partners and the staff on the "goodness of climate" of the firms is another example of the 'old school and new school' trying to work together. Will the gap be greater in the late 90's, particularly with the speed of technological progress, and the nature of education of architects.

The next study should reveal some very interesting results when considered in the light of this reported research and in the light of changes in society , the profession and the education of architects.

Already reports on competency standards for architects [8], the RIBA strategic study of the Professionas [9] and a special report for the NZIA [10] have indicated that changes should have taken place and should take place in the items measured in this research.

6 References

1. Carr-Saunders, A.M. and Wilson, P.A. (1964) *The Professions*, Frank Cass and Co. Ltd, London.

2. Ibid p.185

3. Mackinnon, D.W. (1970) The Personality Correlates of Creativity, a Study of American Architects in *Creativity*, Penguin Books, London.

4. More, D.M. and Kohn, N. (Jnr) (1960) Some motives for entering dentistry. The *American Journal of Sociology*, Vol 66, No. 1.

5. Litwin, G.M. and Stringer, R.A. (Jnr) (1968). *Motivation and Organisational Climate*, Harvard University, Graduate School of Business, Boston.

6. Pace, C.R. in *Organisational Climate, Explorations of a Concept,* by Taguri, R. and Litwin, G.W. 1968, Harvard University, Graduate School of Business, Boston.

7. Frederiksen, N. (1966). *Some effects of organisational climates on administrative performance*, Research Memorandum R.M. 66-21, Educational Testing Service.

8. R.A.I.A. 1994 Competency Standards for Architects.

9. R.I.B.A. (1993-95) *Strategic Study of the Profession*, RIBA Publications, London.

10. UniServices (1995) *NZIA 1993 Delphi Exercise Report*, UniServices, Auckland.

APPENDIX

Table 5 Composite average need satisfaction results N=26

	A	B	C	D	E	F	G
Dimension	Social economic status	Security	Social internal	Esteem and power	Self-Actualisation	Personal direction	Professional attachment
Range	1-7	1-7	1-7	1-7	1-7	1-7	1-7
Range achieved	3.5-70	2-6.25	4.5-6.5	4.5-7	3-6.75	4.87-7	4.41-7
Mean	5.3	5.24	5.55	5.7	5.45	5.84	5.54
Standard deviation	.78	.86	.53	.67	.77	.55	.67
95% Confidence limits on mean	±.27	±.29	±.18	±.23	±.26	±.19	±.23

Table 6 Composite climate results N = 26

Dimension	Authority	Standardisation and formalisation	Ambient personality system	Nature of work	Interpersonal relations	Innovation and creativity	External environment	Professional presence
Scale range	0 - 1	0 - 1	0 - 1	0 - 1	0 - 1	0 - 1	0 - 1	0 - 1
Range achieved	.22 -.86	.20 - .90	0 - 1.00	.22-.73	.29-1.0	.05-.80	.10-.75	0-1.00
Mean score	.62	.60	.62	.49	.71	.47	.39.	.53
Standard deviation	.17	.17	.27	.18	.18	.19	.17	.30
95% Confidence limits on mean	±.06	±.06	±.09	±.06	±.06	±.06	±.06	±.10

MULTI-CHARACTER MODEL OF THE CONSTRUCTION PROJECT INITIATION PROCESS

Källe Kahkönen
VTT – Building Technology, Finland

Summary

In spite of its widely agreed significance, the construction project initiation process is still a relatively new research topic. Our learning continuum along research and development efforts has just begun, and a lot of important challenges face researchers in forming a basis for improvements in the current practice of companies. VTT Building Technology is carrying out a two-year study of the project initiation process. The study, which is called *Preplanner,* is a joint effort involving VTT Building Technology and five companies representing owners of heavy industry. This paper presents the findings of sub-project of Preplanner study. The sub-project focused on the modelling of the project initiation process. A *multi-character* model of the project initiation process was obtained as the main result. The model provides a balanced representation of various characteristics of the project initiation process. The feedback obtained from company personnel suggests that, compared with the other identified models, this model provides improved capabilities to explain the nature of the project initiation process.

Sommaire

Malgré son importance largement reconnue, la méthode de lancement d'un project de construction est encore un sujet de recherche relativement nouveau. Noitre projet d'études continues dans le domaine de la recherche et du développement vient à peine de commencer et les chercheurs se trouvent en face d'un défi important pour trouver une mise au point afin d'amélioner le méthodes courantes utilisées par les compagnies. Le Département Construction du VTT (Centre de recherche technique de Finlande) est en train d`éffectuer un project de recherche réunit les efforts de cinq compagnie représantant l'industrie lourde. Cet

The Organization and Management of Construction: Shaping theory and practice (Volume One). Edited by D.A. Langford and A. Retik. Published in 1996 by E & FN Spon. ISBN 0 419 22230 8.

article nous relate les découvertes d`un sous-projet, faisant partle de la recherche Preplanner. Ce sous-projet est concentré sur l'établissement d'une méthode de lancement d'un project. Les avis donnés par les compagnies utilisatrices laissent supposer que cd modèle, comparé à d'autres modèles reconnus, permet de mieux expliquer la nature de la méthode de lancement d'un projet de construction.

Keywords: construction projects, industrial plants, project definition, early project stages, project planning

1 Introduction

Project initiation is defined here to mean the process prior to final investment decision-making. The main objective of this process is to maximize the chance of successful project realisation. First, this is achieved by producing strategic information for the owner, resulting in *sufficient understanding*. Second, an important issue within the project initiation is the necessary *project implementation arrangements*. Examples of these arrangements are the establishment of the core project team and negotiations with main equipment suppliers and contractors. In practice, the project initiation culminates in a number of choices and decisions of strategic importance. These define the project concept, select alternatives, define the technical content and the scope of a project, determine its financial and commercial requirements, etc.

Nowadays, it is widely agreed that the project initiation process has great potential to improve significantly the success of the whole project. However, the project management concept itself is traditionally too often understood and applied in a way that does not introduce a systematic and well-structured project initiation process. The findings of earlier research carried out in Australia and USA show that the current practice is very uneven.

Compared with the later stages of project management, it seems that the basic nature of the project initiation process is poorly modelled and understood, leading inevitably to unsatisfactory practical implementation. Within the project initiation process one can often encounter unclear or conflicting objectives, high levels of uncertainty relating to most estimates, communication problems between individuals, unrealistic opinions and a lack of creativeness, flexibility and consensus between various parties. In summary, project initiation can be characterised as a learning process where those involved are trying to achieve a shared opinion and understanding about the problem and its possible solutions.

The objective of this paper is to present the results of the two-year study on the project initiation process. In particular, a project initiation process model is presented. This model aims to reveal some important characteristics of the nature of the project initiation process, and to form the basis for a project initiation guidebook and other training material.

2 Background to the study

It seems that, due to some historical preoccupations regarding project planning and management, the project initiation process has only gained a rather small professional role. Historically, project management and planning have been very much *project-realisation-oriented.* This means that we presuppose that an *already* existing project demands our attention in terms of its realisation. The project formation process and its necessary decision-making has not aroused so much interest. In the first half of 1990s, however, awareness of the importance of project initiation has increased continuously in the research community and companies.

The project initiation process is still a relatively new research topic. At present, the well-known project planning theories and techniques provide only a very limited understanding of this process.

Laufer [1] presents a summary of the essentials of project initiation and concludes by presenting some evolving principles for effective project initiation. Based on his experience of construction projects, Morris [2,3] has found that most of the factors explaining the success of construction projects stem from project initiation.

In the USA a research group has been working on this topic for several years. After a number of research reports addressing some specific topics, their research programme has resulted in a project initiation handbook [4]. Furthermore, in Australia work led by the Construction Development Agency has also led to the publication of a practical guidebook aimed at improving the quality of project initiation in companies [5].

Apparently, our learning continuum along research and development efforts has just begun, and a lot of important challenges face researchers. Examples of research problems to be solved in the near future are

1. Measurement of effective and sufficient project initiation;
2. Effects of characteristics of projects on the project initiation process;
3. Enhancement of current project initiation process models towards multi-character models (for example, this can mean putting together chronological phases, project initiation stages, topics and its cyclic nature); and
4. Appropriate decision-making techniques.

3 Preplanner - research project

VTT Building Technology is currently carrying out a study of the project initiation process. The study, which is called *Preplanner*, is a joint effort involving the participation of VTT Building Building Technology and five companies representing owners of heavy industry (paper and pulp, energy, mining and chemical industries).

It is acknowledged that advanced project management comprises two main components: i) knowledge about methods and tools, and ii) individuals' personal experience of live projects. The main objective of this project is to improve understanding about the project initiation process and develop practical tools. The companies involved have provided six live projects, and the experiences of the senior project managers and and the companies' chief executives have been collected. These experiences form an important starting point for the various sub-projects of this research effort. The topics of the sub-projects are 1) Project initiation model; 2) Decision-making techniques; 3) Value management; and 4)

Training package and its computerised version. This paper presents the findings of the project initiation model sub-project.

4 Project initiation process as a part of the project life-cycle

The coverage of the sufficient project initiation process over various phases preceding construction can vary substantially between different projects. In certain cases owners need a rather detailed project manual before the final investment decision can be made. The level of detail depends upon the project uncertainties and the risks caused by them. Even basic engineering or detailed design of some key equipment item might be necessary if unproven technology is to be applied (Figure 1). Additionally, with regard to overseas projects, adequate project initiation might require careful examination of local conditions, leading to the identification and definition of a reliable procurement strategy. In some situations the necessary earthworks have been started or almost finished before the final investment decision. This "jumping-the-gun" phenomenon is a consequence of the need to meet to very tight project completion times.

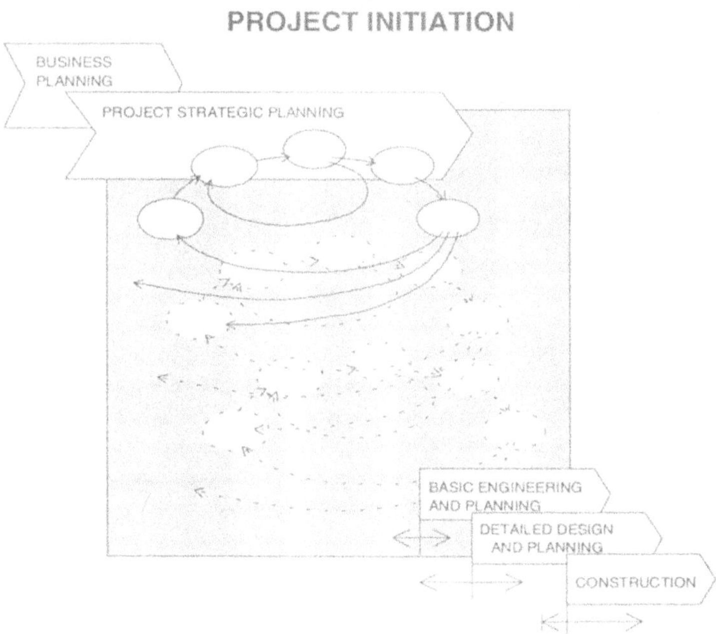

Fig. 1. Project initiation process as a part of the project life-cycle.

However, awareness and understanding of project risks seems to be surprisingly uneven, even though this is a major project characteristic affecting project initiation. Large companies are quite often equipped with best practice, but the problems lie in the small and medium sized enterprises. What seems to be needed are practical rules and guidelines expressing the role and coverage of the project initiation process with respect to the project life-cycle.

Knowledge acquired from experienced project managers suggests that in a *typical* project: i) the cost of project initiation is 3-5% of the cost of project planning and design; and ii) the cost of completing individual project initiation stages increases exponentially compared with the earlier less-detailed stage (completion of a single project initiation stage costs ten times more that the completion of the preceding stage, see section 9 of this paper).

5 Multi-character model

It is likely that the multifarious nature of the project initiation process often makes its understanding and successful implementation difficult. For example, it can be almost impossible to verify the suggested project initiation approach since the people involved do not share a common understanding about the process in question. What is needed is a simple and flexible model capable of explaining the dynamic nature of project initiation and its various characteristics.

Figure 2 shows a project initiation model called the *multi-character model.* The model comprises a conceptual classification which is used to explain the nature of project initiation.

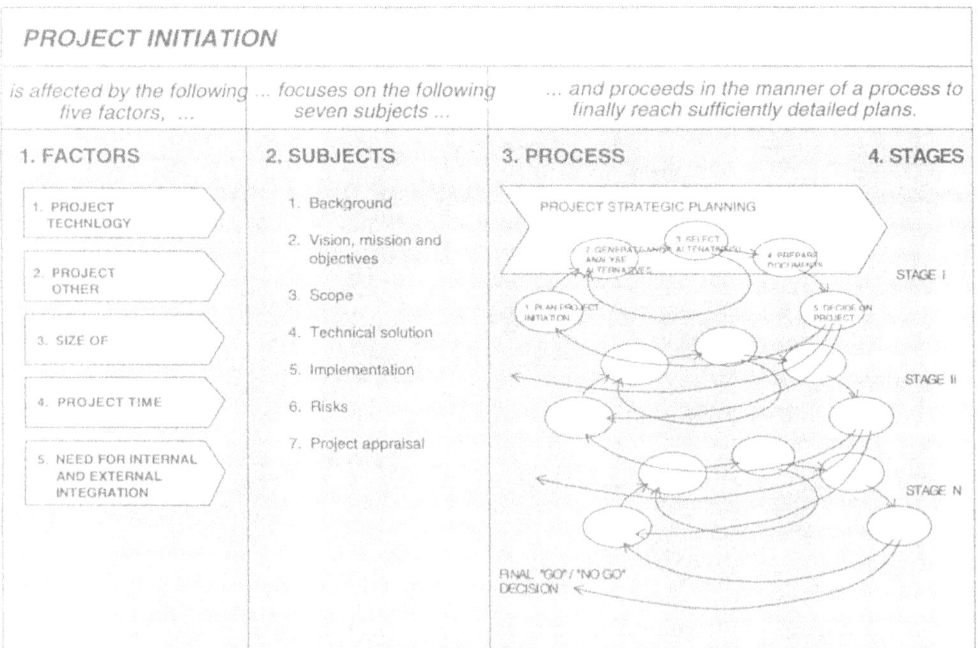

Fig. 2. Multi-character model of the project initiation process.

The classification includes four groups of viewpoints and characteristics, which are:

1. FACTORS affecting the coverage and activities of project initiation.
2. SUBJECTS on which the work focuses within each project initiation stage.
3. PROCESS by which project initiation completes each stage and proceeds to the next more-detailed stage.
4. STAGES of the project initiation process express the level of detail of the information and understanding to be gained.

An important finding is that these four groups together form a basis to better explain and understand the nature of the project initiation process. One group can provide only a limited view. Each group is briefly presented in the following sections.

6 Project characteristics affecting the project initiation process

Five project characteristics have been identified as having a direct effect on the extensiveness, methods, level of detail and completion time of project initiation. These are:

1. Riskiness of technology.
2. Riskiness of other success factors.
3. Project size.
4. Project time.
5. Need for internal and external integration.

The first two groups refer to the overall riskiness of the project initiative. It was found that the technology risk is still often underestimated and thus it often causes the most severe problems. The rules, such as '*in the case of development projects the time required is often four times more than the first estimate*' [6], clearly suggest that the need for careful and detailed project initiation is too often ignored. Several authors have suggested that the role of technology risk is decreasing, but it is apparent that technology is still the most important risk - at least in the case of industrial plant projects to be carried out by the small and medium sized companies and in the case of very big projects often having even some industry-wide or national importance.

The other four characteristics are also important, requiring attention and action during project initiation.

7 Aspects of project initiation

Project initiation activities focus on seven main subjects:

1. Background
2. Vision, mission and objectives
3. Scope
4. Technical solution

5. Implementation
6. Risks
7. Project appraisal

It is important to understand that the people involved are supposed to improve understanding with regard to all subjects within each project initiation stage.

8 Nature and repetitiveness of the project initiation process

Experienced project managers have found that the same type of working and activities are encountered many times during the project initiation process. in particular, a lot of work is being done all the time in order to generate, analyse and decide on alternative technical, project implementation or financial alternatives. This process is composed of the five phases shown in Figure 3. These five phases comprise a single project initiation stage.

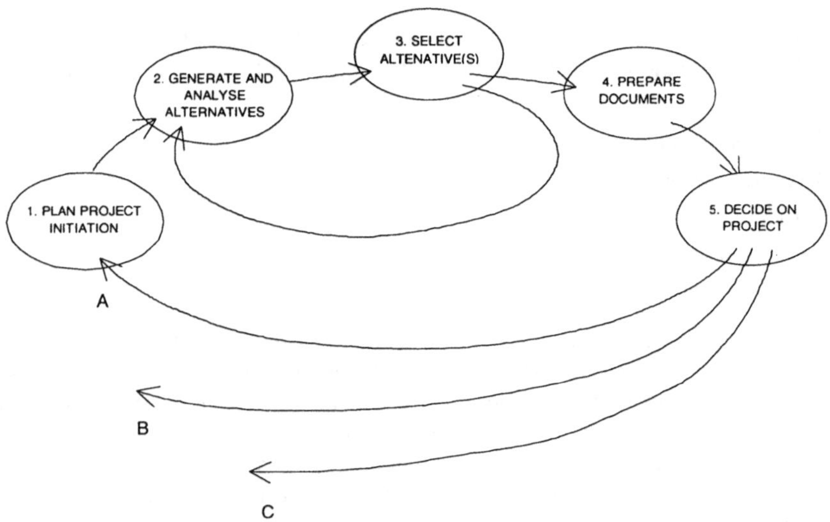

A = MAYBE (the same project initiation stage will be carried out again but with a different option, assumption or working method)
B = GO or NO GO (project realisation starts or the project is cancelled)
C = MAYBE (move to the next project initiation stage)

Fig. 3. Internal repetitiveness of the project initiation process is composed of five phases.

9 Project initiation stages

Different project initiation stages are used to address and define the level of detail of the results expected from the project initiation process. As shown in the previous section, the completion of any project initiation stage requires chief executives to decide whether or not to conduct further studies, halt project preparations or start project implementation. Various project initiation stages are an important means ofing understand and managing the sufficiency of the project preparations made to datye, particularly with respect of the project uncertainties, which are supposed to give way to reliable estimates when project initiation proceeds to the more detailed stages. The main project initiation stages are:

1. Project strategic planning
2. Pre-feasibility study
3. Feasibility study
4. Front-end engineering
5. Basic engineering

Project initiation is very much a question of co-operation between representatives of owner company and consultants (or other suitable organisation), who are often in charge of organising the work, carrying out the main activities and preparing the documentation. The people from the owner company are primarily responsible for tasks falling under the heading *project strategic planning* (this is not clearly a project initiation stage, but rather the initial input from the owner). Project strategic planning starts from quite general statements about the project mission and goals, gradually evolving into more detailed evaluation and selection of project implementation alternatives, and finally becoming increasingly concerned with technical solutions. Likewise, the same type of change takes place during the stages (2-5) performed by consultants and/or by a specific team (Figure 4).

10 Discussion

Awareness of the significance of project initiation seems to be increasing amongst project practitioners. Company people and particularly those involved in the Preplanner research effort would like to see a major shift away from the current very uneven practice towards a systematic project initiation process, which would clearly increase the possibilities for successful industrial construction projects. Obviously, this can be achieved with the aid of practical tools which could be straightforwardly applied by practitioners. These tools must be holistic, flexible and simple in order to be of sufficient value.

However, it is author's opinion that our current theories and especially our models about the project initiation process do not provide a coherent basis for the different tools required. Theories that can be used as a starting point when such models are developed - for example, soft systems engineering by Checkland [7] - are not very well known even amongst researchers. Thus the need for new improved project initiation models is an important issue.

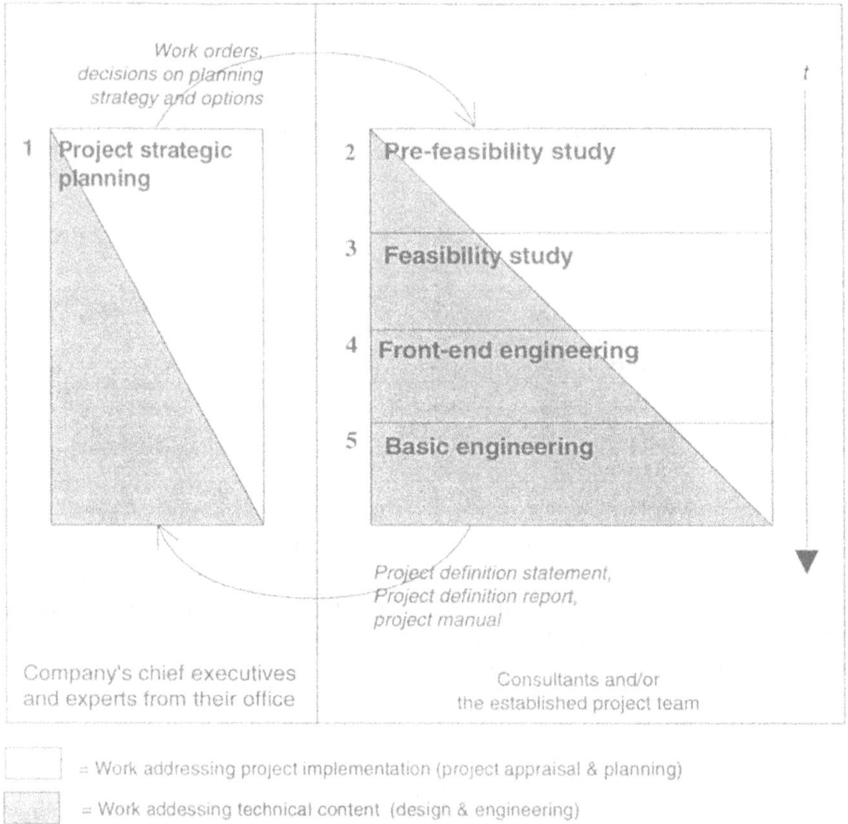

Fig. 4. Stages of the project initiation process and the relationship between them.

11 Conclusions

A new project initiation model called the *multi-character project initiation model* was presented in this paper. It is considered that this model provides an improved representation of the nature and characteristics of the project initiation process compared with the models represented in the literature so far. This statement is based on the feedback obtained from companies involved in the Preplanner research project.

The model provides a starting point for the development of project initiation guidebook and other tools.

Acknowledgements

I gratefully acknowledge Mr. Seppo Hänninen (Outokumpu Engineering Services Ltd.), Mrs. Eeva-Liisa Lehikoinen (JS Terasto Ltd.), Mr. Markku Levonen (IVO International Ltd.), Mr. Markku Metsäranta (Jaakko Pöyry Ltd.) and Mr. Mauri Toivanen (Teollisuuden Voima Ltd.) for their support and constructive critisism during my work.

12 References

1. Laufer, A. (1987) Essentials of Project Planning: Owner's Perspective, *Journal of Management in Engineering*, 6(2), pp. 162-176.
2. Morris, P.W.G. (1988) Initiating Major Projects - the Unperceived Role of Project Management, *Proceedings of the 9th INTERNET World Congress on Project Management*, September 4th-9th, Glasgow, UK, Volume 2, pp. 801-813.
3. Morris, P.W.G. (1994) *The Management of Projects*, Thomas Telford Services Ltd, London, United Kingdom.
4. Construction Industry Institute - CII. (1995) *Pre-Project Planning Handbook*, Special Publication 39-2, University of Texas at Austin, USA.
5. CIDA - Construction Industry Development Agency. (1994) *Project Initiation Guide*, Sydney, Australia.
6. Kharbanda, O.P. and Stallworthy, E.A. (1983) *How to Learn from Project Disasters*, Gower Publishing Company Ltd., Aldershot, Hampshire, United Kingdom.
7. Checkland, P. (1981) *Systems Thinking, Systems Practice*, John Wiley & Sons, Chichester, United Kingdom.

Part III
The firm and its resources

1 The firm and the human resource

UNDERSTANDING AND AFFECTING CHANGE

Gary R. Smith
Department of Civil and Environmental Engineering,
Pennsylvania State University, Pennsylvania, USA

Abstract
This paper explores the primary models of change management that have been found to be successful in a variety of industrial settings. Resistance to change is discussed as it relates to implementation of new management or technological systems. Construction firms do not generally maintain a consistent process for implementation of new ideas, even if they recognize the need for change. Because of their inconsistency in identifying and implementing new technologies, many valuable innovations are disposed of before their full capacity is realized.

Research into the implementation of management and technology systems provided the opportunity to explore the more fundamental issues surrounding change and change management. This revealed that change management strategies, total quality management and successful corporation attributes have arrived at similar strategies which suggests there is a fundamental set of attributes companies should strive to acquire.
Key Words: Innovation, Implementation, Change

The Organization and Management of Construction: Shaping theory and practice (Volume One).
Edited by D.A. Langford and A. Retik. Published in 1996 by E & FN Spon. ISBN 0 419 22230 8.

Sommaire

Le présent papier explore les premiers modèles de la gestions du changement qui se sont avérés efficaces dans plusieurs contextes industriels. La résistance au changement y est discutée au regard de ce qu'elle est inhérente à la mise en place de nouvelles gestions ou de nouveaux systèmes technologiques. Les entreprises du bâtiment ne maintiennent pas, d'une façon générale, de processus consistant pour l'introduction de nouvelles idées, même si elles reconnaissent la nécessité du changement. Et c'est cette inconsistance-même dans l'identification et la mise en place de nouvelles technologies qui fait que de nombreuses innovations de valeur disparaissent avant d'avoir atteint leur pleine capacité.

Les recherches menées sur la mise en place de systèmes de gestion et technologiques ont été l'occasion d'explorer les questions qui sont au coeur du changement et de la gestion de celui-ci. Cette exploration a révélé que les stratégies de gestion du changement, la gestion de la qualité totale et les atouts de succès de l'entreprise tendent à faire apparaître le même type de stratégies, ce qui laisse penser qu'il existe un ensemble fondamental d'attributs que toute entreprise se devrait de chercher à acquérir.

1 Introduction

No single phenomena will play as demanding a role in determining the success or failure of construction firms in the future as their response to change.[1] While the construction industry is typically cast as a slow changing and low technology industry, many sweeping advances have overtaken the industry. The introduction of laser controls for equipment, automated measurements, robotics, electronic design and drafting capabilities, advances in materials, and many other examples can be cited to illustrate that a slow moving and low technology image is misleading. In addition to technology changes, many types of management change can be affecting the workplace at any particular point in time. Consider changes such as those:

- in the work environment (regulations, unions)

- in organization priorities (market vs. technology driven, quality vs. price)
- in organizational structures (functional vs. matrix, partnering, team building)
- in the way work is done (self inspection, self management)
- in personnel policies (innovation rewards, increasing creativity, incentives)
- in corporate culture (eliminating corporate traditions, total quality management)

The list should not imply that implementation of new ideas has become easier for any construction organization or that all changes are good and successful. Overall, even when organizations recognize the need for change, they seldom have an organized approach to implementation. Change is a continuous process, and should not be treated as a project that is assigned to a project manager who is between projects. The purpose of this paper is to summarize information on those strategies or processes that have been successful in sustaining organizational change, and the factors to consider in planning and controlling resistance to change.

2 What Constitutes Change?

Routine changes are minor variations of existing systems or technologies. Standard operating procedures often change between projects and these changes create few implementation problems. Radical changes involve many systems or technologies that have not been used before. More complex management and training issues result from radical changes. Nord and Tucker identified three levels of implementation development in their work which are appropriate to classifying changes [2]:

Level 1: Initiation: no solution to a problem is known to exist. Complete research and development is required. There are no guaranteed results

Level 2: Adaptation: A few prototypes exist which may work, but lack packaging for immediate implementation. More development is needed for implementation.

Level 3: Borrowing: Many well packaged innovations exist and a company can copy one and apply it with little modification.

Many concepts that are successful in other organizations will fail in a new setting. The particular characteristics of a firm will determine how they implement new ideas. A mixture of adaptation and borrowing is evident in many firms. As evidenced by traditionally low levels of research funding, construction firms prefer to operate at the borrowing level, and occasionally they will budget funds for implementation at the adaptation level. The low-tech image of the industry gains support from the more consistent Level 3 implementation efforts.

3 What Creates The Need For Change?

Beckhard and Harris identified several issues that create the need for change [1]. Worldwide competition, technology, and economic variations cause companies to change their basic organization or processes. A major cause of recent changes affecting firms has been changes in regulations. Hazardous waste and safety are two obvious examples of regulations requiring changes in company operations. A new market niche or a new technology may pressure the company to change its mission statement to conform with the market or technology. Over time companies need to adjust the way they do business to cope with these and other changes.

Steele structured the forces of change as demand-induced and supply-pushed [3]. Demand induced innovation is change generated outside company improvement efforts. The need for change, in this case, is first recognized by others. Once recognized, demand induced changes are often easier to implement because of competitive forces. Steele indicates that this form of change is the largest source of change in US industry. Beckhard and Harris also state that "forces requiring change in large systems today tend to originate outside the organization [1]."

Supply pushed innovation results from innovations from within the organization, whether a need for change was recognized or not. Supply pushed changes, while not the majority of changes, are likely to account for major changes. However, this type of change also carries with it the greatest risk of failure. Supply pushed innovations will exhibit higher resistance from employees than the demand-induced changes.

4 Resistance To Change

Resistance to change generally carries a negative connotation. Those who resist change are bad or uncooperative people, while those who accept change are good or desirable people. Clearly, this is the perspective of those driving the change. Resistance to change is natural and desirable for many reasons. People change when there is a sufficient reason to do so [4]. If no apparent benefit is clearly demonstrated, why should they change? In many cases a known result, even if it is an unfavorable result, can be more acceptable to an employee than not knowing the potential outcome.

There needs to be some resistance to change. Otherwise both good and bad changes may be implemented before proper consideration can be made regarding the overall impact of a change. Because some resistance to change is desirable and predictable, plans to address this resistance are needed when the decision has been made to carry a plan forward.

4.1 Common Sources Of Resistance

Resistance may be formed from what people perceive they are losing not by what they will gain. When the change involves organizational restructuring, patterns of communication and socialization among the workforce will be affected. Some sense of insecurity may occur when someone's job location or pattern changes. Managers should evaluate what will be lost as well as gained when evaluating change.

Lack of ownership in the idea may create limited resistance. If the idea comes from a competitor, as discussed earlier, the resistance may be somewhat lower. In the construction industry where borrowing and adaptation are common methods of innovation, the lack of ownership is a weak form of resistance. Managers often complain about changes that only increase the amount of paperwork in an already paperwork-burdened environment. This may be legitimate resistance where the goal of the paperwork is unclear or it seems an unnecessary collection of data.

Large changes experience boredom and lack of interest due to a perceived lack of progress. The lack of progress is often associated with the time it takes to inform and train personnel in the new system. Resistance, in the form of apathy, is very difficult to change once the attitude is accepted.

Lack of top management support is evident where changes are demanded yet little in terms of resources or assistance in achieving new goals is provided. In many cases, previous goals remain unchanged and new goals are added to the requirements. A common problem with proposed changes is when current procedures have no known flaw, therefore, the proposed change seems to be for sake of change.

4.2 No Resistance

Harvey discussed four possible problems associated with no resistance to change [4].

1. It was no real change. The organization is already doing it in some fashion.
2. People in the organization are dead from the neck up.
3. Everyone thinks alike. No one knows if change is good or bad since there are no conflicting viewpoints.
4. Resistance exists, but no one will say so.

The last category is called "stealth resistance" because of the apparent lack of resistance. Stealth resistance often remains dormant throughout much of the planning process. Once the implementation process starts, the resistance appears, often from unanticipated groups. Unfortunately, many commitments have been made already. Thus, stealth resistance may be the most dangerous, financially.

5 Strategies to Implement Change

5.1 Change Strategies

Power strategies imply that changes are forced onto the employees. Action to stop the implementation is overcome through the personal or positional power of the group implementing the change. This type of change can create a lack of trust in management, confusion in the workplace, and possible loss of talented people who do not desire to work under the new scheme or to cooperate with the implementation. Explicit and implicit coercion are often used in conjunction with power strategies. Coercion ultimately results in a "do it or else option" that permits employees to choose a personal course of action. Manipulation is a covert attempt to influence others through a conscious structuring of events or a selective use of information. If the manipulation is "discovered" by the workforce, trust is virtually destroyed.

Rational strategies require a time investment for developing information related to the change. The information is used for presentation to the affected groups in the organization. The basic philosophy of the rational model is that the work force will cooperate with a change once they understand the current situation and have time to consider what the future will hold if they do not change. One approach to sharing this information is to hold an organization wide confrontation meeting to explain the need for the change and the facts supporting the need. The specific plan for implementing the plan may also be presented or the details may be withheld pending employee reaction to the general information. Small group meetings or one-on-one meetings are generally treated as educational intervention (education and communication) techniques and are particularly good if resistance is due only to a lack of information.

Strategies that follow a **normative approach** involve employees in planning the proposed change and seek their input into many parts of the implementation. Participation should involve potential resistors in some aspect of the proposed system design and implementation. Lawrence states that participation, to be of value, must be based on a search for ideas that are seen as truly relevant to the change under consideration [5]. Otherwise, involving resistors will appear to be an attempt to silence them through the committee.

5.2 Pilot Project Implementation

Pilot projects are similar to experiments. Some experiments fail and should be tried again, but failures in management systems are difficult to clean-up. In general, if it is not successful the first time, it is unlikely to get a second opportunity.

The following steps for pilot projects are suggested.

- Start Early: maximum benefits result when planned early in the project
- Commitment: establish executive sponsorship
- Organize the Team
- Project Execution Plan: Identify concepts, recognize barriers, identify differences from normal procedures, develop goals, define measurable criteria, estimate costs, define schedules, define progress reporting
- Execution: distribute the project execution plan, designate responsibility to team members, monitor actions of the team, create open lines of communication, establish routine, continually monitor
- Project Close Out: analyze the results with a specific effort to quantify the

benefit to cost ratio. Document lessons learned, make recommendations.

6 Project Level Change Constraints

A critical assumption, made by most manufacturing oriented studies, is that a manager operates within a time frame sufficiently long enough that the time function can effectively be ignored as a variable. A construction manager, in contrast, has a comparably short cycle in which to implement changes. When a change is planned at the project level, time becomes a key parameter in planning for the change. The project team's primary task is the completion of the project within a time and budget developed according to the project needs. The team's focus is usually not on the implementation needs of a management system being changed. Project personnel are not focused on implementation of a new management scheme or the introduction of new technology, and project managers are typically rewarded by "on time" and "under budget" measures. New technology and management schemes may prevent accomplishment of these activities in the planned fashion.

7 Project Findings

7.1 Project Description

The objective of the investigative research was to determine the level of implementation of research developed by the Construction Industry Institute at the project level. Twenty-two projects were visited, and 61 interviews were conducted. The owner, engineer, and constructor project managers for each project were interviewed for each project. Prior to the interviews, the participants were surveyed with respect to their company practices for implementing new management or technology systems. Forty-one surveys were completed and returned. The projects were primarily for the construction of process systems and industrial facilities whose values ranged from US$2 million to US$600 million. Rather than present survey statistics, the findings are reported as consensus views of the participants.

7.2 Resistance At The Project Manager Level

Little resistance to new ideas and systems was evident at the project level. Most project managers felt free to implement new ideas at their own discretion. However, most wanted cost-benefit or some form of guarantee that the system would work. Therefore, discretion should be defined as within budget parameters and schedules. None made allowances for implementation within project budgets. A few projects received home office support when the change involved revisions to computerized reporting and new computing systems. A persistent desire to have all materials at the Level 3 innovation level was noted at most project sites.

The basic problems noted most frequently by the project participants were that the criteria for a successful project did not include implementation of research or any other new system, unless an unmistakable benefit could be identified. However, when asked if they tracked cost-benefit on systems they had implemented in the past, most indicated that they did not track implementation information in that form. Favorable results were often attributed to poor initial estimates and not a result of systems improvement.

Workforce training was observed to be lacking at many levels. The average training time was less than two weeks per year and often the material was not coordinated with any particular implementation plan.

7.3 Strategies

Few companies had strategies or plans in place to permit integration of new ideas and technology into the company. However, from an implementation standpoint, the companies which had implemented more of the targeted research materials, were also involved in total quality management (TQM) processes. Not all were fully integrated into TQM, and several had quit using a formal improvement program, but many retained key features in an informal continuous improvement program.

8 Plan For Implementation

The development of one implementation strategy is recommended. The choice of strategy depends on how drastic the change may be, the speed of implementation desired, the amount of preplanning to be performed, and the level of involvement by others. The last two are prime contributors to the most common mistake in implementation, a lack of resources. Often the implementation plan moves too fast with too few people involved to effectively communicate the reasons for the change.

Planning to implement a change requires some preliminary investigation for complete development of a change strategy. The first stage in planning requires that the support needed to implement the plan be defined. The next important phase is commitment planning. Commitment planning is intended to identify those target individuals and groups whose commitment is needed and what form of resistance they might present. Through this process a target group is defined that represents the critical mass needed to ensure the effectiveness of a change. The focus of the planning effort is on those key elements or personnel in the organization that are necessary for the implementation.

8.1 Plan Elements

Key personnel have three choices as they consider a new plan presented to them, assuming they do not openly resist the plan. Their choices are to let it happen, to help it happen, or to make it happen. Key personnel can be informally evaluated for their current position on this scale. This will help determine the strategy needed to move the "let it happen" group into helping or making the change happen. The plan should cover the major elements of:

- Identifying the commitment of the critical mass
- Create a monitoring process to measure resistance and identify potential problems
- Establish an educational intervention program for the group whose support cannot be immediately counted upon.

- Resistance management would include, identifying resistors and identifying appropriate strategies for managing the resistance.

When assessing various organization positions with respect to willingness to change, relevant subsystems should be identified and assessed for their readiness and capability to implement the change. Readiness is the motive, aim, and willingness to implement the change. Capability is the power, influence, and authority to allocate appropriate resources. Capability also implies the subgroup possesses the information and skills necessary to carry out proposed implementation tasks.

When viewed from a project perspective, any change will require a champion that possesses either personal or positional power and is willing to use the power to benefit the project [6]. The champion may use their power to recognize that they do more than just aid the project. They often go well beyond the expected and traditional job responsibilities. It is important to identify and encourage the emergence of champions. Those individuals who are willing to take a leadership role in the change are key components in successful implementation. Encouragement of champions can be assisted when the corporate culture climate rewards risk takers (innovation is not without risk).

It is also important that a champion is free to act when needed. A champion should not be tied concurrently to traditional and innovative processes. If necessary, use multiple champions. Some champions perform well at the early investigation and planning stages (the "big picture"), while others may be more adept at implementing the change (nuts and bolts implementation).

9 How Is Change Management Different From TQM Or Continuous Improvement?

From a holistic view, total quality or continuous improvement concepts and philosophies are similar to successful change management strategies. The adoption and action of Deming's 14 TQM points are a signal that the company management intends to stay in business and aims to protect their investors and jobs [7]. To improve continuously, change must be a continuous process.

Several of Deming's points relate directly to the importance of change, leadership of change, education and training, trust, teamwork, while several other key management traits are covered in the others. In essence, the Deming method is a detailed normative strategy for implementing routine and radical changes under the auspices of quality improvement. If the points that include cost, product, and quality issues are removed, the framework of a normative strategy is clearly demonstrated. These points, paraphrased, are listed below:

- Adopt the new philosophy, take on leadership for change.
- Institute training on the job.
- Institute leadership. Supervision of management is in need of overhaul as well as supervision of production workers.
- Drive out fear, so that everyone may work effectively for the company.

- Breakdown barriers between departments. People in research, design, sales, and production must work as a team, to foresee problems of production and in use that may be countered with the product or service.
- Eliminate slogans, exhortations, and targets for the work force asking for zero defects and new levels of productivity. Such exhortations only create adversarial relationships, as the bulk of the causes of low productivity belong to the system and thus lie beyond the power of the workforce.
- (a) Remove barriers that rob the hourly worker of the right to pride of workmanship. (b) Remove barriers that rob people in management and engineering of their right to pride in workmanship.
- Institute a vigorous program of education and self- improvement.
- Put everyone in the company to work to accomplish the transformation. The transformation is everyone's job.

The features of a successful organization discussed earlier are also visible in Deming's 14 points. It would be expected, therefore, that companies with a TQM approach to business operations and projects would have fewer difficulties in identifying technology and changes important to them and successfully implementing the improvement.

10 Successful Organizations

Assuming that successful organizations are those able to change and include new technology and management schemes into their business practices, the following traits are important to note [1].

1. An effective organization is purposeful and goal oriented.
2. Work is organized, resources allocated, and decisions made are defined by the work requirements, not by authority or power requirements.
3. Decisions are made based on sources of information rather than position in the hierarchy.
4. Reward systems are related to the task, not the status of the performer.
5. Communication is relatively open.
6. Inappropriate internal competition is minimized.
7. Conflict is managed, not suppressed or avoided.
8. The organization is viewed as an open system.
9. Each individual's identity, integrity, and freedom is supported.
10. The organization operates in a learning mode.

These also indicate strong support for a normative strategy and reflect many points similar to Deming's quality philosophy.

11 Summary

Successful organizations, total quality management principles, and successful chang strategies have many common features. Each of these stressed goals or goal setting, ope communication, trust, conflict management, decisions made on the basis of informatio not position, and the importance of continuous learning. These elements, onc established, would eliminate many of the points of resistance currently encountered b firms. Given that change will be continuous, a process that allows continuou modification and learning is likely to ensure continued success by an organization.

12 References

1. Beckhard, Richard and Reuben T. Harris, Organizational Transitions, 2nd. Edition, Addison Wesley Publishing Company, Reading MA 1987

2. Nord, Walter R. and Sharon Tucker, Implementing Routine and Radical Innovations, Lexington Books, D.C. Heath and Company, Lexington, 1987

3. Steele, Lowell W., Innovation in Big Business, Elsevier North Holland Publishing Company, New York, 1975

4. Harvey, Thomas R., CHECKLIST FOR CHANGE: A Pragmatic Approach to Creating and Controlling Change, Allyn and Bacon, Boston, 1990

5. Lawrence, Paul; "How to deal with Resistance to Change," Management of Change, Harvard Business Review, 1991

6. Pinto, Jeffery K. and Dennis P. Slevin, "The Project Champion: Key To Implementation Success," Project Management Journal, The Project Management Institute, Vol. XX, No. 4, December 1989, pp. 15-20.

7. Deming, W. Edwards; Out of the Crisis, Massachusetts Institute of Technology, Center for Advanced Engineering Study, 1986

8. Rydz, John S., Managing Innovation: From the Executive Suite to the Shop Floor, Ballinger Publishing Company, Cambridge, 1986

9. Tatum, Clyde B., "Innovation on the Construction Project: A Process View," Project Management Journal, Project Management Institute, Vol. XVIII, No. 5, December 1987 pp. 57-67

ENHANCING LEARNING IN CONSTRUCTION PROJECTS

D. Boyd

Faculty of the Built Environment, University of Central England, Birmingham, UK

A. Robson

Department of Building Finance, Birmingham City Council, Birmingham, UK

Abstract

The performance of the UK Construction Industry has been perceived as bad for decades. Many technical and managerial solutions have been tried and have failed. It is contended in this paper that many of the problems within the construction industry arise because of its barriers to learning. This paper investigates the enhancement of learning within projects through the use of two semi-formal techniques: personal project diaries and debriefing, as part of making construction into a learning industry. These techniques were used both as tools to facilitate learning by individuals and as research tools to determine learning needs and barriers. The format of these diaries and their use by a small group of participants in two case study projects is described. The critical incidents recorded in the weekly diaries were used as the focus of the structured debriefing at monthly intervals. The debriefing process used the technique of cognitive mapping which allowed the researcher to capture a shared understanding of the events. This debriefing made the learning from the task and process more explicit in a way that could be useful to the participant and, because it was explicit, to their organisation. The paper concludes that coupling personal project diaries with structured debriefing can deliver enhanced individual learning and that, in the right organisational circumstances, this learning can be captured and could be used by the organisation.

Sommaire

C'est l'affirmation de cette communication que pleusieurs des problèmes qui concerne l'industrie du bâtiment brittanique ont leur origine en les obstacles contre l'apprentisage. Cette communication examine comment il est possible d'augmenter l'apprentisage pendent les des enterprises par le moyen de deux techniques: journaux privés de l'enterprise et le 'de-briefing' - de gagner l'information des participants après coup. Le format des ces journaux et leur usages par une groupage des participants penant deux enterprises est décrit. Les moments critiques qui sont rapporté dans les journaux hebdomadaire a eté utiliser comme foyer du 'de-briefing' à l'intervaux mensuels. Le processus de 'de-briefing' se servera de la tracerie des connaisances (en anglais *cognitive mapping*), qui permettre au chercheur d'attraper le comprehesion communal après coup. Cette 'de-briefing' a fait plus evident l'apprentisage de la tâche et le processus dans une façon utile a la fois aux participants et a leur organisation. La communication conclure que la couplage des journaux prives avec un'de-briefing' formel peut livrer l'augmentation de l'apprentisage individu, et que, en les circonstances organisationelle exactes, cette apprentisage peut être attrapé et utilisé par l'organisation.

1 Introduction

The UK Construction Industry is perceived both externally and internally, as having a poor record of performance which has national consequences [1]. Past studies have attributed this in terms of: under capitalisation [2], fragmented organisational structure [3], and inadequate communications [4]. It is contended in this paper that many of the problems within the construction industry arise because of its barriers to learning. The research reported here has investigated the use of two semi-formal techniques: learning logs/diaries and debriefing, to enhance learning within the industry. This paper focusses on the justification and application of these techniques. This is part of a wider

The Organization and Management of Construction: Shaping theory and practice (Volume One). Edited by D.A. Langford and A. Retik. Published in 1996 by E & FN Spon. ISBN 0 419 22230 8.

research project which is advancing the concept of "construction as a learning industry", exploring what the idea of the learning organisation [5] might mean in the contruction industry.

The industry has again been under scrutiny by a number of bodies: Latham [1], Technology Foresight [6], and the SERC [7]. All of them see the industry as problematic and unsuccessful. This present time is seen as the opportunity to 'sort it out' otherwise foreign competition might herald its demise like other heavy industries such as shipbuilding or steel. Suggested solutions tend to be technical, formal and managerial: design procedures or better contracts; making construction more like manufacturing, new technology or business process approach. However, it has to be noted that in the past such solutions have not met their full expectation. Systems building failed in the 70's and the popularity of quality assurance and even project management as panaceas appears to be waning. Thus, it has been suggested that this 'search for certainty' [8] is a recurring myth which is doomed to failure.

It would be difficult to disagree with the notion that new techniques and processes can help the industry. But, a different industry cannot just be set up; the current industry needs to be changed. Even in the implementation of relatively minor changes such as the introduction of IT in offices, it is now acknowledged that better solutions result from working with the staff in creating the solutions and implementing them. There is a need to engage with the industry as it is, in order to enhance its strengths and to support its weaknesses. Such an engagement can be accomplished through the idea of the learning organisation. In 1986 Geoffery Holland, then Director of the Manpower Services Commission, called for a new management development initiative in the UK: "If we are to survive as individuals or as companies or as a country we must create a tradition of 'learning companies' " [9]. This has been acknowledge for the construction industry "An organisations only sustainable competitive advantage lies in its ability to learn faster that its competitors .... to produce world class construction" [6]. The need for learning is even more important for Groak's [10] alternative view of the industry as a field of projects surviving in a turbulent environment with well developed complex adaptive features: "knowledge for action and organisational learning take their rightful place". It is contended that this can be seen to occur within informal systems set up through individual relationships between people, alongside individual's strong motivation to complete the project [8]. These informal systems, which involve coping skills of people to deal with the unexpected and value conflicts, are often employed to expedite projects that would otherwise come to a halt if the formal systems were relied upon [3]. However, the informal system cannot be formally acknowledged as it conflicts with the organisational procedures and the adversarial formal relationships [8]. Learning between projects is then held back because no discussion can take place on what really happened in a project.

Revans [11] in his description of learning organisations focuses upon the empowerment of individuals struggling with intractable problems to combine work and learning. This sets out to seek improvements from within, indeed from the common task, as it acknowledges that the daily round offers constant learning opportunities and that the quality of such learning is often reflected by the morale of the organisation. Thus a definition of the learning organisations as applied to the construction industry would be an industry that acknowledges the informal systems employed by individuals at project level to deal with the complexity of problems on site which arise through the interaction of human and technical elements. The promotion of learning and learning networks as one of their energies of change in the industry has been proposed officially: "Education and training should cater for peoples need to learn, pass on knowledge and adapt to change effectively throughout their careers." [6].

2 Methodology

The research reported here is based on Kolb's [12] model of learning. This states that true learning can only take place if experience is put under reflection. The strategy considered here to improve learning in the construction industry is for participants to keep diaries in order to record experiences making them available for reflection, in the first place, to the individual and secondly to the industry. By backing this up with debriefing by the researcher then this can help the individual to reflect on their experiences and it also provides more impetus for the individual to maintain their diary . Such a semi-formal and structured arrangement is required to counteract the crisis nature of the industry where there is 'never enough time' because of commercial pressures.

The research had three objectives in exploring the aim of enhancing learning in the industry [13]. Firstly to analyse whether the formalised techniques of diaries and debriefing can help to enhance the learning of individuals on construction projects. Secondly to determine whether the formalised techniques of diaries and debriefing can obtain information on the realities of managing construction projects. Thirdly, whether the formalised techniques of diaries and debriefing can be used as a research tool to identify the barriers to achieving all round individual learning. The basic approach was to test these techniques with people working on two case studies of projects that were on site.

The first objective was tested within the case studies directly by interpretting the diary and debriefing responses. In many ways this is intimately connected with the second objective as it assumes that it is possible to make explicit true information about a project and one's role in it. This is a phenomonological position in that it assumes people construct valid meaning to the world around them [14]. Thus, the information is subjective and value laden but this is all that can be known about a social world. In addition the approach was analysed against Kolb's [12] definition of learning both to provide a language to describe results and to explain the results obtained. It is stressed that this first objective was focussing on individual's learning, thus although a totally explicit explanation might not be available, learning might still be achieved. It is only when diaries and debriefing are required for the second and third objectives does the reality and understandability of an explanation become critical. These objectives might be able to provide direction to the next levels of learning that of the project group and ultimately the industry.

In the second and third objectives it is not just the perceptions of what happened that are important but also the feelings and emotions. These dimensions demonstrate individuals' values and their conflicts which guide actions and prevent learning. By making these dimensions have importance in the diaries then subjects are more likely to try to surface them even if this is not normal in their practice. They then provide a point of empathy between the researcher and subject which allows a more in-depth study.

In these objectives, the influence of the researcher needs to be acknowledged. Gummeson [15] indicates one limitation of case study research as the researcher's Pre-understanding. Pre-understanding is defined as peoples knowledge, insights and experience before they engage with a research project. The more consideration that is given to the input, preunderstanding, the greater the validity of the output, understanding. Much effort was expended in setting down the researcher's pre-understandings before embarking on the case study [13]. The researchers knowledge of the construction industry as pre-understanding is important as it allows the researcher to more quickly understand the subject. Gummeson [15] also sees access to cases as problematic. In this research, physical access is essential and explicit, however, access to the mind of a subject is more difficult. Here the importance of developing a relationship of trust and confidence between researcher and the subject who completed the diaries and debriefing was never underestimated and was at the forefront of all approaches.

The research is of an action research character [16] where change is a specific objective. The researcher is not neutral or detached but works with the subjects to implement specific change [17]. In this case the change is the learning of individuals. However, additional change may take place in the short term through modifying the individuals approach to the project that they are working on. This give rise to the possibility of a positive feedback in that the project learning environment may change, making actions and learning occur differently than on a normal project. Thus, the ability to find information on the reality of managing projects might be compromised.

Two case studies were chosen for this pilot research study. This research only considered the construction phase as the critical transformation point where concepts become practice. In order to ensure access, both cases had people who were familiar to the researchers. This has the potential to influence responses towards what the researchers wanted, but this risk was thought less important than having better access and a quicker establishment of trust. Some subjects were unfamiliar to the researchers and so partially overcame the risk of familiarity. In addition, the researchers were not involved as professionals on either of the projects which might break trust or present a conflict of interest [13]. Three subjects were asked to participate in each case study. These subjects were selected to span design, contracting management and site supervision levels in order to investigate differences in perceptions that might occur.

3 Learning a definition for the Construction Industry

The notions of learning are central to the way we, as individuals, respond to the world. Most learning theories adopt a cyclical model with four distinct stages. Kolb's Learning Cycle [12] presents the stages as concrete experience, reflective observation, abstract conceptualisation and active experimentation. Effective or 'all round' learning requires the full journey around the cycle. However the cycle describes an ideal learning pattern and Kolb's insight was that individuals differ in their interest in learning form the different stages of the cycle. Kolb [12] and Honey and Mumford [26] established that individuals evolve skills in one area of the cycle more than others (called their learning style) and so they have the propensity to short-circuit the cycle. Short-circuiting the cycle results in failed or insufficient learning and will cause people to act set ways.

At work, individuals tend to come together in, or are recruited into, organisations which present environments characterised by the opportunity to use a particular learning style. There is therefore a tendency for people with the same learning styles to associate and create organisational environments with particular styles of operation. These ideas of learning environments have been developed by Kolb into a cycle of complex environments which effects the lifelong learning and adaptation of individuals in fundamental ways and involve typical responses to problems [19]. Boyd and Wild [16] considered the barriers to both practice and academia achieving all round learning, i.e. learning from all four segments in the experiential learning cycle. They demonstrate that construction practice is concerned with 'getting the job done', finding pragmatic solutions to problems and only addressing problems as they arise. The expectations of the organisational field mirrors the learning style preferences of the individuals working together in the organisation and generates structural and cultural barriers to learning [19]. In order to achieve all round learning in the construction industry it is necessary to close the gap in the learning cycle between the behaviourally and symbolically complex environments. Thus, it is not sufficient to have an experience in order to learn as without reflecting upon the experience it may be quickly forgotten or its learning potential lost. Schon [20] suggested that this gap can be closed by the 'reflective practitioner' and Boyd and Wild [16] suggested action research.

Argyris and Schon [21] have shown that most organisational learning is single loop, or error detection and correction. They saw only isolated examples of double loop learning which changes current operating assumptions, norms and values and which involves deeper enquiry, questioning and, inevitably , conflict and power struggles. Argyris and Schon argue that organisations can only learn through the learning of individuals and that it is through understanding this process, through duetero learning, that whole learning is brought about. Argyris and Schon suggest that a learning organisation is one which becomes better at double loop learning and handling the conflicts to achieve this. They propose a model of expert facilitated intervention in the reflective practitioner.

Technical skills form the grounding of education in the construction industry. However this approach alone will only serve to maintain the status quo and will not respond to the need for change [22]. Experienced practitioners in the construction industry have a vast amount of knowledge which can be use as a learning base for the industry. This knowledge includes the skills and informal systems employed by individuals for dealing with uncertainty, complexity, instability, uniqueness and value conflict which exists in the industry. The industry needs to harness this knowledge to challenge current norms and achieve continuous improvement. It is only if individuals, and the industry, know what people are taking account of, and what makes people do what they do, that the situation can be improved. The industry needs the skill of reflection in order to achieve learning. By learning through self-development individuals and the industry will be able to form a greater learning base which can be used to promote innovation and change [22].

4 Diaries

Diaries have been used for some time as a tool for research data collection in the social sciences [14]. The criticism of them for this use, in that it modifies behaviour [24], is the very aspect important for this study. Further, Zimmerman and Weider [25] suggest that they are an alternative to participant observation if coupled with interviews, thus, supporting their use for establishing the realities of construction projects. Diaries as learning tools have been suggested by Honey and Mumford [26] and Pedlar Burgoyne and Boydell [27]. Pedlar et al [27] suggest that although there is nothing new in the idea of keeping a diary for reflecting on ones own behaviour, to help make such an exercise more meaningful for self-development the diary should be structured round a simple model of an individual's behaviour in a particular situation. According to this model there are three aspects to this behaviour: (i) the individuals feelings, (ii) the individuals thoughts and ideas, and (iii) the individuals action tendencies.

The two examples of diaries described above had been designed for self-development as exercises in a learning text carried out alone. These used blank pages to act as the diary so as not to constrain personal entries. In the research project described here, the subjects were unfamiliar with the theories of learning, thus leaving blank pages to be filled in was not considered sufficient. The structure of the diary was more rigid and tried to take subjects through an event described (see figure 1).

DIARY ENTRY	LEARNING SUMMARY
Date ___ Time ___	Date ___ Time ___
WHAT HAPPENED?	LEARNING Learning from previous diary entries
MY RESPONSES My thoughts and ideas My feelings My possible actions My chosen actions	Suggestions for improving the operation of the project

Fig. 1. Personal Project Diary Format

Guidance on the information required under each heading was provided as an introduction. Further, in designing a diary for use in the construction industry, it was felt that there might be resistance to self exposure which had to be shared with the researcher and maybe with the different organisations involved in the case study projects. The diary takes stages one and two of Pedlar et al [27] journal, firstly asking for a description of events which will provide information on the realities of what happened. It then goes on to ask for an analysis of an individuals thoughts/ideas, feelings and possible actions. Finally it asks the subject to identify the chosen action and analysis the reasons for this. It is here that the diaries can deliver information on the barriers to learning and the theories-in-action employed by individuals. After approximately four diary entries a second step is added that of looking for learning from the previous diary entries and an action plan or suggestion for improvements to the project or future projects. It is here that evidence can be gained on the enhancement of learning.

The frequency of entries requested for a diary entry were: once a week on a Friday to prompt a review of the weeks events, and approximately every four weeks for a learning summary. As guidance it was anticipated that each week this would take half an hour for a diary entry. Weekly intervals would give an opportunity to obtain a longitudinal glimpse at the project over three months. In order to prompt participants into the discipline of keeping the diary a suggested timetable of dates for entries was included whilst leaving it open for participants to insert the actual date and time of each entry. These diaries were then be used as a basis for debriefing exercises.

5 Debriefing

Debriefing was considered to compliment the diaries in two ways: firstly to clarify the meaning of the words to the researcher and secondly to help the subject reflect more holistically on the events. The focussing on an event in this case is similar to the 'critical incident' technique [28] which has been used to establish deep causes of events. It was felt that subjects did not or could not describe the full set of consideration in their diary and that this could be teased out through the facilitation of the researcher. The subject was asked to clarify their assumptions and then to explore their response to the situation. There is a need to fix a chronological order as many participants become so involved in their part they become unaware of the total sequence of events. This involvement can be very emotional and it is important to release these feelings in a constructive way to establish the assumptions and preconceptions subjects brought to the event.

It was considered that there were still some things which the subjects would not be able to put into words, in particular the tacit knowledge [29] and theories-in-use [20]. This information makes an event unique for an individual and is an excellent site for issues that can provide double loop learning. To overcome this language and concept barrier, the technique of cognitive mapping was used [30]. Cognitive mapping is a technique which aims to visualise an individuals ideas, beliefs, values and attitudes through creating a pictorial map of how the subject frames a problem (e.g. figure 2). This helps in exploring the real meaning behind the event as well as ensuring that this meaning is understood by the researcher.

As the subject explains a diary entry, the researcher builds up a view of the subjects world. The researcher then encapsulates the event or issue in a word or short phrase which is written at the centre of a large sheet of paper. The subject is asked to think of satisfactory alternatives to this event and these are recorded on the paper. The next step is to develop ideas through addressing the questions "why does this matter to you"? or "why are you worried about this"?. As the ideas are expressed they are recorded and are linked back to the alternatives and to one another to explore how ideas and actions are inter-dependant. To help explore issues and to develop a shared

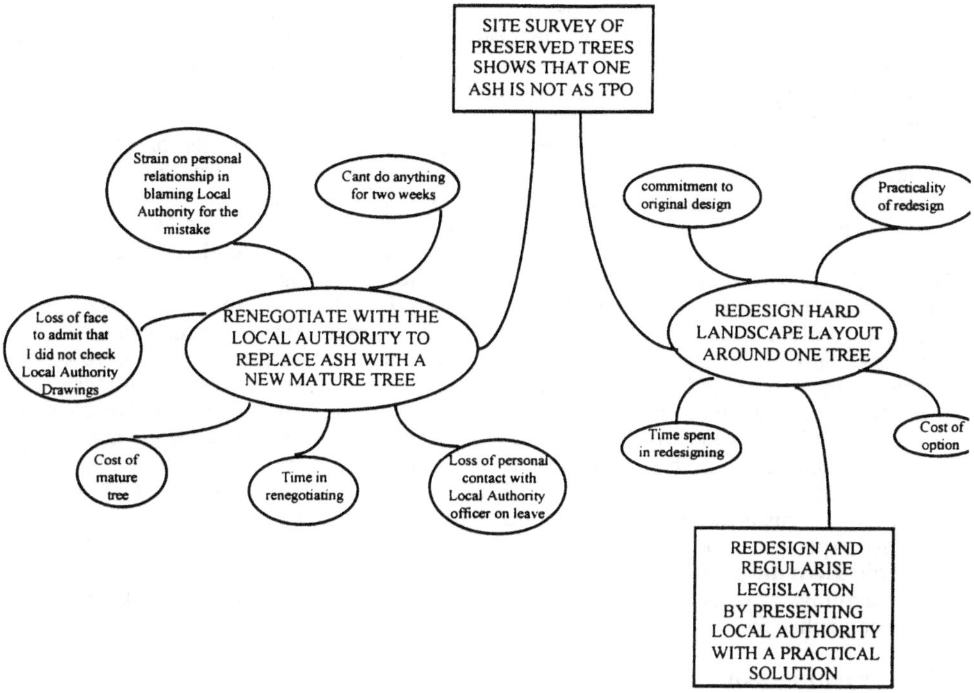

Fig. 2. Example Cognitive Map (psychological opposites omitted)

understanding between subject and researcher, the meaning of ideas/issues are explored through asking subjects for the psychological opposites [30]. The resulting pictorial model is not only evidence of careful listening but an explicit presentation of the meaning of the problem to the particular individual. Depending on an individuals' disciplinary background, organisational role, past experience and socio-political interests they will frame problems in different ways. This technique allows them to find their learning from an event in the way best understood by them. The researcher is able to share this learning through the pictorial representation and might be able to help transfer this to the industry.

6 Case Studies

The first case study project was the construction of new student accommodation to the value of £5.5M under a design and build contract. The project was on two adjacent sites in an inner city area and involved the construction of thirteen, three and four storey blocks. The research was carried out towards the end of the 51 week contract period. When the research began the contract was on programme and there was a general feeling that time and cost pressures had been controlled to a compressed programme. This case study examined learning within a large multi-national contracting organisation with the research subjects being the Site Manager and the Design Manager. The third research subject was the Architect from a small local practice.

The second case study project was the construction of a 2140 m^2 office development to the value of £1M under a traditional form of contract. The site was in an inner city area and the project involved three blocks, two new build and one refurbishment where planning consent was subject to listed building consent to retain the facade. The research was carried out in the middle of the 50 week contract period.

When the research began the contract was behind programme on the refurbishment element. There was a general feeling that time and cost pressures and a dominant client on the project were hampering progress on site. The research subjects were the site manager belonging to a small local contracting firm, the project architect and the project quantity surveyor.

The case studies were analysed through the series of events based on the perceptions of individual research subjects. Research subjects sometimes used the same event which provided an interesting comparison of different perceptions. By analysing the information given under **What Happened ?** a description of the events from an individual's perspective and the antecedents to the events were determined. These descriptions provided information about the realities of managing construction projects and the informal system employed. By analysing **The Responses,** information was gathered on feelings, and on the process of problem solving including the evaluation of options. This allowed an understanding of the theories-of-action which the research subjects employed and some insight into their values, assumptions, attitudes, and beliefs. The evidence of theories-in-use included the presentation of alternative courses of action which were not followed, and imaginative experiments during the debriefing which indicated what the individual would do under various circumstances. The evidence of the governing variables will come from graphical representation of tacit knowledge. This information allowed the identification of barriers to all round learning due to experience, personal development, or the reluctance to change. From the periodic **Learning Summaries**, it was determined whether single loop or double loop learning occurred. This was then used as evidence to examine whether learning had been enhanced.

7 Learning Analysis

The diaries of the research have produced a large amount of information which is still being analysed [13]. All subjects demonstrated a wealth of experience, however, the diary entries showed that the problems in day-to-day operations on site were resolved pragmatically due to time and commercial pressures. This demonstrated only single loop learning where subjects created new techniques for coping e.g.'covering the losses' due to poor performance of sub contractors or inadequate design information. In addition, the suggestions for improvement at a individual level illustrated single loop learning, e.g. to make subcontractors feel part of the team, or to generate design solutions on site. However there was also a strong common thread in the feelings of "frustration", "disappointment", "I've been here before" and that "this is typical of the industry in recession" illustrating that the values of individuals were at odds with the environment in which they worked.

In the learning summaries the research subjects set boundaries around the recurrent themes from their diary entries to come up with definable problems. The summaries provided the opportunity to reflect outside their professional and organisational roles and to consider the consequences of daily decision making on the project as a whole. The summaries identify double loop learning, for instance, in their concern for the reasons for accepting lowest tenders which led to poor performance of subcontractors on site, or for the reasons behind inadequate design information. From the learning summaries the boundary between learning in a behaviourally complex environment, typical of the environment of construction practice and focused on results, can be crossed to a symbolically complex environment and focused upon consequences [16]. This facilitates all round learning for individuals. These steps, towards looking below the surface of everyday problem solving on site, gives an indication that individuals at a project level have the ability through self-development to initiate change from their experience. The frustration felt at project level when the same problems occur again and again appears on the surface to be outweighed by the buzz of crisis management and the satisfaction of achieving results and seeing progress on site against the odds.

The case studies and subsequent analysis have shown that construction practice on site restricts the learning process to experiencing and acting and that project teams are generally forced to focus on short term results, due to the structural and cultural pressures required by the system in which they operate [19]. The research has started to examine the barriers to learning, which will form the subject of a future paper. It is believed that at an individual level, the diaries and debriefing have begun to attack both the situational barrier to learning by presenting the opportunity for reflection through the discipline of completing the diary and providing time for debriefing, and the intellectual barrier to learning by making valid the concepts of all learning to the individuals. Diaries need to be continued at a group or team level in order to attack some of the further barriers to learning identified, for example, the reduction of a blame culture. The research subjects were not willing to share their individual diary entries with others, but were happy to take items recorded in the learning summaries to a group for suggestions for improvements to the project, or to the organisation for suggestions for improvement. But feedback needs to be encouraged and learning to achieve improvement. Thus, this formal approach to the informal process of learning through experience and self-development could prove of long term benefit and ultimately result in change and innovation in the industry.

8 Conclusion

This paper has contended that many of the problems within the construction industry arise because of its barriers to learning. The research investigated the enhancement of learning within projects through the use of two semi-formal techniques: personal project diaries and debriefing. The most prominent problem seemed to be a lack of time formally set aside to engage with double loop learning through commercial pressures. Short term problem solving was practised by all the research subjects due to pressures of the organisational environment and the construction programme. The thoughts and feelings of the research subjects when approaching day to day problems were of frustration and exasperation at the underlying causes of the problems faced, but in the daily action they had to put aside these feelings and find pragmatic solutions. This is indicative of Kolb's [18] behaviourally complex environment and it was concluded that individuals in the industry adopt a problem solving approach which is focused on results not consequences [16]. Both case studies illustrated how problems were resolved through informal systems which allowed progress on site to work around the value conflicts between individuals and organisations. It was felt that individuals wanted to change the cultural core of the industry to alleviate the commercial pressures. Thus, at an individual level, double loop learning was demonstrated, and so learning was enhanced by the diaries and debriefing. However, in order to move this work on, the barriers to learning that are fundamental to the organisation and operation of the industry need to be bridged.

References

1. Latham M.(1994), *Constructing the Team*, HMSO
2. Hillebrandt P.M.(1984), *Analysis of the British Construction Industry*, Macmillan
3. Higgin G. and Jessop N.(1965), *Communications in the Building Industry*, Tavistock
4. Banwell H. (1964), *The Planning and Management of Contracts for Building and Civil Engineering Work*, HMSO
5. Pedlar M., Burgoyne J., and Boydell T. (1991), *The Learning Company: a strategy for sustainable development* , McGraw-Hill
6. HMSO (1995),*Technology Foresight: progress through partnership*. Office of Scence and Technology, HMSO, London
7. SERC (1994), *Innovative Manufacturing Initiative*, Science and Engineering Research Council, Swindon

8. Boyd D. and Wild A. (1993a), 'Innovation and Learning in Construction Project Management', *9th ARCOM conference Sept 1993*, Ed. R Eastham and R.M. Skitmore. ISBN 0951906070
9. Holland G.(1986), Excellence in Industry: Developing Managers - A New Approach, speech at Dorchester Hotel, London, 1.2.86, Manpower Services Commission
10. Groak S.(1994), Is Construction an Industry?, *Construction Management and Economics*, Vol.12, pp 287-293
11. Revans R.W. (1982), *'The Origins and Growth of Action Learning'*, Chartwell Bratt
12. Kolb D.A., (1984) *'Experiential Learning'*, Prentice Hall
13. Robson A. (1995) *Learning in Construction Projects*, MSc Construction Project Management dissertation, University of Central England, Birmingham
14. Easterby-Smith M., Thrope R. and Lowe A. (1991), *Management research: an introduction*, Sage, London
15. Gummeson, E. (1991) *Qualitative Methods in Management Research*, Sage, London
16. Boyd D.and Wild A.(1994), Action Research and the Engagement of Construction Education and Practice,*10th ARCOM conference Sept 1994*, Ed. R.M. Skitmore.
17. Whyte W.F. (Ed)(1991), *Participative Action Research*, Sage
18. Kolb et al (1984), *Organisational Psychology*, 4th ed., Prentice Hall
19. Richardson J. and Bennett B.(1984), Applying Learning Technique to On-the-Job Development, *J. of European Industrial Training*, Vol 8
20. Schon D., (1985) *'Educating the Reflective Practitioner'* , Jossey Bass
21. Argyris C. and Schon D.,(1974), "*Theory in Practice, increasing personal effectiveness*", Jossey Bass
22. Boyd D. and Wild A.(1993b), 'Education and Training for Quality', in proceedings of conference: *What is the future for CONSTRUCTION?*, University of Central England, 29th October 1993, Ed D. Boyd, ISBN 1 869954 28 9
23. Lait R. (1993), *'Human Resource Management as a Source of Criticism of Present Approaches in Balfour Beatty*, MSc Construction Project Management dissertation, University of Central England, Birmingham
24. Oppenheim, A.N.(1966) *Questionaire Design and Attitude Measurement*, Heinemann, London
25. Zimmerman D.A. and Weider D.L. (1977) The Diary Interview Method, *Urban Life*, 5(4), January, pp479-499
26. Honey P. and Mumford A. (1986), *'The Manual of Learning Styles'* self published
27. Pedlar, M., Burgoyne J.and Boydell T. (1986) *A Manager's Guide to Self development*, McGraw-Hill, London
28. Flanagan J.C.(1954), The Critical Incident Technique, *Psychological Bulletin*, 1, pp 327-358
29. Polanyi M. (1957), *'The Tacit Dimension'*, Wiley
30. Eden C.(1988), Cognitive Mapping: A Review, *European J. of Operational Research*, No.26, pp1-13

THE MANAGEMENT OF PUBLIC RELATIONS AS A BUSINESS PROCESS IN CONSTRUCTION

C.N. Preece and K. Moodley
Department of Civil Engineering, University of Leeds, Leeds, UK

Abstract
Organisations in the construction industry are faced with a number of challenges. How can they respond to greater client and public concern about the environment and demonstrate their corporate and social responsibility?

Construction organisations need to plan and co-ordinate their corporate and marketing communications, moving beyond a reactive and crisis management mode to a more proactive stance in addressing the wide network of stakeholders ranging from employees, investors and customers, to potential recruits, local communities and the general public

This paper identifies the issues in the management of public relations and builds a framework which integrates corporate, public relations and marketing communications in construction. The benefit of adopting this is that organisations will be able to better strategically plan their overall image and communicate with an increasingly diverse network of audiences.
Keywords: Public relations; construction industry, construction firms, communications, corporate image

Sommaire

Les entreprises du batement et travaux publics (BTP) doivent faire face a un certain nombre de challenges. Comment peuvent elles faire face a une demande plus importante au niveau du respect de l'environement de le part de leurs clients et du public? Et commentpeuvent elles demontrer leur responsabilites sociales et corporatives?

Les entreprises du BTP ont besoin de planifer et de coordiner leurs declarations au niveau de leur marketing et leur corporations afin de passer d'un mode de gestion reactif ou de aise a une attitude plus proactive - aci en s'adressant a un reseau etentu d'uinvestisseurs qui comprendrait aussi bien des employes, des clients, des investisseurs ou des organisations telles que des communes et le grand public.

Cet article pose la question de la gestion des relations publiques - Il met en place un schema de gestion qui complendiait les corporations, les relations publiques et le marketing daus le monde du BTP.

The Organization and Management of Construction: Shaping theory and practice (Volume One).
Edited by D.A. Langford and A. Retik. Published in 1996 by E & FN Spon. ISBN 0 419 22230 8.

Il est a souliaiter que de L' emploi d'une telle gestion resultena une meilleure planification strategique de la part des entreprises du BTP et que ceci leur permettia une meilleure communication avec une audience toujours plus etendue et diverse.

Mots cles: Relatipns publiques, BTP (Batiment et travaux publics), entrprises du BTP, communication (Moyen de..), image des corporations.

1 Introduction

Recent reports to the construction industry have emphasised the need for greater attention to be paid to its image (1, 2). Construction organisations; contractors and professional consultancies alike, need to consider their own response to increased competition for contracts, more demanding and sophisticated clients and a general public which may at best be uninformed, and at worst hostile to the industries activities.

The 1990's has seen increased action by environmental and other groups, most notably the anti-roads lobby. There is evidence of a movement particularly amongst the larger organisations in construction, to develop public relations management in an attempt to improve communications with a wide variety of different stakeholders, and to develop their relationships with local communities through community policies (3,4,5,6).

There has been little attention paid in construction to the management of this business function, and the practice and processes involved in developing an overall communications strategy. This paper will seek to define public relations in terms of the construction industry, and develop an integrated approach to internal and external communications.

2 Defining Public Relations As A Management Function

Public relations consists of a plethora of activities. A generic definition provided by the Institute of Public Relations states:
'Public relations is the planned and sustained effort to establish and maintain goodwill and mutual understanding between an organisation and its publics."(7).

The plural word 'publics' refers to the fact that every company has several different groups with which it wants to communicate. For instance it's own management and staff, customers and clients, the general public, government etc. Each one of these groupings may be known as a public.

Public relations as a business process can be used to create a positive image and identity for the organisation. It can support marketing in a highly competitive environment and improve communications with management and staff resulting in improved morale. The public relations function may be involved in a diverse range of corporate and marketing related activities including producing press releases, the annual report, marketing brochures and internal newspapers, journals and other material. In publicly quoted firms, the function may be responsible for investor relations. As a response to the need to improve an organisations corporate image with the general public it may be involved with educational and careers activities and sponsorship of community activities.

3 The Role Of Public Relations In The Construction Industry

Public relations is being propelled by rapid changes in technology and the way business is conducted. The after-effects of streamlining, downsizing, and other business strategies that have eroded employee trust and loyalty, present communicators with opportunities and challenges in media relations, employee communication, strategic communication planning, investor relations, marketing, community relations, and related areas (8).

As mentioned earlier, with the exception of a number of articles in professional journals in the mid 1980's (9,10,11), public relations as a business process and a function of management has not been given attention in construction. This section will examine the role of public relations in the construction industry under five broad headings; PR in support of marketing, investor relations, internal communications and the general image of the construction industry.

3.1 PR In Support Of Marketing & Sales

Despite the importance of the construction industry to the economy, there is little empirical data on the PR strategies used by firms to communicate business messages to clients. As indicated in recent research into the promotional strategies of major UK construction firms (12,13) there is a need for organisations to develop and communicate more distinctive and differentiating messages; as a means of educating potential customers about the firm and to retain existing clients.

Public relations activity is much misunderstood, and in the context of the construction industry it may be better known as 'press relations'. Media relations is one of the most important tasks facing any PR department. In commercial terms, the specialist press such as *New Civil Engineer*, *Building*, and *Construction News* can raise the profile of the firm and its activities with those who influence the client; architects, quantity surveyors and engineers. General business magazines and newspapers may be targeted to reach client organisations in specific industries.

Many clients have been found to be concerned with bad reflective image and are seeking a responsible approach to operations on site and minimal disruption and damage to the natural environment (14). The business case for organisations in construction to develop community relations plans in addition to environmental policies is becoming clearer.

Press and media relations has an increasingly significant role to play in support of the marketing of contractual services in construction in communicating with clients about the company and creating a positive business climate before the use of more direct promotional techniques i.e. promotional brochures, direct mail and pre-qualification team presentations and interviews.

3.2 Investor Relations & Enhancing Corporate Image With The City.

There are several ways of achieving improved relations with investors and enhancing corporate image with city analysts.

The legal obligation, the corporate annual report, is the biggest single communication and requires considerable effort on behalf of the most senior directors and management of the firm. In a survey in the US more than 51% of institutional investor respondents ranked the annual report as the most useful and informative information source within a company's financial program, more important than any investor relations officer, the Chief Executive Officer and electronic information databases such as Dow Jones (15).

Interim and final results are also seized by financial journalists and relations with the 'money' papers such as the *Financial Times*. and the financial section of the other broadsheets is essential

The cornerstones of an investor relations (IR) program for publicly quoted organisations are valuation, disclosure, shareholder relations, and capital formation (16). Valuation primarily encompasses activities designed to ensure that the price of the stock reflects realistic prospects for the company and reasonable assessments of valuation by the marketplace. Disclosure is disseminating information considered material to the investment process. The purpose is to ensure fair and equal announcement of both good and bad news, as well as developments that might have a material effect on the investment decision. Shareholder relations encompasses the care and feeding or a satisfied shareholder base. The capital formation aspect is ensuring that there is a market for the company's securities and that capital is available at reasonable, competitive costs.

Recent research which analysed annual reports in the construction industry, identified that many could be improved to provide financial and other information in a more user-friendly manner (17). In construction the strategy may be to publicise the most profitable divisions of the firm rather than those with the most turnover.

3.3 Improving Internal Communications

Public relations has an internal function within the organisation. A firm needs to communicate with its own management and staff, and indeed all those who combine to produce the products and services offered by the organisation. Public relations may be involved in producing a variety of in-house literature including journals and newspapers or organising internal events such as management conferences to discuss future strategy or announce major changes to the organisation.

It is important that employees have a favourable image of their company because they are important ambassadors in their communities. Organisations are often criticised for their lack of social responsibility and respect for the individual. The U S Air Force for example has had serious community relations problem, particularly with recent base closures. In addressing this problem, Air Force Public Affairs have attempted to make base newspapers more relevant, focusing on those issues that are most important to people - career development, pay, benefits, and the future of the company (18) .

Much of the construction industry is service based and heavily reliant on people. The variety of different managers, designers and engineers, across functions and within and between organisations and projects need to be involved in the culture, mission and objectives of the firm or practice. Improving the image of the organisation can only be effective if all those whom the firm relies on in delivering the services and products to the customer understand what is expected of them in terms of performance.

3.4 Responding To Environmental & Other Lobbying Groups

In construction there are many possible crisis situations, most of which are applicable in any business, for instance the unexpected death of a senior executive, or a high-level fraud inquiry. Two cases that are particularly industry specific are the unexpected failure of a structure or building leading to damage to property or human life, and the growing problem of direct action in protest at environmental damage; the Twyford Dam being a most recent case..

Crisis management deals with issues that are big and dramatic enough to rapidly damage a company's image. Good crisis management must include methods of avoiding a crisis developing, as well as how to react in the worst case scenario. The PR function needs to identify likely risk areas, recommend changes to procedure to reduce these identified risks and finally prepare how to react swiftly and effectively should the preventative moves fail.

3.5 Improving The Overall Image Of The Construction Industry

The construction industry has had a very poor image with a wide section of the general public, customers and clients and the City. This image is partly shaped by the major players, the construction firms and major consultants who are judged largely on how they execute their activities with as little disruption to normality for the population as possible.

Public relations which is aimed at informing and educating the wider society about the construction industry's practices would do much to improve the perceptions of many. Following from the Latham Report (19), the professional institutions such as the Chartered Institute of Building, Chartered Institute of Civil Engineers, Royal Institute of Chartered surveyors and the Royal Institute of British Architects together with the construction firms and professional consultancies need to consider their response to more demanding and critical clients and increased competition in the industry.

4 A Strategic Framework Of Public Relations In The Construction Industry

This section will synthesis the previous discussion in the form of a basic framework which integrates public relations and marketing communications in the construction industry.

Fig 1 shows diagramatically the development of this model of public relations which is based on a literature review and exploratory research with public relations practitioners in a small number of major construction organisations (N=12). The intention is that this should be the basis of further and more detailed work which would test the validity of the model in the construction industry.

The model is presented in five phases in the development of public relations strategies. Each phase is split under two headings; internal and external considerations. This relates to the fact that the management of public relations as a business process is concerned with communications with internal audiences i.e. management and staff within the organisation, and external audiences i.e. clients, government, shareholders and investors, local communities, the media and the general public.

4.1 Clear & Measurable PR Objectives

Internal communications would seem to be an essential part of any public relations strategy. The management and staff of the firm or practice are essential in providing a quality service to the client or customer. Public relations can be used to communicate the mission and objectives of the firm or practice through a variety of different media; from the company notice board and newspaper in smaller firms, to the management conference and house journals in the medium to large organisations.

A further objective must be to build a positive working environment with an organisations subcontractors and suppliers who will have either a positive or negative impact on the overall image of the firm with its many audiences. Issues concerning quality and performance expectations need to be communicated with all those involved in providing the service to the client.

Public relations should have a separate and defined role within the construction organisation. It should, however, be complimentary to the marketing function. Construction organisations need to adopt both a pro-active and re-active strategy to enable not only initiation of action in creating a favourable business climate for the firm or practice, but to be able to deal efficiently with crisis situations.

4.2 PR Strategies

A comprehensive public relations program educates, informs, explains, and persuades. It includes a wide range of powerful tools that are used throughout the year, rather than just an occasional press release. In addition, it must support the organisation's corporate and marketing objectives. A multidimensional program should support any advertising, literature, and more direct marketing and sales activities (20).

4.3 Analysis Of PR Environment

It is important for public relations practitioners to have an early focus on issues as a means of shaping public opinion. Research plays an important role in the early identification of possible public relations issues and problems. Acting in a pre-emptive manner to shape public opinion about a project before it is formed can save time and money. One U S organisation that has used this method successfully is Arkla Inc. By employing one of the most fundamental of research methods - face-to-face meetings and interviews - the company worked with environmental groups to build an interstate pipeline system (21). This may seem to contrast with the reactions of major construction firms who may have taken a more heavy handed and reactionary view to environmental direct action groups.

4.4 Choices Of Strategy

One of the key choices in the provision of an overall commitment to public relations is whether to develop an in-house public relations function, or to contract out PR work to agencies or consultants.

Important issues concerning conflict which may arise between a firm and an agent or consultant have been identified in terms of concerns over knowing each other's businesses, contributing to a consistent communication flow, finances, and 'chemistry' (22). Organisations may decide to contract out specialised work such as investor relations or political lobbying where particular skills and experience are necessary.

4.5 Implementation Of PR Strategies

Implementation of PR strategies will include active promotion of the need for a professional approach to the top management of the firm. Senior management, possibly across functional or geographical boundaries will need to be engaged in the strategies for them to be effective. The decision will need to be taken as to whether to develop an in-house PR team or to select outside PR consultants or agencies to undertake the work. Managerial, executive and mechanical functions would need to be defined.

There would appear to be an attitude within the construction industry that PR is free advertising. This would indicate a need for more formal training of those charged with the implementation of public relations within construction organisations. Firms need to employ those with both experience and formal education and to actively encourage the formal training of existing personnel.

4.6 Feedback

Using research, practitioners can meet the growing need to justify the cost of public relations programmes, and prove results. The key is to make research part of planning a campaign strategy. During the late 1980's a number of medium to large UK construction organisations were found to be in the process of conducting corporate image research to identify how they were being perceived by their target clients and others (23) Public relations researchers provide quantitative and qualitative measures of what the media is saying about a client's products and issues. Research can be applied to assess what influence, if any, the PR techniques and media applied have on peoples' attitudes, perceptions and actions towards the organisation (24)

It would seem to be most important to feedback results to the senior management of the organisation who may not be convinced of the need for investment in the communication function.

5 Conclusions And Recommendations

Public relations is an essential part of the business process, particularly in the dynamic economic climate of the construction industry. At a time of increased competition and ever more demanding and critical clients, the business case for developing a positive public relations profile for the organisation is clear.

Public relations can improve the perceptions of a wide and diverse range of internal and external audiences and thus create a positive general and business environment for the firm. A public relations strategy in the commercial sense needs to address both internal and external audiences. For a construction firm or practice this means communicating with its management and staff, helping to improve morale, retain skilled personnel and maintain commitment to the overall objectives of the organisation

CLEAR & MEASURABLE PR OBJECTIVES

Internal & External

Build a positive corporate image and identity * Communicate mission, objectives and strategy of organisation * Create a client oriented culture * Improve morale and retention of management and staff * Change perceptions in the minds of target audiences * Support marketing efforts * Improve relations with customers, investors, local communities * Tackle crisis situations

PR STRATEGIES & PROCESSES

Internal & External

Employee Relations * Investor Relations * Media Relations * Community Relations * Crisis & Issue Management

ANALYSIS OF PR ENVIRONMENT

Internal & External

Clients * Employees * Suppliers & Subcontractors * Shareholders * Westminster, Brussels & the Local Council * Local Community * Professional Bodies * Print & Broadcast Media

CHOICES OF STRATEGY & TECHNIQUES

Internal & External

Choices between managing PR in-house or through outside consultants or agents * Different techniques to be used * Newspapers, House Journals and Management Conferences * News Releases * Press Conferences, Press Briefings & Photo calls * Feature Articles * One-to-one Press Contact * Exhibitions & Speaker Opportunities * Sponsorship * Corporate Hospitality * Advertising * Corporate Identity

IMPLEMENTATION OF PR STRATEGIES

Internal & External

Actively promote PR to the top management of the firm * Engage senior managers in PR activities * Development of in-house PR team * Selection of outside PR consultant or agency * Establishment of managerial, executive and mechanical functions * PR management development and training

FEEDBACK

Internal & External

Actively seek feedback on the effects of PR on improved client, investor, employee and community relationships and commercial success * Ensure results are fed to the senior management of the firm

Fig 1 A Framework For The Development Of A Model Of Public Relations In The Construction Industry

From an external stand point, the construction business needs to continually communicate with its clients and customers, financial investors, subcontractors and suppliers, the media, central and local government. Construction is a highly transient industry. Local community relations would seem to be especially important at the level of building and civil engineering projects on a national and increasingly international level.

In recent years public relations has developed into a distinct management function within the organisation structures of many medium to large construction organisations. If managed effectively, and with the active support of top management, public relations can develop and sustain a competitive corporate image and identity for the construction business

Organisations need to view there own PR activities, not only as a business process which will enhance their own corporate image, but which will improve the wider image of the industry. A combined cross-industry effort may be pursued to improve perceptions of construction and those who work in it. Public relations has a serious lobbying role. Professional institutes and construction organisations need an overall strategy in promoting the industry's cause to those influential in local, central and European government, the print and broadcast media and the City. An overall public relations strategy in the industry would also target the younger generation as a critical audience, through educational establishments i.e. schools, colleges and universities (25).

This paper has set out to outline the development of a model of public relations which takes an holistic approach to corporate and marketing communications in the construction industry. The next stage will be to test the validity of this model in the industry.

6 References

1. Centre for Strategic Studies in Construction (1989) *Investing in Building 2001*. Reading University.
2. Latham Report (1994) *Constructing The Team*, HMSO.
3. Costain Group Publications - *Costain and Young People Joint Venture*. Costain Group Plc.
4. Costain Group Publications - *Community Involvement Policy*. Costain Group Plc.
5. John Laing Group *Annual Reports* 1986 -1994
6. George Wimpey PLC *Annual Reports* 1989-1994
7. Green S G (1994) *Winning PR tactics; Effective techniques to boost your sales*. The Institute of Management Foundation, Pitman Publishing p 7.
8. Clemons J G (1994) A communication office in the year 2010. *Communication World* Vol: 11 Iss: 1 Jan/Feb p: 42-43
9. Kreamer P 91985) A strong identity creates a memorable image. *Building Technology & Management*. Chartered Institute of Building. December. pp 20-21.
10. Leech A (1981) PR begins at home. *Building Technology & Management* Chartered Institute Of Building. February. p 24.
11. Rutland P (1985) Presenting the company image. Building Technology & Management. Chartered Institute of Building. October. pp 13-14.
12. Preece C.N (1994) Promoting construction for competitive advantage. *Chartered Builder* July/August, Chartered Institute of Building, UK.
13. Male S & Preece C (1994) Promotional strategies in the marketing of contractual services within UK construction firms. *A J Etkin International Seminar on Strategic Planning in Construction Firms*, Haifa, Israel, June.
14. Preece C N (1993) *The management of promotion for competitive advantage in UK construction firms*. Unpublished PhD. Heriot-Watt University, Edinburgh, UK
15. Hutchins H R (1994) Annual reports (... who reads them?) *Communication World* Vol: 11 Iss: 9 October p: 18-21
16. Brown L F (1995) Getting a grip on investor relations Directors & Boards Vol: 19 Iss: 2 Winter 1995 p: 44-46

17. Preece C N (1993) op.cit
18. Dalton H J (1992) The importance of internal information during organisational change. *Executive Speeches* Vol: 7 Iss: 3 December 1992/January 1993 p: 10-15
19. Latham (1994) op.cit
20. Marken, G A (1993) Effective public relations ... more than a few news releases Public Relations Quarterly Vol: 38 Iss: 4 Winter 1993-1994 p: 47-48
21. Hart M A (1993) Early focus on issues shapes public opinion. *Public Relations Quarterly* Vol: 38 Iss: 4 Winter 1993-1994 p: 43-44
22. Bourland P G (1993) The nature of conflict in firm-client relations: A content analysis of Public Relations Journal, 1980-89 *Public Relations Review* Vol: 19 Iss: 4 Date: Winter 1993 p: 385-398
23. Preece (1993) op.cit
24. Hauss D (1993) Measuring the impact of public relations. Public Relations Journal Vol: 49 Iss: 2 Feb 1993 p: 14-15
25. Preece C & Moodley K (1995) A public relations strategy for the construction industry. *Chartered Builder* CIOB September.

HUMAN RESOURCE DEVELOPMENT IN LARGE CONSTRUCTION COMPANIES

M.R. Hancock, C.K. Yap and D.S. Root
Construction Study Unit, University of Bath, Bath, UK

ABSTRACT
This paper is concerned with the incidence of human resource development (HRD) within large construction companies.

Organisations that undertake HRD comprehend that their human resources are an asset rather than simply an expense. Furthermore, these firms believe that a sense of commitment towards employees signals to employees that they are regarded as important to the company's future success.

The hypotheses of the study are that large construction companies understand the basic concept of HRD, the practice is common and that HRD contributes to the enhancement of managerial performance.

The practice of HRD encompasses the activities of training, management development and appraisal. Analysis of the data suggests that large construction companies which undertake HRD understand the basic concept but that its practice is not common within large construction companies. Nevertheless, those that do practice HRD perceive that it can contribute to managerial performance.

The study suggests that proper understanding and practise of HRD can be a contributing determinant for competitive advantage in large construction companies.

Keywords: Human Resource Development, Strategic Planning

SOMMAIRE
Cette étude se rapporte avec l'incidence du développement de la resource humaine (DRH) dans les companies de la construction grandes.

Les organisations qui se chargent du DRH comprennent que ses resources humaines sont une valeur, plutôt qu'une dépense. Bien plus, cettes firmes croient qu'un sentiment d'engagement envers les employés signe aux employés qu'ils sont considérés comme important au succès ultérieure de la compagnie.

The Organization and Management of Construction: Shaping theory and practice (Volume One).
Edited by D.A. Langford and A. Retik. Published in 1996 by E & FN Spon. ISBN 0 419 22230 8.

Les hypothèses de cette étude sont que les compagnies de la construction grandes comprennent le concept fondamental du DRH, la pratique est courante et que le DRH aide au rehaussement du fonctionnement directorial.

La pratique du DRH entoure les activités de l'education, du développement directional et l'évaluation. L'analyse de la donnée suggére que les compagnie de la construction grandes qui se chargent du DRH comprennent le concept fondamental, mais que sa pratique n'est pas courante en dedans des compagnies de la construction grandes. Néonmoins, celles qui mettent en pratique le DRH percevoient qu'il peut contribuer au fonctionnement directional.

L'étude suggére qu'un entendement approprié et la pratique du DRH peut être un déterminant contribuant pour l'avantage concurrentiel dans les compagnies de la construction grandes.

Mots-clefs: Le Développement de la Resource Humaine, La Planification Stratégique.

Introduction

Human Resources Management (HRM) has received much attention in the general management literature during the 1980's and 1990's. A number of writers suggest that the reason for this lies in there being a direct correlation between strategic HRM and economic success. Porter (1985) believes that HRM can help a firm obtain competitive advantage. Schuler and Macmillan (1984:p242) make a similar point; that "effectively managing human resources gives benefits which include greater profitability." Arguably, economic gain is the primary objective of any organisation going into business (even charitable organisations. have to generate 'profit' in order to sustain their activities). This is not a new idea for businesses or for management commentators: The Classical School stressed the importance of 'scientific management' for improving efficiency and effectiveness with the sole intention of increasing profit. The research reported here considers only one aspect of HRM: the development of human resources

Mangham and Silver (1986) conclude that with the exception of a few large organisations, minimum attention has been given to the systematic development of managers. They indicate that companies regard 'learning from doing' (rather than actual management training) as critical to the development of managerial competence. Consequently, most managers have to develop themselves, struggling along as their career progresses with little or no opportunity to reflect on what they have learned. Experience indicates that the construction industry tends to "make or break" its managers by hurling them in at the deep end, letting them learn as they progress (sometimes with the consequence of major mistakes), rather than developing them intelligently. If this idea has any foundation, then there will be consequences for large construction firms because:

1) Large organisations tend towards formality and inflexibility with a plethora of rules and regulations likely to impact upon the efficiency of managers.
2) Within large organisations it becomes difficult to control everything from the top without depriving junior managers of authority and autonomy.
3) Large organisations usually undertake a variety of types of work which may create an 'ability gap' which the organisation must fill with good calibre managers.

Hypothesis

Given the aforesaid and the importance of large construction firms both to the industry and the general economy, the following three hypotheses were raised:

1) Human resource development (HRD) is a well-understood concept within large construction companies.
2) HRD programmes are common amongst large construction companies.
3) Large construction companies consider the provision of HRD programmes to be a positive aid to the enhancement of managerial performance.

Methodology

A literature review considers the relationship between strategic planning, HRM and HRD. For the empirical work, a postal questionnaire survey was adopted enabling a wide geographical area to be covered. Questionnaires were sent to human resources / personnel managers considered to be the most appropriate respondents.

The survey was confined to large construction organisations (determined by number of employees rather than annual turnover) a figure of greater than 115 employees was chosen...

a) to concur with statistical data available from the Department of Environment on Housing and Construction Statistics.

b) to achieve an adequate sample population for the survey to be carried out and,

c) because it is debatable whether construction organisations below the chosen size would be adequate. Given the scale of the work and number of people employed, organisations in this group are unlikely to possess the corporate or organisational sophistication, capacities, capabilities and geographical spread of operation, etc. to reveal the type of information that would satisfy the purpose of this study.

Hypothesis testing was carried out using a combination of the Mann-Whitney U Test (hypothesis 1) and Chi Square Test (hypotheses 2 & 3).

Strategic Planning and HRM

Strategic planning has been identified as the management of change (Fellows et al, 1983) and change has both direct and indirect effects on the resources of the organization. Despite the recent doubt cast on the efficacy of strategic planning due to the impossibility of controlling the firm's environment (Mintzberg 1994 a & b) there is evidence (Hussey 1982, Thune & House 1970) that, in general, organizations who undertake strategic planning are more successful than those who do not. In a direct response to Mintzberg's claim, Ansoff (1994) asserts that strategic planning has transformed itself into different varieties as a response to environmental challenges. Whatever the arguments, it seems reasonable to suggest that organizations that do not systematically analyse the environment, organisation, job and individual, are more likely to fail to anticipate how, when and where the people will be needed. Fellows et al (1983) recognise that attempts within the construction industry to plan manpower have been generally fruitless as there are too many variables for the goals to be realised. This is echoed in Langford, Hancock et al (1995) who claim that the number of variables demonstrates that building is rather less in control of its environment than most other industries. Key variables, identified by Hillebrandt and Cannon (1990:p137) are :-

1) "the finite construction period of each project;
2) the wide geographical spread of location of projects and especially that of large ones which can be undertaken only by major firms;
3) the uneven requirement for specific skills over the life of the project;
4) the wide diversity of skills required is such that any one contractor may not be able to supply all of them;
5) and the demand for any particular type of works fluctuates over time."

Because of these variables, construction firms find it difficult to anticipate workload, which leads to difficulties in the long term employment/deployment of manpower. All of these variables affect the planning of manpower within the industry and go some way towards explaining the current employment trend away from direct labour to self-employed and labour only sub-contracting: manpower planning has been confined/restricted to the management level of most construction organisations. This idea is supported by Hillebrandt and Cannon (1989) who claim that contracting firms are basically selling their management skills. The quality of management personnel is therefore key and requires the organisation to invest in HRD.

HRD is an area of HRM which itself has been defined by Armstrong (1991:p81) as "concerned with the development and implementation of people strategies and ensuring that the culture, values and structure of the organisation and the quality, motivation and commitment of its members contribute fully to the achievement of its goals."

The employment trend referred to above demonstrates the means by which organisations are trying to enhance productivity in the face of environmental influences. Firms hope that improved productivity will lead to a 'competitive edge' and thus enlarge their economic gain. It has been implied by writers such as Hussey (1982), Ouchi (1981), and Peters and Waterman (1982), that resources such as finance, plant and material are equal to all businesses; the only element that can make a difference is the human resource. Many organisations are now adopting the concept of HRM to try and gain that competitive advantage.

Strategic human resource development

The process of HRD starts with the strategic plans of the organisation which define where the business is going and the type of resources required to get there. These strategies are then translated by manpower planning into more specific definitions of how many and what sort of people will be needed in the future. Simultaneously, the business plan defines what levels of performance are required to achieve these objectives.

HRD must be specific and organised, in the sense that there are objectives, a process of learning, and provisions for evaluation. As HRD encompasses the activities of training and management development and performance appraisal, it should be regarded as central to anything that can be sensibly termed HRM.

If the term HRM is to be taken as something more than 'a new label for an old bottle of wine', then the terminology must consider the employees (present and future) of the organisation to be a resource, as opposed to commodity or cost. This implies investment therein whilst management implies that strategies aimed at the motivation, development and deployment of this resource and its associated investment will be directed in such a way as to maximise its potential. HRD plays a vital role here. Writers have adopted a various perspectives in considering the subject.

Nadler and Wiggs (1986) view HRD as a comprehensive learning system for the release of the organisation's human potentials - a system that includes both learning experiences (classroom, mediated, simulated) and on-the-job experiences that are keyed to the organisation's reason for existence (profit, survival, service, etc.). The organisation's HRD should aim to maximise the employee's effectiveness on the current job; facilitate the individual's mobility to the next assignment; and increase the employee's commitment to the organisation through learning and opportunities for personal growth and change.

Keep (in Storey, 1989) stated that companies that, for whatever reasons, are inclined to treat their employees simply as a cost or commodity, and who hence fail to invest in training and development activity cannot meaningfully be said to be practising HRM. He further argued that there is a solid economic rationale that is likely to bind the incidence of training and development to wider HRM policies. If a company has invested in training and development of its workforce, it then makes sense to develop policies that will help to retain these employees and to motivate and develop them in such a way as to put their skills to best use, thereby maximising the return on investment. Furthermore, training and development activities imply attempts to motivate and involve the workforce. Thus, Keep implied that a failure to treat expenditure on training and management development as a necessary investment could rapidly undermine the credibility of an organisation's attempt to adopt HRM practises.

Hall (in Fombrun et al:1984), defines strategic HRD as the identification of needed skills and active management learning for the long-range future in relation to explicit corporate and business strategies. He argues that the element most often missing is the linking of development needs and activities to an explicit organisational mission and strategy. Investment in training and development may be considerable, but many organisations rarely examine how training and development can most effectively promote organisational objectives, or how development activities should be altered in the light of the business plans. He further argues that the focus is frequently on immediate skill requirements in new or present assignments rather than on future projected requirements (i.e. operational rather than strategic planning).

Lastly, there is often an inadequate amount of planning devoted to the identification of future skill needs compared to the growth of those skills. To be strategic about development means to analyse future business opportunities and plans and to think objectively about future employee skills that will be necessary to implement these plans.

For an organisation to understand and be more effective about HRD, various problems and issues need to be resolved at the strategic level. This requires a clear understanding and commitment from top management that HRD provides valuable resources to the organisation. The success of HRD programmes depends on many factors at organisational level and the clearer the organisation is about development policies, the more effective control can be exercised over employee's growth.

Management training and development are considered a strong area for effective HRD. Recent investigations have sought to establish the level of commitment to management training and development within British companies (MSC, 1985;

NEDO/MSC,1984; Coopers and Lybrand, 1985; Lovey, 1991). These investigations reveal that whilst training and management development are undertaken by companies this is rarely adequate.

Training and management development has also been investigated within a construction industry context (Williams, 1987; Binder Hamlyn/UMIST, 1988; Mphake, 1988) and reveals that an increasing level of management and supervisory training was undertaken during the boom period of the 1980s.

Another important aspect of HRD programmes is in the area of performance appraisal. Appraisal can be used as an instrument for a systematic approach of evaluating staff in respect of the work and comparing it with a pre-determined target. Appraisal is considered a necessary element because it makes explicit what constitutes effective and efficient behaviour of employees: this is critical to the implementation of any strategic plan. It is also considered as the foundation of the training and development process.

Investigations into appraisal trends have been carried out to determine the extent of the practice (Gill, Ungerson and Thakur, 1973; Gill, 1977; Long, 1986). These reveal that the majority of companies use appraisal schemes with the intention of evaluating developmental needs and to improve current performance. These findings concur with a survey conducted in the construction industry (Aboobucker, 1987) which also revealed that the majority of users are contractors rather than consultants.

Survey, results & analysis

Sample & response rate

112 firms were chosen from listings in the New Civil Engineer's Contractor File of 1993 and 1994 and the Contract Journal 1994. This represents about 13% of the companies in the 'large' category. A total of 47 replies was received representing about 42% of the sample. From these 47 replies, 30 (26.8%) were usable and the rest were letters regretting inability to assist.

Current practices of large construction companies.

As a prelude to the main hypothesis testing, the data collected was used to indicate trends within large firms, related to the existence of strategic HR policies; training management development programmes and performance appraisal.

1. Strategic human resource policies

Respondents were asked whether their company implements any strategic human resource policies. 17 respondents (56.7%) have a strategic human resource policy within their company. This is encouraging in that the linking of human resource policy to a strategic plan shows that companies appreciate the benefits of 'humans' as assets.

2. Training and management development programmes

Respondents were asked whether their company implements any comprehensive training and management development programmes. 20 respondents (66.7%) indicated that their company has a comprehensive training and management development programme.

This result is interesting as earlier construction specific research by Williams (1987), Binder Hamlyn/UMIST (1988) and Mphake (1988) stresses the need for more awareness of the benefits training and management development programmes can bring to the company; and emphasise the need to increase this function within the company.

Although a "like to like" comparison could not be made with the earlier research (due to bias in the sample group), in general, the overall trend seems to show an improvement. More construction companies appear to be undertaking training and management development programmes than before.

Of the 20 companies that implemented comprehensive training and management development programmes, 19 commented favourably towards the programmes with 1 expressing no opinion. This seems to show an almost unanimous belief that comprehensive training and management development programmes are positive in their contributions to the company, especially in the area of enhancing managerial performance. Even for those 10 companies that do not have a programme, 5 felt that it could benefit the company: 3 expressed no opinion and 2 saw no potential benefit.

3. Performance appraisal system

Respondents were asked whether their company has a performance appraisal system. 26 respondents (86.7%) replied positively. This is in line with surveys undertaken by Gill, Ungerson and Thakur (1973) and Gill (1977) which reveal that appraisal has become a well-established feature of the British personnel function. Construction specific research by Aboobucker (1987) supports this.

Of these 26, 21 commented positively on the system, 2 expressed no opinion and 3 considered it to be not effective. 20% (5 out of 26) of the respondents do not think or have no opinion as to whether their appraisal system is effective. Latham (in Fombrum et al:1984) points out that for appraisal to be viewed positively, it must be an instrument to sustain effective and/or efficient job performance rather than a reward/punishment instrument.

Nevertheless, the result shows nearly 80% (21 out of 26) of those companies having an appraisal system are positive towards the contribution the system can bring to the company. This may be attributed to the fact that the trend in performance appraisal systems is moving towards more openness of reviews with an increase in career counselling (Long, 1986).

The practice of HRD in large construction firms

The foregoing are all indicators of current practice and firms have not been considered to be practising HRD unless all three factors are in place.

53.6% of respondents (15 Nr) confirmed that a strategic human resource policy; comprehensive training and management development programmes; and performance appraisal system existed in their firms. This indicates that more than half of the large construction firms surveyed have HRD programmes.

Of the 15 companies that practise HRD, 13 commented favourably towards it with 2 expressing either dissatisfaction or no opinion on training and development or the appraisal system. This shows that nearly 87% (13 out of 15) of the respondents are positive that HRD contributes to effectiveness in the company.

Statistical analysis of hypotheses

Following the establishment of the simple trends shown above, formal statistical testing of the hypotheses was undertaken in order to confirm that the patterns observed in the data were not merely chance occurrences and are 'robust' in the sense that they would be replicated should the research project be repeated (Kinnear and Gray:1994).

Hypothesis 1 - HRD is a well-understood concept within large construction companies.
In order to test the above hypothesis, various statements were made within the questionnaire for respondents to 'voice' their opinion. These statements were designed to test the respondent's conceptual understanding of HRD by their strength of answer from disagreement to strongly agree. The hypothesis was sub-divided into 5 sub-hypotheses (within which both null and alternative hypotheses were set). In each case the Mann-Whitney U Test was applied and in each case the corrected 2-tailed p-value was smaller than the significant p-value (at the 1% significance level in all but the first sub-hypothesis which was significant at 5%). As a result the sub-hypotheses accepted were:
Sub-hypothesis 1.1 - Human resources are an asset to the organisation rather than an expense.
Sub-hypothesis 1.2 - Strategic Human Resource Planning has been successful in linking the management of human resources to the organisation's overall strategic goals and objectives.
Sub-hypothesis 1.3 - Humans are the most important resource for a construction company (compared to finance, material and machinery).
Sub-hypothesis 1.4 - HRM helps the organisation to optimise return on its investment in people.
Sub-hypothesis 1.5 - A proactive HR policy is essential for the effective and efficient running of the organisation.

Hypothesis 2 - HRD programmes are common amongst large construction companies.
Null and alternative hypothesis were set and the Chi-square test carried out to compare the observed frequency with the expected to see if there was a significant relationship between them. The result showed the calculated chi-square value to be smaller than the tabulated critical value. Accordingly we have accepted the null hypothesis and conclude that HRD programmes are not common in large construction companies.

Hypothesis 3 - Large construction companies consider the provision of HRD programmes to be a positive aid to the enhancement of managerial performance.
Once again the chi-square test was applied to null and alternative hypothesis. In this case the calculated chi-square value was greater than the tabulated critical value at the 1% significant level and therefore we rejected the null hypothesis and accepted the alternative hypothesis. This result suggests that the provision of HRD programmes in large construction companies has a significant relationship with the enhancement of managerial performance.

Conclusion

Results of the hypothesis tests indicate that large construction firms understand the basic concept of HRD. This is a good sign as the literature suggests that companies that practise HRM exist in a continuum. Companies demonstrating an understanding of the concept are at the 'strong' end of the continuum and have a distinct approach to labour management where strategic interventions are designed to elicit commitment, development and full utilisation of resources.

Having established that companies understand the basic concept behind it, hypothesis 2 sought to determine the extent of this practice within large construction companies. Results show that there is no significant incidence of HRD in large construction companies. From a sample of 28 companies tested, 15 practise HRD (representing 53.6% of the valid total). As the test indicates that no significant relationship exists, the sample is not a representation of the population. This is mainly because the sample group is taken from the top ranking construction organisations and it can be concluded that these organisations are already high achievers and thus more advanced in their management technique. It is proper therefore to show that the practice of these groups of companies cannot be a representation of the population on the whole. If the sample was taken on a random basis, the result might have shown a significant relationship.

Hypothesis 3 was designed to test whether those companies that undertake HRD consider it to be an enhancement to managerial performance. The test shows that there is a significant relationship between large construction companies which practise HRD and the enhancement of managerial performance. Although no hard data is gathered to prove this relationship and the relationship is based on the perception of the one answering the questionnaire, nevertheless, a strong significance (beyond the 1 per cent level) indicates that HRD programmes have been viewed as a positive aid to the enhancement of managerial performance. The purpose of HRD programmes is to create conditions whereby potential of employees can be utilised to the fullest and thus contribute to the success of the organisation.

Bibliography

Aboobucker,C.M.M. (1987) *Appraisal of Staff Performance in the Construction Industry* - Unpublished MSc Thesis, Brunel University.

Ansoff,H.I. (1994) Comments on Henry Mintzberg's Rethinking Strategic Planning *Long Range Planning Journal,* Vol. 27, No 3.

Armstrong,M. (1991) *A Handbook of Personnel Management Practice (4th Ed)* Kogan Page, London.

Binder Hamlyn/UMIST (1988) *Survey on Supervisory and Management Training Needs in the UK Construction Industry* CITB, London.

Coopers & Lybrand Associates (1985) A Challenge to Complacency: Changing attitudes to training. *Report to the Manpower Services Commission and the National Economic Development Office* Coopers and Lybrand

Department of Environment *Housing and Construction Statistics of 1980-1990, Great Britain* HMSO, London.

Fellows,R. et al (1983) *Construction Management in Practice* Construction Press, Essex.

Fombrun,C. et al (1984) *Strategic Human Resource Management* Wiley, N.Y.

Gill,D., Ungerson,B. &Thakur,M. (1973) *Performance Appraisal in Perspective: A Survey of Current Practice* Institute of Personnel Management, London.

Gill,D. (1977) *Appraising Performance: Present Trends and the Next Decade* Institute of Personnel Management London.

Hillebrandt,P.M. & Cannon,J. (1989) *The Management of Construction Firm (Aspects of Theory)* Macmillan, London.

Hillebrandt,P.M. & Cannon,J. (1990) *The Modern Construction Firm* Macmillan, London.

Hussey,D. (1982) *Corporate Planning (Theory & Practice) 2nd Ed.* Pergamon Press, Oxford.

Kinnear,P.R & Gray,C.D. (1994) *SPSS for Window Made Simple* Lawrence Erlbaum Associate, Hore.

Langford,D., Hancock,M.R., Fellows,R. & Gale,A.W. (1995) *Human Resource Management in Construction* Longman Scientific & Technical, Essex.

Long,P. (1986) *Performance Appraisal Revisited* Institute of Personnel Management, London.

Lovey,L. (1991) *Management Training and Development in Large UK Business Organisations: A Survey* Harbridge House, London.

Mangham,I.L. & Silver,M.S.(1986) *Pilot survey on management training: Commissioned by the Economic and Social Research Council and the Dept. of Trade and Industry* - University of Bath, Bath.

Manpower Services Commission (1985) *Adult Training in Britain* HMSO, London.

Mintzberg,H. (1994a) Rethinking Strategic Planning (Part 1:Pitfalls and Fallacies) *Long Range Planning Journal*, Elsevier Science Ltd, Vol 27, No 3.

Mintzberg,H. (1994b) Rethinking Strategic Planning (Part 2:New Roles for Planners) *Long Range Planning Journal*, Elsevier Science Ltd, Vol 27, No 3.

Mphake,J. (1988) *Construction Companies Investment in Management Development* Unpublished MSc Thesis, University of Bath.

Nadler,L. & Wiggs,G.D. (1986) *Managing Human Resource Development* Jossey-Bass Publishers, London.

National Economic Development Office / Manpower Services Commission (1984) *Competence and competition* NEDO, London.

Ouchi,W.G. (1981) *Theory Z (How American Business can meet the Japanese Challenge* Addison-Wesley, London.

Peters,T. & Waterman,R. (1982) *In Search of Excellence* Harper & Row, N.Y..

Porter,M. (1985) *Competitive Advantage* Free Press,N.Y.

Schuler,R. & Macmillan,S. (1984) Gaining Competitive Advantage through HRM Practices *Human Resource Management Journal*, March, p241-55.

Storey,J. & Sisson,K. (1993) *Managing Human Resources and Industrial Relations* Open University Press, Buckingham.

Thune,S.S. & House,R.J. (1970) Where Long-Range Planning Pays Off *Business Horizon*, August.

Williams,R. (1987) *Management Training for Civil Engineers in Mid-Career* Unpublished MSc Thesis, University of Bath.

CHARACTERISTICS OF WINNING CONSTRUCTION MANAGEMENT TEAMS

D.H.T. Walker
Department of Building and Construction Economics, Royal Melbourne Institute of Technology, Melbourne, Australia

Abstract

Good management of construction time is closely associated with sound management practices. Recent research undertaken on a representative sample of 33 construction projects in Melbourne, Australia has revealed interesting insights into the reason why some buildings are constructed more quickly than others. One hundred and two factors which may affect construction time performance were analysed in the study using regression analysis, one way analysis of variance (ANOVA) and correlation analysis.

Characteristics of 'winning' construction teams are discussed from the perspective of the manager of the construction team and construction time performance. Interesting and valuable insights can be gained from this work. Important construction team leader characteristics are discussed which can be useful, in terms of construction time performance, to select an appropriate leader of the construction team.

Sommaire

La bonne exploitation du temps de construction est alliée fort avec la direction solide. Les recherches recents sur une section réprésentative de 33 projets de construction à Melbourne, Australie, nous ont permis de comprendre pourquoi quelques bâtiments sont construits en plus de vitesse que des autres. Cent deux facteurs liés avec le façon d'accomplir la construction ont été analysé dans l'étude, en employent l'analyse regression, l'analyse variance (ANOVA) et l'analyse correlation.

Les charactères des équipes de construction qui ont de succés sont discutés de la part du cadre de l'equipe, ainsi qu'en vue de l'exploitation du temps de construction. Cet oevure offre des perspicacités avantageux et interessants, en tenant compte du charactère du cadre d'une equipe de construction, etant utile en fonction de chiosir le cadre juste pour l'equipe.
Keywords: Construction Time Performance, Leadership, Construction Management.

1 Introduction

Construction time performance has been identified along with cost and quality as one of three crucial success factors for a construction project (PRD 1991a, p32). A number of researchers have investigated the role and influence of the construction leadership team in achieving timely construction and meeting the client's needs and have concluded that the construction management team plays a pivotal role (Ireland 1983;

The Organization and Management of Construction: Shaping theory and practice (Volume One). Edited by D.A. Langford and A. Retik. Published in 1996 by E & FN Spon. ISBN 0 419 22230 8.

Sidwell 1982; Morris 1983; Walker 1989). Case studies of best practice reveals interesting, though indicative, evidence taken from small samples of projects (BCA 1993; CIDA 1993; CII 1995). These add weight of evidence to support conclusions drawn from earlier studies (NEDO 1988) which advanced preferred characteristics of construction management teams to achieve sound construction time performance. Many leading academics have contributed to the debate and this has lead to the view that in order to perform well, winning construction management teams must engage in effective planning and communication with other teams involved with construction.

With the benefit of a substantial body of knowledge and empirical evidence being available in this area of investigation from the late 1970's and early 1980's, a detailed study was undertaken from 1990 to 1993 on a representative sample of Australian projects undertaken in Melbourne (Walker 1994a). General results have been reported elsewhere (Walker 1995) and more specific aspects have also been reported upon (Walker 1994b; Walker 1994c).

Sumarised specific findings are reported in this paper on an investigation into construction time performance. The work represents an in-depth study of 33 construction projects of mixed type of end-use, of scope (measured in 1990 Australian dollars) from $4 million to $45 million constructed and completed during the boom-bust period of 1987 to 1993. The findings reported in this paper relate specifically to the construction management team factors.

Conclusions flowing from the research provide useful best practice characteristics of leaders of the construction process. The findings suggest that winning construction teams are sophisticated and competent in the discharge of their duties. Managerial performance of the construction team was found to be the most significant factor, the interaction between client's representative and the construction management team was also found to be pivotal in facilitating good construction time performance. The value of the research reported upon in this paper lies in the weight of evidence that supports conclusions reached.

The paper is structured with a brief outline of research findings followed by more detailed discussion of the evidence. It is considered beyond the scope of this paper to discuss the methodology developed for the study. The paper concludes with recommendations.

2 Research Findings

The thesis tested by the research (Walker 1994a) was that *'variance between actual performance and trend line performance can be substantially explained by managerial effectiveness of the project team in response to challenges posed by factors outside the control of the construction management team.'*

More specifically, four principal hypotheses are tested by this work:

client's representative team's management effectiveness -

P_1-H_0 *that construction time performance IS NOT significantly affected by the management effectiveness of the client's representative;*

P_1-H_1 *that construction time performance IS significantly affected by the management effectiveness of the client's representative;*

construction management team's effectiveness -

P_2-H_0 *that construction time performance IS NOT significantly affected by the management effectiveness of construction management teams;*

P_2-H_1 *that construction time performance IS significantly affected by the management effectiveness of construction management teams;*

Design teams' effectiveness -

P_3-H_0 *that construction time performance IS NOT significantly affected by design team management effectiveness;*

P_3-H_1 *that construction time performance IS significantly affected by design team management effectiveness;*

Project challenges -

P_4-H_0 *that construction time performance IS NOT significantly affected by a small number of challenges posed by factors outside the control of the construction management team;*

P_4-H_1 *that construction time performance IS significantly affected by a small number of challenges posed by factors outside the control of the construction management team.*

Investigation of evidence to support or reject the principal hypotheses involved testing 102 sub-hypotheses because the principal aim of the work was to investigate the reason why some projects are built more quickly than others by establishing a 'league table' of factors for the sample group to also indicate the strength of each factor tested.

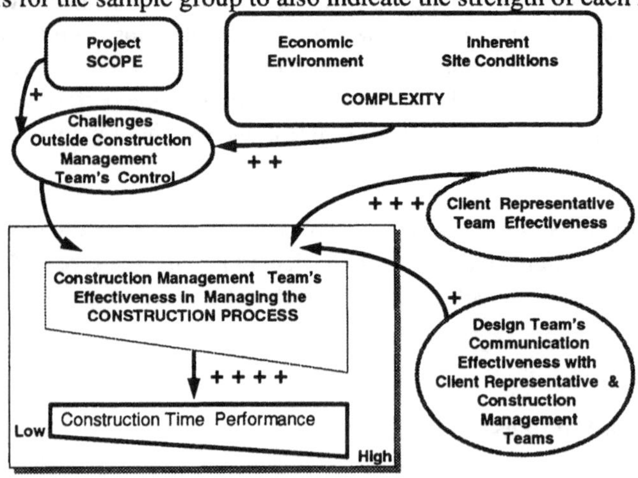

Figure 1 Model of Factors Affecting Construction Time Performance

The '+' symbols provide ordinal scales of significance. A '+++' significance is not three time more significant than a '+' influence. Significance means likelihood of the result NOT happening by chance.

A model, based upon the literature and findings from a pilot study, was refined and expanded through knowledge gained in analysing results from this 33 case study research (Walker 1994a). Figure 1 helps explain the strength of relationship between factors affecting construction time performance. It expresses two ideas. construction time performance is directly affected by the construction management team's managerial performance. This performance is affected by challenges outside the construction management team's control, the client representative's team effectiveness and the design team's communication effectiveness.

This model places responsibility for good construction time performance firmly with the construction management team. Berkeley *et al.* (1991, p5) state that *'the fact that it may be difficult to control the occurrence of risk does not absolve the manager from the need to anticipate the risk, and thus find ways of lessening the impact of its occurrence on the achievement of project goals. This may involve ensuring that the planning for the project is flexible in the face of the occurrence of the risk and/or that subsequent contingency planning can be made in the event of its occurrence.* Their view that construction time performance is substantially in the hands of those capable of dealing with risk supports the conclusions of this research. It is proposed that this group is generally the construction management team.

The model indicates that the most significant cluster of factors relate to the construction management team's effectiveness in managing the construction process which forms the filter through which construction time performance is determined. The construction management team is in direct control of construction activities and it is this team which can best formulate plans to control and mitigate risks to achieve satisfactory construction time performance. High construction management team performance results in high construction time performance and low construction management team performance results in low construction time performance.

No specific team structure is proposed, however, the results strongly support the idea that team member's qualities are crucial. A well functioning construction management team will be able to identify risks and plan to mitigate detrimental impact upon construction time performance, be able to effectively communicate using appropriate media and have the capacity to develop sound inter- and intra-team synergies. Though the construction management team's management style was not identified as a factor affecting construction time performance, it is moderately correlated to construction management team management related factors that do affect construction time performance. The model in figure 1 indicates that high construction management team effectiveness leads to high construction time performance.

The model, illustrated in figure 2, explains the relationship between construction management and client representative team effectiveness and other factors acting on these two teams. The construction management team's effectiveness in managing the construction process is affected by developing an organisational structure that is best suited to meet the challenges and risks accepted (Morris 1994, p219).

This model helps explain the elements and their relative significance and impact upon contributing to the construction management team effectiveness in managing the construction process.

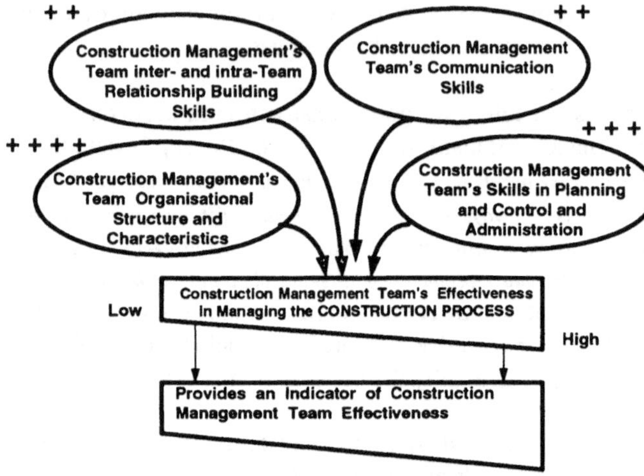

Figure 2 Indicator of Construction Management Team Effectiveness
The '+' symbols provide ordinal scales of significance. A '+++' significance is not three time more significant than a '+' influence. Significance means likelihood of the result NOT happening by chance.

3 Discussion

Table 1, The ANOVA test (at the 95% confidence level), reveals that important summary and detailed construction management related factors affect construction time performance. Six of the seven factors affecting construction time performance with a significance level of less that 0.01 are construction management team related. This demonstrates overwhelming support for the conclusion that construction time performance is most significantly affected by construction management team performance. The 0.01 significance level means that this result may occur by chance in one in one hundred cases.

Table 1 illustrates the sub-hypotheses tested and their ANOVA significance level for summary construction management variables and table 2 for detailed variables. The column headed 'HY' is the tested sub-hypothesis. In the short-hand description of the sub-hypothesis the term HI+ refers to a high or greater level measure, AVG+ refers to average or better, SH+ refers to slighly high or higher, CM refers to the construction management team and CR refers to the client's representative team. The third column indicates the significance level of the conclusion being due to chance, thus a 0.001

result (for the first row in table 1) indicates that 1 out of 1,000 cases may be chance results. The 0.95 confidence level expects a significance result of 1 - 0.95 or 0.05, thus results less than 0.05 are considered significant.

HY	Sub-hypothesis (Short-hand) Description - In Significance Order	Sig.
62	HI+ CM's organisational structure to manage risk	0.0010
56	AVG+ effectiveness of the construction team's planning	0.0188
87	SH+ CM's communication management to facilitate decision making	0.0258
61	HI+ effectiveness of the construction team's monitoring and control	0.0269
100	HI+ effectiveness of the CM team in managing the construction process	0.0273
49	HI+ impact of CR/CM working relationship	0.0468

Table 1 - Summary construction management Sub-hypotheses Supporting P_2-H_1

HY	Sub-hypothesis (Short-hand) Description - In Significance Order	Sig.
58	HI+ CM's planning - responding to problems or opportunities	0.0016
59	SH+ CM's effectively coordinating resources	0.0028
60	SH+ CM's developing an organisational structure to maintain workflow	0.0032
93	HI+ decision making communication within the CM team	0.0094
96	SH+ CM decision making, communicating and actioning	0.0113
84	SH+ key sub-contractor's task-oriented management style	0.0156
9	5% (Est.+EOT) workdays being within 5% of actual duration	0.0161
52	SH+ CM's forecasting planning data	0.0275
57	HI+ CM's monitoring and updating plans to reflect work status	0.0305
53	AVG+ CM's analysing construction methods	0.0315
90	SH+ CM and design team communication effectiveness for decision making	0.0448
99	AVG+ CM's effective use of information technologies	0.0452

Table 2 - Detailed construction management Sub-hypotheses Supporting P_2-H_1

The 18 reasons highlighted in tables 1 and 2 indicate why construction time performance is affected by client representative factors. Reasons fall into three broad groups, with HY-100 being a summary sub-hypothesis, which describe good construction management practice:

- HY-9, 52, 53, 56, 57, 58 and 61, construction management planning management and control procedures;
- HY-49, 59, 60, 62 and 84, construction management organisational structure, coordination and management style;
- HY-87, 90, 93, 96 and 99, construction management communication skills.

Results from the sub-hypothesis 9 test indicates that construction time performance is dependant upon 'the building duration time being within 5% of estimated duration adjusted for approved extention of time (EOT)'. The only significant association with any of the factors compared in the analysis was a moderate correlation with impact of client representative/construction management working relationship. This may indicate that projects constructed ahead of approved revised schedule tend to demonstrate better construction time performance than others. It may also mean that in those cases the estimated duration was generous and could be reduced, that the client representative was generous in approving EOT, or some combination of both. It is probably more realistic to assume, given that projects are awarded under market conditions, that the estimated duration was reasonable. The most likely explanation for these results is that this measure indicates good time management on the part of the construction team and that the client representative granted reasonable EOT claims.

Sub-hypothesis 56 is a summary of sub-hypotheses 52 to 55 of which sub-hypothesis 52 'CM's forecasting planning data' and sub-hypothesis 53 'CM's analysing construction methods' are indicated to significantly impact upon construction time performance. Sub-hypothesis 61 is a summary of sub-hypotheses 56 to 60 and each of these impact upon construction time performance, i.e. sub-hypothesis 57 'CM's monitoring and updating plans to reflect work status', sub-hypothesis 58 'CM's planning - responding to problems or opportunities', sub-hypothesis 59 'CM's effectively coordinating resources' and sub-hypothesis 60 'CM's developing an organisational structure to maintain workflow'. These hypotheses generally have high correlation with each other and with team communication variables. Test results for sub-hypothesis 56 indicate moderate correlation with a number of client representative factors. This supports the hypothesis that a sophisticated client representative, who can facilitate an environment in which a construction management team plans and controls well, can achieve high construction time performance.

Correlation results for sub-hypothesis 61 also indicate high correlation with factors that do not directly affect construction time performance such as 'CM's management's systems and procedures' and 'CM's effectiveness in team management to achieve synergy'. Management style has been shown to have moderate influence on 'effectiveness of the CM's monitoring and control', particularly 'CM's people-orientated management style', 'task-orientation of the CM's management style' and flexibility of both the client representative and the CM's management style. The general picture emerging from the Spearman Rank Correlation analysis is that CM performance is linked, and is probably assisted, by positive client representative sophistication and managerial effectiveness factors.

The importance of team relationships, organisational structure and management style is indicated through the results of those sub-hypotheses tested and found to significantly affect construction time performance, notably sub-hypotheses 49, 59 60, 62 and 84. These pertain to a well organised, well managed and cooperative construction management team.

Sub-hypothesis 49 test results for 'impact of CR/CM working relationship' suggests that a better quality relationships between the client representative and construction management team leads to better construction time performance. Spearman Rank Correlation results support this, indicating high correlation between this factor and 'confidence of the CM and design team in the CR' and also with 'CM's communication management to facilitate decision making'. Most of the factors affecting construction time performance have moderate correlation with this factor. There is also high correlation between this factor and 'the CR's ability to quickly make authoritative decisions'. The weight of evidence suggests that the client representative's influence upon this factor is strong with elements of good management performance being linked to a good construction management/client representative working relationship, which is also shown to influence good construction time performance. The importance of positive construction management/client representative working relationship is supported by this factor's moderate NEGATIVE correlation between 'direct use of power in the CR's management style'. This indicates that a power rather than cooperative approach results in lower construction time performance.

Test results for sub-hypothesis 59, 'CM's effectively coordinating resources', highlights the importance of good construction management managerial practice. It is not surprising that Spearman Rank Correlation analysis reveals very strong association between this variable and 'CM's planning', 'responding to problems or opportunities' and 'effectiveness of the construction team's monitoring and control'. The high correlation between the variables 'CM management systems and procedures' and 'CM's analysing work sequencing to achieve and maintain workflow' supports the overall assessment of good team management skills reflected in its association with 'CM's effectiveness in team management to achieve synergy'.

Results from testing sub-hypothesis 60 revealed a strong correlation between 'CM's developing an organisational structure to maintain workflow' with the summary CM management variable. The particularly strong association with this variable and 'effectiveness of the construction team's monitoring and control' confirms an expectation that strong evidence could be found to link the CM's organisational skills and good managerial practice. The moderate correlation between this variable and with construction management and client representative management style suggests a construction management 'hands-on' and involved but team-oriented style is associated with positive construction time performance results.

Sub-hypothesis 62 test results clearly demonstrate a strong link between the construction management team structure, its quality of personnel and staffing level being appropriate for the management task, and good construction time performance. Correlation analysis strongly links this factor with 'effectiveness of the construction team's monitoring and control' and 'CM's analysing work sequencing to achieve and maintain workflow'. A strong association exists between this variable and 'the construction team's confidence in the CR's contribution'. This association, together with moderate correlation with the client representative's ability to contribute both construction and design ideas, indicates that an involved and sophisticated client representative assists the construction management team to achieve management effectiveness. The moderate association of this variable with 'stability of client/client representative objectives' and 'clarity of communication of client/CR objectives' supports this view.

The only management style factor shown to directly impact upon construction time performance is the 'key sub-contractor's task-orientated management style'. This factor has moderate association with a number of construction management planning, control and communication variables that directly impact upon construction time performance. Associations with factors not directly affecting construction time performance include the management styles of other team members. The picture emerging from association with both people-oriented and task-oriented managements styles of the client representative and construction management suggests a task-oriented and people-oriented management style encourages key sub-contractors to be results-oriented, using people's skills to overcome task related problems in a flexible manner.

Construction management communication skills is an important element for good construction time performance. Sub-hypothesis 87 test results provides clear evidence of construction management's communication management facilitating decision making affecting construction time performance. This variable has strong correlation with impact of 'CR/CM working relationship', 'CM's effectively coordinating resources', 'effectiveness of the construction team's monitoring and control', and 'CM decision making, communicating and actioning'. This evidence links communications management with planning and control and developing sound client representative/construction management relationships as well as the effectiveness of decision making. The moderate association indicated of this factor with others not directly affecting construction time performance indicates that client representative sophistication and management style also plays a part. The moderate negative correlation between this variable and direct use of power in the client representative's management style supports the contention that a confrontationalist management style has negative impact upon construction time performance.

Test results for sub-hypothesis 90 'CM and design team communication effectiveness for decision making', sub-hypothesis 93 'decision making communication within the construction management team', and sub-hypothesis 96 'CM decision making, communicating and actioning' support the evidence that general communications and decision support affects construction time performance. The argument is further strengthened by construction management and design team communication effectiveness being moderately associated with a design team people-oriented management style. construction management decision making effectiveness is moderately linked to both 'planning and control' and 'general communication performance'. construction management decision making performance is also moderately associated with a task-oriented client representative and construction management and people-oriented management styles. 'CM decision making, communicating and actioning' is highly correlated with construction management

planning and control and is also highly correlated with 'CM's effectiveness in team management to achieve synergy' and communication management to facilitate decision making.
Sub-hypothesis 99 'construction management's effective use of information technologies' test result also suggests that good construction management communication skills in using decision support technology assist the construction management to become a more effective manager.

4 Conclusions

The ANOVA and Spearman Rank Correlation analysis has provided valuable statistical evidence which helps explain why good construction time performance is significantly affected by construction management team performance. The construction management's professional approach to planning and control contributes to good construction management team performance. This approach is evidenced by the construction management team's:

- *initially thinking through the plan and defining the issues, planning construction methods effectively, and putting in place a coordination strategy that allows for adequate monitoring and control;*
- *communication both with internal and external team helps to facilitate decision making, decision communication and decision action;*
- *adoption of a management style that maximises cooperation but still maintains a focus on results, this does not mean that task-orientation and flexibility or people-orientation are mutually exclusive.*

The useful application of this work is that it proves, within the statistical constraints and sample definition, that good construction time performance has strong association with good construction management management performance and that the client representative has a strong influence on providing an infrastructure of cooperation, communication and commitment which significantly affects construction management performance. This work, however, draws to our attention the critical importance that leadership on the part of the client representative has upon construction management performance.

In the past some client representative's have been able to influence client's into believing that a confrontationalist management style, direct use of power and an unyielding results-oriented approach to management produces sound construction time performance outcomes. It is always difficult *post-hoc* to fairly allocate praise and blame for construction time performance results because issues get muddied by details, personality factors, and the time taken to design and build. This work has provided a clear picture of client representative and construction management best practice for achieving good construction time performance and as such can form a sound basis for developing benchmarks for achieving these best practices.

5 Recommendations

The work reported upon here is restricted to construction management teams and the research investigated wider issues which included the impact of the client's representative team. The following summary statements and recommendations are proposed. There are a number of lessons to be learned from the research findings which construction companies and building clients could implement to their advantage.

1 Results indicate that a well organised and functioning construction management team can offer much to the achievement of construction time performance. This is exemplified by effective planning and communication for decision making. A construction team's flexibility in responding to problems and opportunities is based upon sophisticated planning for control where plans are seen as a focus for action rather than static or rigid commands or procedures. This implies that effective use of planning tools should be encouraged for simulation analysis to predict a range of possible futures of which a 'preferred future' can be selected. Close involvement of team—participants in the construction planning and implementation of work is needed to achieve this. This needs to be integrated into the monitoring cycle, checking reality against the 'preferred future' of the plan. If construction teams can see plans as 'preferred futures' which are subject to impact of reality and unexpected events, then they may be more prepared to see plans as goals to be constantly aimed for rather than predictions in which guilt may be attached to those who prepared them or those who 'failed' to realise them.

- *It is proposed that construction companies should ensure that team members, charged with defining plans and those closely involved with carrying out the work, be trained in the use of planning techniques, idea generation and problem solving. This will provide a better chance for enhancing a team's responsiveness to changed environments and circumstances.*

2 The research results also draw clear conclusions about a favourable nature of the client and construction team relationship. Research evidence presented in this thesis suggests that where a good relationship exists there is a better chance of good construction time performance.

- *It is proposed that the client's representative team should gain a good understanding of the project's complexity and constraints to assist the development of confidence of the construction management team in the client's representative team. This can be achieved through mechanisms such as partnering (Patching 1994) or other models which better integrate the client's representative and design team with the construction management team in problem solving and investigation of strategies for construction. These have been documented as being successful in many cases in Japan and the USA (Bennett 1991; Walker 1989).*

3 The research results give clear indication of the need for competent team members. This may seem obvious and the evidence provided confirms this requirement to achieve high construction time performance. According to Bennett (1991, p254) competent people require not only a good technical and general education but also practical skills in problem solving and analytical thinking.

- *It is proposed that the construction industry invests in its future by promoting life-long learning. This can be achieved in a number of ways. Formal training at trade, undergraduate, and post graduate levels can be, and in many cases are, encouraged. The methodology developed for this thesis also provides a useful framework for project team members to learn from experience.*

This short and by no means exhaustive list, explores how industry can adapt to improve construction time performance. It attempts to link what has been learned from the research to provide a few practical and useful recommendation for industry to improve construction time performance.

6 References

Bennett J. 1991, *International Construction Project Management General Theory and Practice*, Butterworth Heinemann, Oxford.

Berkeley D., Humphreys P.C. & R.D. Thomas, *'Project risk action management'*, Construction Management and Economics, vol. 9, pp. 3-17.

Business Council of Australia 1993a, *Case Studies of Project Implimentation in the Building Construction Industry,* BCA, Melbourne.

Construction Industry Development Agency (CIDA) 1993, *'Project performance update - a report on the time and cost performance of Australian building projects completed 1988-1993'*, CIDA, Sydney.

Construction Industry Institute (CII) 1995, *Winning Teams,* CII, Australia.

Ireland V. 1983, *The Role of Managerial Actions in the Cost, Time and Quality Performance of High Rise Commercial Building Projects*, PhD thesis, University of Sydney, N.S.W., 1983.

Morris P.W.G. 1983, *'Project management organization'*, Construction Papers, vol. 2, no. 3, pp. 5-18.

Morris P.W.G. 1994, *The Management of Projects A New Model*, Thomas Telford, London.

NEDO Construction Industry Sector Group 1988, *Faster Building for Commerce,* Report prepared for the National Economic Development Office (NEDO) Commercial Building Steering Group, London.

Patching A. 1994, *Partnering and Personal Skills for Project Management Mastery*, APA, Sydney.

PRD 1991a, *'Productivity and the Australian construction industry'* A report prepared by Policy and Research Divisions (PRD), Royal Commission into Productivity in the Building Industry in NSW.

Sidwell A.C. 1982, *'A critical study of project team organisational forms within the building process'*. PhD thesis, Department of Construction and Environmental Health, University of Aston in Birmingham.

Walker A. 1989, *Project Management in Construction*, 2nd edn., BSP Professional Books, London.

Walker D.H.T. 1994a, *An Investigation into Factors that Determine Building Construction Time Performance*, PhD thesis, Royal Melbourne Institute of Technology, Victoria 1994.

Walker D.H.T. 1994b, *'Risk identification and developing benchmark measures of construction time performance'*, Proceeding of INTERNET '94, 12th World Congress On Project Management, Oslo, June 1994, vol. 2, pp.41-47.

Walker D.H.T. 1994c, *'Procurement systems and construction time performance'*, Proceedings of CIB W-92 East Meets West conference, Hong Kong, December 1994, pp. 343-352

Walker D.H.T. 1995, *'An investigation into construction time performance'*, Construction Management and Economics, vol. 13, no 3, 265-274, Spons, UK

ORGANISATION BEHAVIOUR IN A TRANSITIONAL ECONOMY – ROMANIA

C. Oltean-Dumbrava

School of Construction and Environment, University of Abertay, Dundee, UK

Abstract

The decision to change a country from a command economy to a market economy impacts especially upon how companies are subsequently managed during the period of transition and afterwards. Previously large construction companies in Romania functioned knowing that contracts would be awarded because competition did not exist. However, with the advent of a market economy, it has been necessary for these large organisations to be divided into small or medium sized enterprises, and some new privately operated companies have also been created. The concept of generating working capital to operate such companies, providing new and additional plant and equipment, and the need to develop commercial business practices for such matters as tendering and marketing, has created major challenges within these companies during the transition. The paper will identify these challenges and problems from a Romanian perspective and discuss how these are being resolved.

Keywords: Transition to a market economy, organisation behaviour, privatisation, generation of capital.

Sommaire

La decision de changer un pays d'une economie centralisée a une economie du marché affecte surtout la maniere dans laquelle les entreprises sont gerées pendant et après la période de transition. Avant, les grandes entreprises de construction functionnaient sachant qu'ils allaient avoir les contracts pursque la concurrence n'existait pas. Avec, l'introduction de l'economie du marché, il est devenu necessaire pour ces grandes entreprises de se diviser en petites et en moyennes entreprises. Des nouvelles entreprises privées sont aussi créées. Le concept de génération du capital de fonctionnement, la necessité d'acquérir des nouvels materials et de développer la strategie commerciale pour des activities telles que l'offre et le marketing presentent un grand defi pour ces entreprises pendant la periode de transition. Cette article

The Organization and Management of Construction: Shaping theory and practice (Volume One). Edited by D.A. Langford and A. Retik. Published in 1996 by E & FN Spon. ISBN 0 419 22230 8.

identifie ces defis et problemes dans la perspective de Roumanie et discute comment ils sont résolus.

1 Introduction

The major changes which affected Romania, starting in 1990, had a great impact on the economy as a whole, and on the construction industry in particular.

Lack of money from public sources, coupled with a virtually non-existent private sector at the beginning of the 1990, brought in a great turmoil in this sector. The volume of investment reduced from 100% in 1989 to 45.2% in 1992, and the amount of the investment in the GDP from 29.6% to 14.8%. In the same period of time, the amount of public investment in the total investment went from 57.5% in 1989 to 16% in 1991 (See Figure 1).

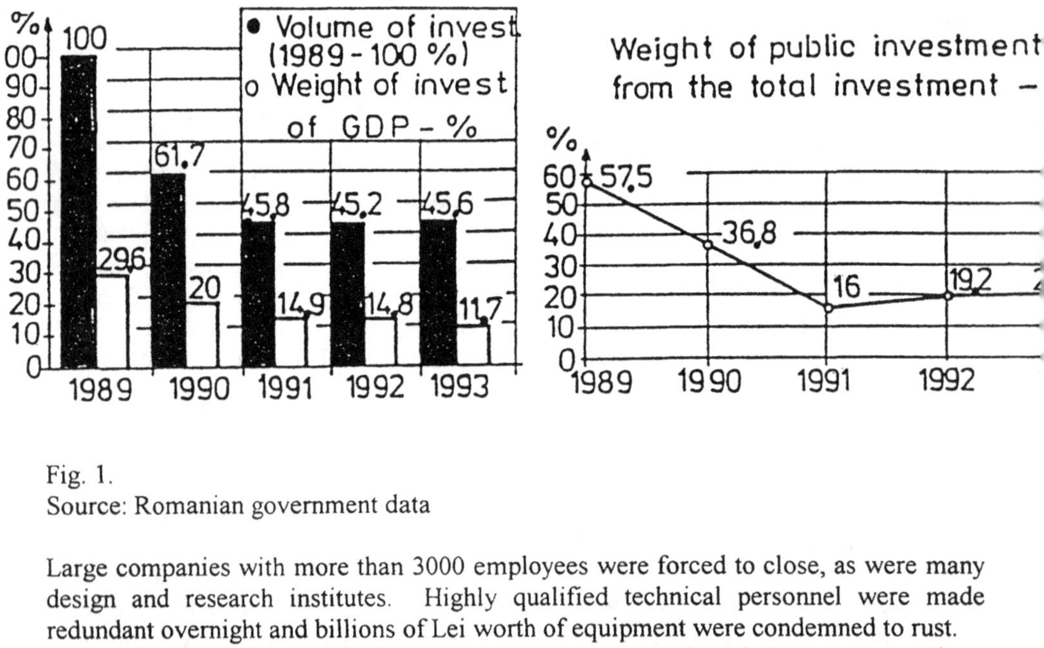

Fig. 1.
Source: Romanian government data

Large companies with more than 3000 employees were forced to close, as were many design and research institutes. Highly qualified technical personnel were made redundant overnight and billions of Lei worth of equipment were condemned to rust.

Faced with all these difficulties, the remaining companies tried to overcome these problems in different ways. Some of them were lucky to find niches abroad and started to subcontract particularly the labour and technical know-how in the Ukraine, Germany and the Middle East. Some others split into smaller companies which were more flexible and thus more able to cope with a rapidly changing environment, and address their activities to the emerging private sector.

2 The formation of the private sector in the construction industry

The majority of the population could not even consider starting a business. They had to struggle for their life in a dramatically changing economy, where jobs for life were replaced with job insecurity, fixed prices with rapidly growing ones, coupled with a high rate of inflation. Some of them were the first to be made redundant and the only chance they had to earn money was to start their own business. Some others were more fortunate, having relatives abroad, or working abroad, and had saved or borrowed a small sum of money in hard currency. This made a great deal of difference. Others started joint ventures with foreign partners with as little as US $5000. This was the start of the privatisation process, coupled with a legislative framework such as: the Law of Privatisation, the Law of Foreign Investment, the Law on State Owned Enterprise Restructuring and the Land Law, which all tried to encourage local and foreign investment. The state or co-operative property gradually started to change in a great variety of types of property (see Table 1, in which is presented the type of property of the business companies involved in building activities in mid 1993).

Table 1. Type of property of business companies involved in building activities
Source: ARACO Bulletin No 6/1993

No	Type of Property	No of Enterprises	Average No of Employees
1	2	3	4
1	Independent managing companies	4	906
2	State business companies	544	341,640
3	Other state economic enterprises	3	1108
4	State autochthon & foreign capital	5	402
5	State, private autochthon & foreign capital	4	1245
6	State & private autochthon capital	50	21,830
7	State & foreign private capital	2	214
8	General partnership company	149	2690
9	Sleeping partners company	19	297
10	Joint - stock partnership company	2	190
11	Joint - stock company	89	18,669
12	Limited liability groups	2992	128,322
13	Domestic co-operative industry	9	1388
	TOTAL	3872	518,321

The weight of investments of private sector as a percentage of the total investments rose from 1.9 in 1989 to 17.4 in 1993, (see Figure 2) and a large number of small and medium sized companies were formed (see Figure 3). Presently, more than 57% of the companies employ less than 200 employees, compared with only 12% of companies employing more than 200. A large number of companies (32%) were employing temporary labour during sporadic contracts. These companies had not previously employed a permanent staff.

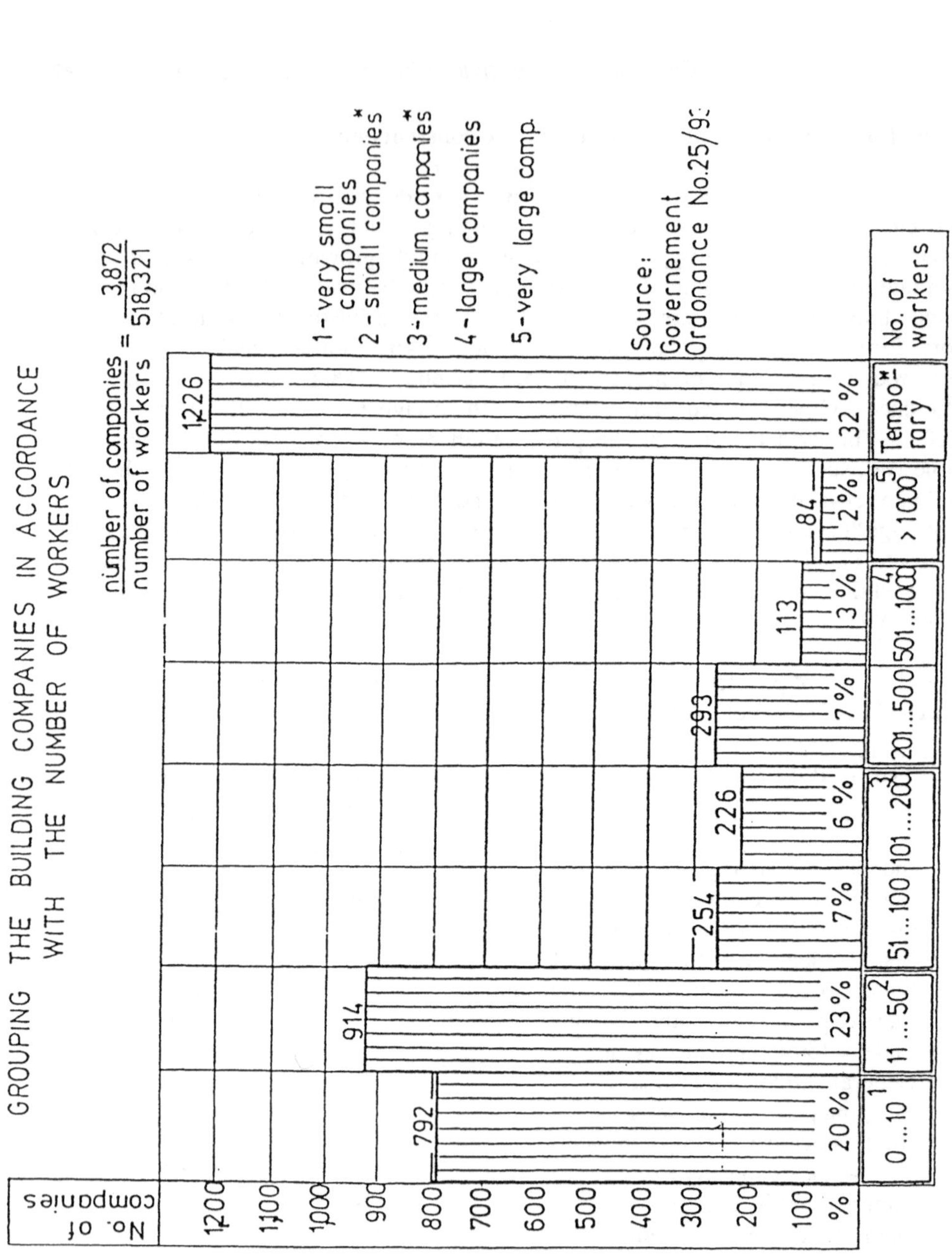

Fig. 3.

SOURCE: BULLETIN ARACO 5/1994

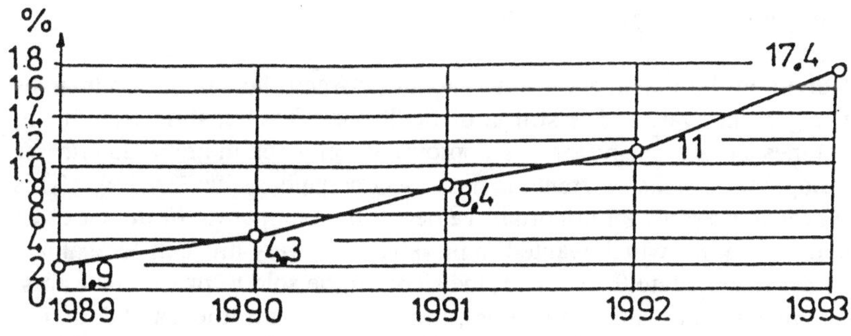

Fig. 2. Weight of investments of private sector in total investments (%)
Source: Romanian government data

The state sold the flats in the blocks to their tenants at very low prices. Thus, by the end of 1992, 6.8 million (88%) dwellings were private property (see Figure 4). On the one hand, the new businesses, including the new banking systems, needed headquarters, and on the other hand, the new owners of flats and houses became more conscious of their values. This contributed to the creation of a new market for the construction industry, formed of new investments, maintenance and repairs of the existing properties.

The large scale privatisation of the industry also required investments in the infrastructure, which offered a market for the larger companies.

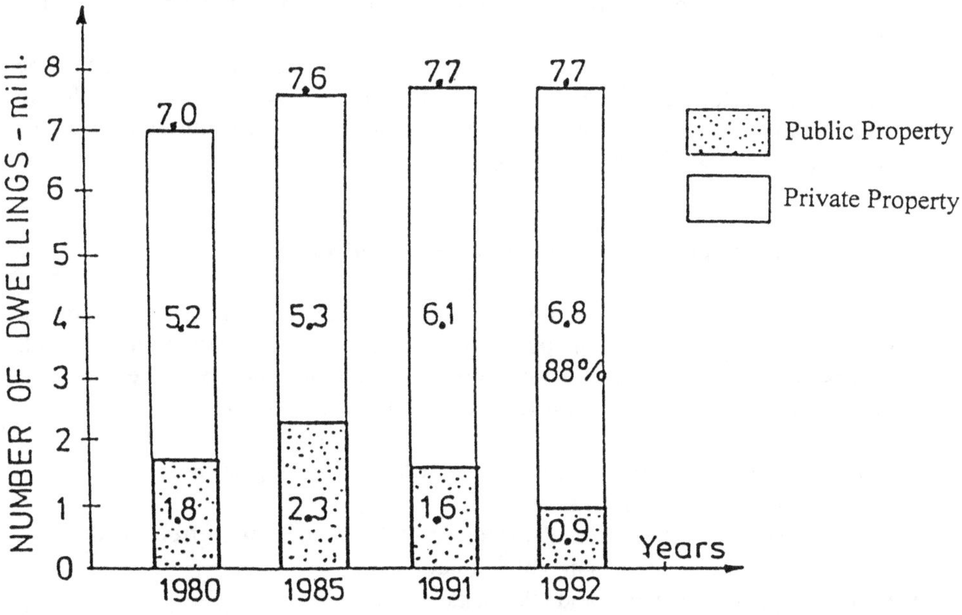

Figure 4. Dwellings stock by type of ownership
Source: Romanian Statistical Yearbook 1993

3 The process of generating working capital

The process of dividing the existing working capital between the smaller companies formed from a larger one was, and is still, a difficult one. The existing equipment, such as tower cranes or big excavators, is appropriate only for major industrial or residential developments. Its current market secondhand value is low. This equipment, however, represents an important value for the company. In the cases of former large state companies which wanted to privatise, it is very difficult to evaluate their capital in order to calculate the value of the shares to be sold to private investors, because of this low market value equipment, which represents the capital of the company. Residual values are high, the writing-off period was fixed to between 10-20 years, and the situation has been exacerbated by high inflation and devaluation of national currency in the years 1990-1993. Thus, this type of investment opportunity is not attractive.

The nature of the market has changed completely. Major industrial, residential or agricultural developments are now hardly undertaken because of lack of funds. The existing built asset has to be maintained in the present form or refurbished to improve its image. This market needs other types of equipment and a change of emphasis towards a better quality, especially for finishings.

Small companies need to invest in tools rather than plant because of the nature of their activity. The supply of tools, plant and equipment certainly exists, and the demand is high. However, the construction companies, no matter how large they are, do not have enough money to buy new equipment in order to replace the existing stock entirely. Sometimes, they have to use what they possess, even if it is not efficient, because they cannot afford to buy or rent what would best suit their needs at the present moment.

The joint venture organisations, in which the foreign partner could contribute to the formation of the working capital by bringing in the necessary equipment and tools, are in a more fortunate situation. They can also afford to import some building materials of high quality, which are in short local supply.

4 The development of the commercial business practices

The general climate of economic competition, typical of the market economy, requires managers to have expertise in techniques and methods suitable to satisfy the demands of clients as far as deadlines, quality and initially agreed budgets, are concerned. The Romanian business culture thus needs to be changed in order to:

- engender trust and respect between national or foreign trading partners
- appreciate the true cost and value of time
- explain the need to take personal responsibility for quick and effective decision making, without reference to some higher authority
- have a willingness to manage and take risks

- understand what profit and competition means (there is a tendency to see competition as something to be "killed off", rather than as a challenge to deliver better price, quality etc)
- understand the importance of marketing and business strategy
- have professional integrity and clarity in providing good quality business information.

A new procurement system had to be put in place and enacted. Thus were formed the new laws on authorising construction work, the decision of the Government of Romania on the approval of a procurement system, tender procedures and public works contracting, and the Law of Quality in the construction industry.

This legal framework tries to create the proper climate for healthy competition. In the public sector, especially in infrastructure activities, major projects were funded by EBRD, the World Bank and other important banks. Thus, there is competition from European and other countries in Romania. This is only one example of many to prove the viability of the existing procurement system which encourages competition.

On 27 May 1995, the Romanian Parliament adopted the draft law to accelerate privatisation. In late May, an IMF team visited Romania, recognised Romania's significant macro-economic achievements, but also highlighted the need for measures to increase structural reform.

Some other positive signs were reported in the first quarter of 1995. For example, the monthly inflation rate was 1.6% in April (year-on-year rate fell from 40.4% in March to 34.6% in April), below the NBR's forecast of 4%. The National Bank of Romania cut the auction credit rate for refinancing to 47.5% in May and they are now coming under increased pressure, from businessmen and Parliamentarians who consider that the rates are over-positive, to make further cuts. Romania's M2 money supply rose by 5.1% in the first quarter of 1995 and the household savings have risen by 20.2% in the same period of time. The industrial production rose by 15.7% (unadjusted) in March 1995 on March 1994. All these are positive signs for achieving a swift transition to a market economy.

5 Conclusion

To change a country from a command economy to a market economy is a difficult process which influences the whole activity of the country and its people, their lives, expectations, and standards of living. This process cannot be smooth; it is a dramatic one in which companies and people have to struggle to survive.

There is much scope, will and determination to make the period of transition as short as possible, but this can only be done with much suffering and sacrifice. High social costs will have to be paid. At the end of this change, it is believed that the lives of the people will be better, and that they will enjoy a higher standard of living.

6 Bibliography

1. ARACO Bulletin No 6/1993.
2. Romanian Statistical Yearbook 1993.
3. ARACO Bulletin No 5/1994.

BUILDING ON FAIR FOOTINGS: IMPROVING EQUAL OPPORTUNITIES IN THE CONSTRUCTION INDUSTRY FOR WOMEN

S. RhysJones

RhysJones Consultants, Greenwich, UK

A.R.J. Dainty and R.H. Neale

Department of Civil and Building Engineering,
Loughborough University of Technology, Loughborough, UK

B.M. Bagilhole

Department of Social Sciences, Loughborough University,
Loughborough, UK

Abstract

This paper reviews the work of the Construction Industry Board Working Group 8, one of the groups set up as a consequence of Sir Michael Latham's review of the construction industry, with a remit to improve equal opportunities for women in the industry. The paper describes the working procedures of the Group, and explains the rationale behind the preliminary recommendations, which had the aim of creating parity of opportunity for women at both the professional and operative levels. These recommendations are set in the context of a changing economic and employment market, and the other Latham proposals and Working Groups. The authors contend that there exists a valid business case for diversification of the construction workforce, and that failure to catch up with other sectors' progress towards good equal opportunities practices could ultimately affect the UK industry's ability to compete in global markets. It is argued that equal opportunities, when seen as an integral part of good management practice, provide the tangible benefits of better levels of staff retention and operating efficiency. The paper presents the findings of research conducted by the Working Group into client attitudes and current equal opportunities practice, and goes on to explain the likely recommendations of the forthcoming final report.

Keywords: Equal opportunities, Latham report, women.

Sommaire

Cet article passe en revue le travail du groupe d'action 8 du conseil du secteur de construction qui est l'un des groupes d'action qui ont été formes, apres la critique de Sir Michael Latham du secteur de construction, en vue d'ameliorer l'egalite de chances professionnelles des femmes dans l'industrie.

L'article decris la procedure de travail du groupe et explique le raisonnement derrière les recommandations preliminaires qui avait pour but de créer une parité de chance pour les femmes au niveau intellectuel ou manuel.

Ces recommandations sont formules dans le but de changer le marché economic et de l'emploi, les autres groupes d'action et le projet de Latham. Les auteurs soutiennent

The Organization and Management of Construction: Shaping theory and practice (Volume One).
Edited by D.A. Langford and A. Retik. Published in 1996 by E & FN Spon. ISBN 0 419 22230 8.

l'idée qu'il y a une raison valable pour la diversification de la main-d'oeuvre de construction et que l'echec de rattraper les autres secteurs dans leurs avance pour une bonne pratique d'galite de chances professionnelles peux affeter la competence du secteur de construction dans le marché global. Il est souvent soutenu que si l'galité de chances professionnelle est integrée dans le systeme de gestion, la perte de personnel peut ètre diminuée at laperformance operationnelle ameliorée. L'article presente les resultats d'une recherche conduite par le groupe d'action concernant l'attitude des clients et les pratiques d'egalite de chances professionnelles et aussi expliquer les recommandations probable du rapport final à venir.

1 Introduction

No major British industry has lower representation of women than the construction industry [1][2]. The industry employs around 1.75 million people, yet less than 10% are women [3], and only 5.72% of members of professional institutions at all grades are women [4]. It is unacceptable that an industry which accounts for approximately 10% of the GDP, and which has such a profound impact on peoples' lives, almost exclusively employs men. Furthermore, recent legal cases have highlighted that the few women that do work in the industry are likely to face discriminatory work practices, and find it more difficult to progress their careers than their male peers [5][6].

In addition to the ethical and moral arguments for an increase in female representation, there are also renewed worries over skills shortages, which are expected to become apparent towards the end of the century as the number of people available for employment continues to fall [7][8]. The CITB predicts that 10% of all new jobs in the industry, including professional, managerial and trade positions, will need to be filled by women by the year 2000 [9]. In the wider context of industry in general, concerns are being expressed about the suitability of the male management style to the modern business. The Institute of Management [10] has recently predicted that the female management style will be more appropriate in the millennium. They cite female attributes such as team-working, consensus management, negotiating, interpersonal skills, and the ability to manage several projects at the same time.

There is, therefore, a need for the construction industry to move away from its current male-dominated workforce, to a more diverse and efficient skills base. An increase in the number of women should benefit the professions by providing new blood and energy [11]. The construction industry must make a choice between remaining isolated as *the* male bastion, maintaining its 'special case' argument and retaining its male culture; or take advantage of the benefits of effective employment of non-traditional groups as many other industries have done already.

If the make-up of the construction work force changes significantly, so the industry must react by increasing the flexibility of its employment practices in an attempt to retain its staff. The Hansard Society commissioned report "Women on Top" [12] suggested that age bars, excessive mobility requirements, informal selection procedures, stereotyped assumptions, unspecified job criteria and the 'old boy' network were all responsible for women's under-achievement in paid work. All of these barriers currently operate in the construction industry, which is a long way behind other sectors, and so putting forward practical recommendations to achieve this aim is the key objective of the working group.

Taking a more positive view, one advantage of being one of the last industries to move towards modern personnel practices, is that the construction industry can benefit from adopting good practices developed by others. It is by adopting good tried and tested human resource policies and practices that the industry can quickly adapt to employing women.

2 Under-representation and under-achievement of women in construction - a statistical overview

The labour market is still horizontally segregated across all industries and services, with women tending to occupy positions in catering, administrative positions and the health service whilst men occupy positions in engineering, manufacture and construction, as shown in Figure 1. Despite the number of women construction professionals rising steadily over the past 20 years, in proportion to the total membership of over 331,000 individuals who belong to a professional institution in membership of the Construction Industry Council, only 9,050 (3.98%) are fully qualified women and 9,478 are students. Overall, women represent only 5.71% of the professions. It is also interesting to note that over half of the women in the industry work in architecture, landscape architecture, interior design or town planning. If these professions are removed from the analysis then only 1.54% of qualified construction professionals who may actually work on site are women [4].

Equally important, however, is the degree to which women that do work in the industry are vertically segregated. Unfortunately, no relevant and reliable vertical segregation data currently exist [13][14]. However, 1991 census data indicates that women are concentrated in clerical and secretarial positions, or other positions not directly involved with the construction process, as shown in Table 1.

Figure 1. Employment by Industry: 1994

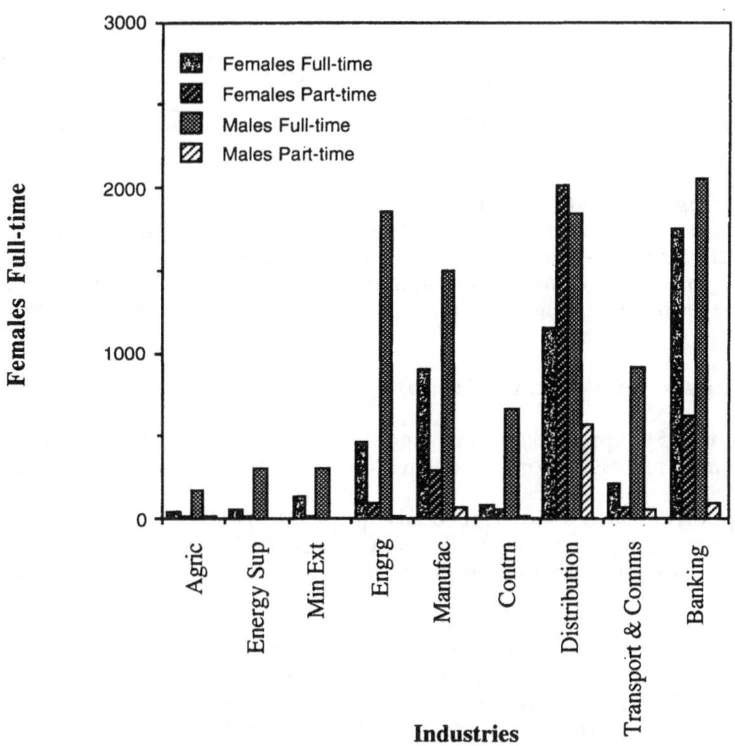

Source: EOC "Some Facts About Women" 1995 after Dept. of Employment 1994.

Table 1. Employment in construction by occupation and sex: 1991 (%)

	Distribution of men %	Distribution of women %	Women's share of total %
Managers and administrators	9.3	15.4	14.0
Professional Occupations	2.9	1.1	3.4
Associate Prof. & Technical	2.9	2.6	8.0
Clerical & Secretarial	1.1	63.0	84.5
Craft & Related	61.2	4.8	0.8
Personal & Protective service	0.2	0.6	22.3
Sales	0.8	4.7	36.7
Pant & Machine Operatives	9.9	1.8	1.8
Other occupations	11.0	5.6	4.8
TOTAL	100.0	100.0	9.0

Source: Court and Moralee [3] 1995 after 1991 Census: Great Britain Economic Activity.

The relatively high proportion of female managers is achieved by their roles as personnel managers and administrative managers. If the positions not directly relating to the construction process are ignored, it can be seen that women generally occupy only the junior positions in construction organisations. These figures indicate that the industry, as well as needing to attract women, also needs to improve women's careers.

3 Background to Working Group 8

Latham's review [15] expressed concern regarding the state of equal opportunities, in particular the under-representation of women both in the trades and in the professions.

> "Equal Opportunities.... be vigorously pursued by the industry, with encouragement from Government. The CIC, CIEC and CLG should produce co-ordinated action plans to promote equal opportunities within the industry and to widen the recruitment base."

The Construction Industry Board, set up to further investigate and implement Latham's recommendations, created a working group (WG8) to examine how equal opportunities could be improved in the industry, and how the recruitment base could be widened. The working group comprises 14 nominated members of the CLG, CIC and CIEC, chaired by Sandi RhysJones and supported in research and analytical tasks by Barbara Bagilhole, Andrew Dainty and Richard Neale of Loughborough University.

Given its original remit, the fact that women represent the single greatest source of potential labour, and the strong evidence that an increase in the employment of women in the industry would lead to less adversarial relationships, an end to skills shortages and an improvement in the image of the industry [16], WG8 concentrated on sexual discrimination and the under-representation of women. Although many of the recommendations and initiatives put forward by the working group will be applicable to other under-represented groups, it is acknowledged that not all will be relevant, and improving opportunities for groups such as ethnic minorities and disabled people demands considerable further investigation and action. The recommendations will, however, apply both to women in the professions and women in the trades. The scope of the working group was, therefore, defined as:

> "To encourage the development of attitudes, practices and physical environments in the industry that neither directly nor indirectly have the effect of placing women at a disadvantage within the construction industry. Also to encourage more women to enter the industry at all levels in all trades and all professions and seek progression within it."

4 Process of producing the report

WG8 has sought to develop a range of practical recommendations that can be easily adopted by the industry at little cost or increase in administrative procedures. They have drawn upon a wide body of existing literature and have consulted experts in the field. Perhaps more importantly, they have continually sought industry opinion from clients, consultants and contractors throughout the study to ensure the general applicability and acceptance of the ideas put forward.

It is important that their recommendations are not seen in isolation from the other working groups. Equal opportunities can be seen as an integral part of the work of all construction firms, and should be built into all recommendations put forward under the Latham review.

Core issues were established by WG8 and investigated by individual members. These were as follows:

- Current practice and experiences within the construction industry at both professional and craft and trade levels;
- Current practice in other sectors, central and local government and other countries;
- European and UK policy and action on equal opportunities;
- Influence and role of education and training;
- Working conditions and how they can be improved;
- Positive action;
- Attitude and influence of customers;
- Innovative construction projects, initiatives, academic studies and statistics concerned with women in construction.

An interim report was published in March 1995, as a discussion document. This explained the views and beliefs of WG8, and the key issues. This report generated many responses from practitioners in the industry, which have been analysed in order to refine the recommendations, and to improve their general applicability.

Also, two questionnaire surveys were conducted to establish the views of clients towards gender issues in the industry, and current practice in the development of equal opportunities policies in consulting and contracting firms. Two "presentation days" in which invited speakers addressed the working group members and invited guests in influential positions within the industry. The speakers at the first prevention day had implemented initiatives to increase the representation of women in the industry, or to improve the working conditions and practices for women. The speakers on second presentation day had initiated programmes in other industries and professions.

All of these data have been handed over to the Loughborough research team who have been commissioned to write the final report, due for publication in early 1996.

5 Principal findings and recommendations

5.1 Recommendations of and reactions to the interim report

The interim report explained WG8's approach to gender issues in the construction industry. Recommendations were that: a code of practice for the formulation of an equal opportunities policy should be formulated; a co-ordinating body be set up to oversee the various initiatives around the country and act as a forum for dissemination of effective ideas; an international body should also be created to collect information on initiatives from around the world; support networks be created to allow women working within the industry to get in touch with each other, learn from each others' experiences and find out about possible job opportunities; data bases of role models and mentors be set up; recommendations on improving working conditions be published; research into standardised job classification models to aid the monitoring process be commissioned; and a review of appropriate apprenticeship programmes and training schemes be conducted.

Analysis has shown that many of the comments and much of the discussion generated by the interim report were contradictory, emphasising the controversial nature of the issues raised, but generally responses from practitioners, and professional institutions were very positive. A common comment regarding attracting women to the industry was that construction suffers from a lack of professional culture, particularly in contracting and outside the professions at an operative level. It has been suggested that improving the standing of the industry, particularly on the contracting side, be offered as a challenge to the professional institutions.

Another popular point was that retaining women should be seen as a priority. Child care, career break schemes, job share and part-time work are all ways in which employers can help women combine work and family responsibilities, and the money saved in retaining their professional staff should more than cover the expense of such flexible working practices. Equal pay was another important point raised as, despite the Equal Pay Act, women may not being paid the same amount as men for doing the same or similar jobs.

The key recommendations of the report were widely welcomed, with some respondents adding that contractors should demonstrate a well-implemented equal opportunities policy as a tender pre-qualification. It was also suggested that an informal annual meeting of all the women in construction groups should be undertaken in place of the umbrella body as similar bodies exist already. The general view was to avoid adding to bureaucratic procedures, but it was acknowledged that a set of guidelines for the effective formulation and implementation of an equal opportunities policy would benefit the industry greatly, particularly if set in a construction specific context and applicable to all sizes of company.

5.2 Research findings

Survey of current construction industry practice - The 20 largest contractors and 20 largest consultants by turnover were asked to submit a copy of their equal opportunities (E.O.) policies. There was a 60% response rate. Some of the companies approached stated that their E.O. policy was private document held by the company and not for release into the public domain. All of the respondents had the minimum of a simple mission statement which stated that the company was committed to equal opportunities. Around half of those surveyed also produced guidelines for its staff on fair recruitment practices including some guidance on advertising. Only one employer, however, had a comprehensive monitoring and review procedure. Most major employers in other sectors now have such procedures, indicating how far behind good practice the industry is at present. This is reflected by construction industry membership of Opportunity 2000, the charity set up by the government to promote the employment of women into the next century. Only two construction firms have joined. The conclusion just be that the construction industry has so far failed to address Equal opportunities Issues.

Survey of Client Attitudes - Ninety five major clients were sent a postal questionnaire requesting information on the sex of their in-house construction related staff, staff from outside agencies working for them and general information about their equal opportunities policy (including whether there had been an increase in the number of women employed since the policy had been in operation). Twenty one responded, a response rate of 22.1%. Although it is difficult to draw firm conclusions from such a small sample, some interesting findings included that clients tended to employ far more women in their construction divisions than construction firms, that they tended to have comprehensive equal opportunities policies and implementation strategies including flexible working practices and that since the formation of their E.O. policy more women had joined the construction section of the business. Overall it seems that clients have a far more positive attitude towards employing women and developing their careers within their organisations.

5.3 Presentation days

The first presentation included speakers involved with proactive initiatives to attract women to the industry and improve the working environment for them. What became clear was that public funds are needed to support positive action initiatives in the industry in the short term. It was felt that this would then achieve a 'critical mass' of women workers so that the industry would be viewed as a standard career for women.

The fact that women's mistakes are more noticed than men's and that they have to work harder for equal recognition were also cited as important barriers to break down. It was suggested that a few well co-ordinated groups looking after women's interests should be set up, although it was acknowledged that many women do not want to join an organisation which implies difference from their male peers. A paradox exists, therefore, in that most women don't want to "look different", but organisations to co-ordinate networking, role models and mentoring are essential if women are to find out about new opportunities and benefit from other's experiences.

The second presentation day concentrated on how other professions had managed to attract and retain women and improve their working practices. In addition to this, other topics included the advantages of diversifying the workforce, improving working conditions and how to succeed as a woman in the industry. Invited guests attended representing the trade press, professional institutions, women's groups, trade institutes and various major construction industry employers. The relevance to the construction industry of the initiatives put forward was established by a lively debate following each pair of speakers.

The central theme running through the speakers' presentations was that change is needed in the industry, and women could provide the innovation and different perspectives needed for progress. To become flexible, the industry first needs to develop a climate that values diversity. Latham's recommendations were seen as very important by a number of the speakers, and it was suggested that women have the ability to integrate as part of the team, look at things in different ways and improve the team. Structural change such as flexible working practices and improvements in working conditions is one thing, but ultimately it was generally agreed that it is cultural change that needs to be fostered. The issue should not be "what can the construction industry do for women", but should be what contribution can women make to the industry.

6 The benefits and case for diversification of the construction workforce

Equal opportunities in construction is not so much a matter of social justice as a vital economic necessity for the industry. Diversity in the work force is now widely recognised as both desirable and inevitable, and can bring many benefits to the working environment. The 'feminization' of the workforce [17] is a continuing and irreversible trend, and therefore women will form a significant and growing proportion of the customers and clients of industry and commerce. The benefits are identified by Howard Davies, Director of the CBI [18] as 'widening not only the talent available to business, but its understanding of the market-place and its links with the community'. The fundamental argument for equal opportunities is the benefit derived from recruiting, retaining and promoting the best people for the job. This will improve the quality of output, lead to an improved customer, enhance organisational flexibility and improve team effectiveness. The following indirect benefits also arise as a consequence of having the best candidates selected, developed and retained; satisfying work environments, improved morale and job satisfaction, improved working relationships between workers, and a better public image.

A significant step is now being made in the evolution of equal opportunities through the concept of 'managing a diverse workforce', which includes women. The main argument is that diversity and differences between people, if managed effectively, add value to any organisation. It is shown that 'harnessing these differences will create a productive environment in which everybody feels valued, where their talents are being fully utilised and in which organisational goals are met' [19]. An effectively implemented equal opportunities policy is fundamental to an organisation committed to total quality management.

7 Working group preliminary recommendations and implementation mechanism

As Latham [15] stated, equal opportunities in the construction industry require an urgent and comprehensive review. Simply promoting the entrance of women into the industry who will leave after 4 or 5 years is a futile exercise. The Working Group contend that creating a working culture accepting of women is the ultimate aim. However, this is a long term objective that will not be realised quickly. In the mean time, therefore, structural changes in the workplace to make it easier for women to develop their careers will begin to promote the type of change needed, for cultural change to follow.

The first presentation day demonstrated that much of what is needed to attract women to the construction industry is already in place, with many women's groups and initiatives already operating effectively in attracting women to construction. The industry should capitalise on this by taking a lead in developing structural policies to ensure equality of opportunity, equal pay, improved working conditions implemented through existing professional institutions, trade organisations and women's support groups.

The industry must begin to learn the value of women's contribution as clients, end-users *and* practitioners. Modern life patterns are very different to those of 20 years ago. Demographic and societal change are defining a new working role for women and men and it is up to the industry to harness this change and take advantage of it.

The economic climate in which the industry operates is currently seen as a constraint on implementing the types of initiatives suggested in this report. However, other industries put a tangible value on retaining their skilled workforce, and promote this by instituting flexible work practices and improving the working environment of their employees. The authors contend that the construction industry should adopt this attitude as in the long term, the investment will be recouped many times by the savings in reduced training costs.

8 Further research

A current problem is that it is hard to assess how well women are progressing in the industry relative to their representation. This is because there currently exists no conformity in job titles or standard career paths to benchmark women's progress against. To overcome this, the Loughborough research team have developed a set of standardised job titles, job descriptions and career paths for all major professions in the industry. These have been built into a questionnaire which is being distributed to the top 500 construction companies. This accurate vertical segregation data should make it possible to establish for the first time the areas of the industry where opportunity exists and the areas where action needs to be taken.

For many of the same reasons that women are important for the construction industry, issues surrounding other under-represented groups such as ethnic minorities and disabled people must be addressed. There needs to be further investigation as part of a more wide ranging review of equal opportunities in the industry.

9 Conclusions.

The role of construction as a major contributor to the fabric of society is undermined by poor image. This hampers the ability of the industry to recruit the best skills, creating a vicious spiral of poor image and poor performance. The process of implementing the recommendations of the Latham Review offers an unprecedented opportunity for change.

Sir Michael Latham calls for fair dealing, trust and teamwork, to improve the way in which business is carried out between client, contractor, specialist sub-contractor and design professional. These are the same factors needed to change the dirty, dangerous and macho culture of the industry to one that will encourage and retain more women, helping to change vicious spiral to virtuous cycle.

In addition to the recommendations outlined in its interim report, the working group on Equal Opportunities called upon its fellow working groups, the industry and its representative institutions and associations to acknowledge the wider application of these three key elements by taking action on:

- fair dealing - in recruitment, equal pay and decent working conditions for all;
- trust - in providing career opportunities and giving responsibility to women and men; and
- teamwork - in rejecting stereotypical attitudes and recognising and utilising the skills and attributes that women can bring to the business of construction.

10 References

1. Gale, A.W. (1991) What is good for women is good for men: theoretical foundations for action research aimed at increasing the number of women in construction management. *Practice Management: New perspectives for the construction professional*, Chapman and Hall, London. pp26-34.
2. Employment Gazette. (1994) *'Labour Force Survey Help line'*, December.
3. Court, G and Moralee, J. (1995) *Balancing the Building Team*, Institute of Employment Studies, Brighton.
4. Construction Industry Council (1995) *CIC Membership Survey 1995.*
5. Architects' Journal (1994) Construction Women Hit Glass Ceiling. *Architects Journal*, 24 Nov 1994. p11.
6. Building (1995) QS Awarded £9000 compensation for sexual discrimination by contractor. *Building*, 20 Jan 1995. p13.
7. Agapiou, A. Price, A.D.F. & McCaffer, R. (1995) Planning Future Construction Skill Requirements: understanding labour resource issues. *Construction Management and Economics,* Vol 13, pp149-161.
8. Ball, M. (1993) Public Capital Expenditure and its Implications foe the Construction Industry: A Research Agenda for Industrial Efficiency, *Birbeck College University of London*, Oct 1993 pp19-23.
9. Opportunity 2000 (1994) 'Did you know....?'. *Opportunity 2000 Newsletter*, Issue Summer 1994, p3.
10. Institute of Management (1994) Management Development to the Millennium, *Institute of Management*, 1994.
11. Greed, C. (1991) *Surveying Sisters: Women in a traditional male profession*, Routledge, London.
12. Hansard Society. (1990) Report of the Hansard Society on women at the top, London: The Hansard Society.
13. Langford, D. Hancock, M. R. Fellows, R. & Gale, A.W. (1995), *Human resources management in construction*, Longman Scientific and Technical, Essex.
14. Gale, A.W. (1994) Women in Construction: An investigation into some aspects of image and knowledge as determinants of the of the under representation of women in construction management in the British construction industry. Ph.D. thesis, University of Bath.
15. Latham, M. (1994) *Constructing the Team*, HMSO, London.
16. Gale, A.W. (1992) The construction industry's male culture must feminize if conflict is to be reduced: The role of education as gatekeeper to a male construction industry, *Proc 1st International Construction Management and Resolution Conference* (Fen, P and Gameson, R eds.), Sept 1992, pp416-427.
17. Hagen, E. and Jenson, J. (1988) Paradoxes and promises: work and politics in the post war years. in *Feminization of the labour force: Paradoxes and promises,* (eds. J.Jenson, E.Hagen, and C.Reddy), Polity Press, Cambridge.
18. EOR (1995) Agenda '95, *Equal Opportunities Review,* No.59, January/February. pp. 12-19.
19. Kandola, P., Fullerton, J. and Ahmed, Y. (1995) Managing diversity: succeeding where equal opportunities has failed, *Equal Opportunities Review,* No. 59, January/February. pp. 31-6.

INDUSTRY AND UNIVERSITY IN A MUTUAL PROJECT OF EVALUATION: A SWEDISH EXAMPLE

Jan Borgbrant and Fredrik Hansson
Department of Construction Management,
School of Civil Engineering, Lund University, Lund, Sweden

1. Summary

The following article accounts for the evaluation of a development project, 3T, carried out at Sweden's largest construction company, Skanska. It describes the methods and results of the evaluation, as well as general ideas concerning future operational development within the two organisations.

The opportunity was given to let the people of the university world and the practical world meet in open discussions with the mutual goals to allow company as well as department to:

1. Lessen the distance between researcher and practitioner
2. Create instructive organisations
3. Increase the knowledge of the production process
4. Analyse the possibilities to improve the process

Thanks to an insightful attitude among the management at Skanska Syd AB, we were given the opportunity to work according to our proposal with continual seminars, where representatives from the company met with the doctorates of the department every three weeks, with approximately seven participants from each group.

In conclusion, some participants of the evaluation make personal reflections concerning this form of evaluation project.

Sommaire

Cet article décrit l'évaluation d'un projet de développement appelé 3T, qui a été conduit dans le plus grande entreprise de construction Suédoise, Skanska AB. Nous décrivons les méthodes d'évalution et les résultats ainsi que la façon dont l'entreprise et l'institution universitaire vont plus loin avec des activités de développement dans les organisations respectives.

Dans ce projet de l'évalution a été donné la possibilité de laisser les personnes des entreprises et de l'université se rencontrer dans des discussions ouvertes avec des objectifs communs au deux partis;

1 Diminuer l'écart entre la recherche et les activités des entreprises.
2 Créer des organisations capable d'apprendre.
3 Améliorer la compréhension du procédé de production
4 Analyser les possibilités d'améliorer le procédé

Grâce à une attitude compréhensive de la direction de Skanska Syd AB, nous avons reçu la confiance pour travailler selon notre proposition avec de fréquent séminaires où les représentants de l'entreprise et les futurs docteurs de l'institution se sont rencontrés toutes les trois semaines, avec environ sept participants dans chaque groupe.

The Organization and Management of Construction: Shaping theory and practice (Volume One).
Edited by D.A. Langford and A. Retik. Published in 1996 by E & FN Spon. ISBN 0 419 22230 8.

En conclusion, quelques personnes qui ont participé à l'évaluation, donnent leurs réflexions personnelles sur cette forme de projet.

2. Background

Since 1991, extensive development work has been pursued within the Skanska concern, Sweden's largest construction company (The annual trade 1994: SEK 32,6 billion, 1995: SEK 38,4 billion), under the name of 3T, which stands for "ThinkTotalTime".

With the 3T-programme, Skanska is hoping to render their enterprise more efficient through a reduction of the time consumption within the construction process. A higher efficiency is wished-for, by means of planning work operations and target practice of material and machine resources more thoroughly. The goal is to work wiser rather than faster. Initiative and new thinking is desirable at all instances. 3T is said to be a process in which all employees take part and continuously develop together.

To support the change in progress, a 3T training programme, aimed at all members of staff, has been planned and to a great extent carried out.

Through Management by Objective (MbO) the company wishes to make better use of the unused potential among the skilled workers. The productivity and competitiveness of the company will be enhanced through commitment and fresh ideas among all employees. Management by Objective is described as "an acknowledgement of the fact that those positioned closest to a task usually know best how to solve it".

3. The Evaluation Commission

Skanska Syd AB (The annual trade 1994: 2 billion Swedish kronor, Employees 1994: 1908):, a subsidiary company within the Skanska Concern, turned to the Department of Construction Management at Lund University, to ask for help with a professional evaluation. A fundamental idea was the achievement of synergy effects between industry and university.

The Department was commissioned to plan and carry out an evaluation. The evaluation work was realised in the form of a series of seminars.

Purpose of the Evaluation

The evaluation had several purposes. The basic purpose was to evaluate the carrying through and the result of the 3T-programme. Besides, the intention was to increase the knowledge of matters of evalutation as well as giving a base for Skanska Syd's and the Department's continued developmental work. This paper emphasises the very evaluation process. Jan Borgbrant and Fredrik Hansson were responsible for the carrying out of the evaluation and the documentation.

Method

The set-up of the evaluation project can, slightly simplified, be described by a three-step-model including the steps Input, Process and Output.

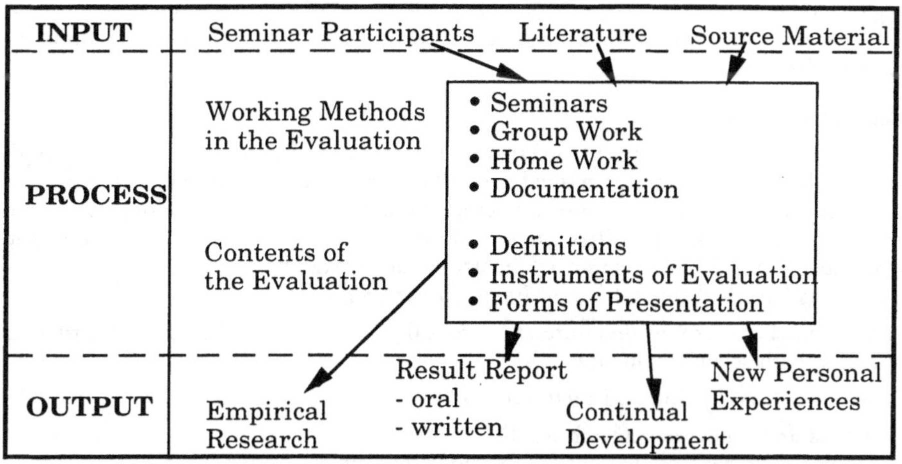

Figure 1 *A three-step evaluation model including the steps INPUT, PROCESS and OUTPUT*

A series of 18 seminars was carried out during the one-and-a-half years in which the evaluation took place. Employees within the company as well as at the Dept. of Construction Management took part in the seminars.

Collection of data has been completed through questionnaires and interviews as well as studies of original sources. Through the questionnaires, a broad inventory of opinions among Skanska employees concerning development work was made. The purpose of the interviews was the acquisition of deeper insight into the level of commitment in single individuals when it comes to the process of change.

4. Working methods in the Evaluation

Theory of Methodical Change

A first step by deep-going changes within an organisation is the participation of all affected parties in the analysis and description of - the CURRENT- SITUATION. The description must include demands from the surrounding world as well as questions that deal with the individual working place and working team. In the second step - the DESIRED-SITUATION, the goal for future activities must be challengingly formulated. It involves thinking outside of established patterns - what at first sight appears unachievable can be formulated as an offensive goal.

The third step - ACTIONS - deals with the strategy of change. Small as well as large improvements, which will lead to the set-up goals, are described in plans of actions. They can concern anything from changes in the organisational structure to individual basic values or technical investments . Managers as well as employees must realise that change is a long-range process.

The fourth step - EVALUATION - thus appears as crucial in the realisation of actual change. It is based upon the following-up and evaluation of every step in the process, as well as the increased insight of what actually supports and undermines the developing process, for individuals as well as working teams and the concern as a whole.

In this four-step strategy, reflection and systematic feed-back play a great part in the implementation of true change. The main hindrance to successful and extensive strategic change is the lack of habit to pause in the daily tasks to exchange thoughts and ideas as to how the process of change really works, and what ought to be done to provide it with new energy.

Series of Seminars

The series of seminars was given a great amount of space. Through the series, a high quality of the final report was wished-for, as well as an open dialogue which would enable continual learning in questions of evaluation. The participating group from the company was made up of skilled workers, foremen, intermediate managers and one representative from the management (departmental manager).
The seminars have had three different items on the agenda:

1. Discussions concerning literature. Theoretical questions were discussed in order to provide high scientific quality.
2. Contents of evaluation and methodical scheme.
3. Analysis and report of research results.

The seminars have been carried out for five hours at a time.

Group Work

During the seminars work in small groups took place. The group work formed the basis for discussions about the forms and contents of the evaluation. The group works were a means of letting the participants exchange personal experiences and make reflections on the goals and methods of the evaluation.

The group work was an excellent opportunity for the making of closer acquaintances, and offered the possibility of understanding each other's professional roles.

Homework/Literature studies

A major part of individual work contributions consisted of literature studies. The literature was dealt with at each seminar session and provided a basis for research-methodical discussions. On several occasions, literature readings proved supportive in the practical evaluation work.

Documentation

During each seminar, notes were taken. The participants in the series of seminars had full access to all material produced by the "evaluation researchers". By making all work material available to every seminar attendee, the quality of seminar discussions was increased. The opinions of each participant have influenced the choice of produced material, such as questionnaires, interview foundations, result summaries and final reports.

The evaluation focused on four areas. To make sure the data compilation covered the four levels individual, group, organisation and surrounding world, the matrix in figure 2 was constructed. This matrix was used to establish whether our questions within the evaluation instrument were complete. At one seminar, great amounts of time were spent correcting imbalances in the data compilation.

Themes / Level of organization	Work Environment	Leadership	Change Strategy/3T	Construction Management
Environment	1A	1B	1C	1D
Organization	2A	2B	2C	2D
Group	3A	3B	3C	3D
Individual	4A	4B	4C	4D

Figure 2 Matrix for the analysis of the data compilation instruments

Questionnaire

The most important compilation of data for the evaluation turned out to be the questionnaire.

6 Do you get response to your ideas:	Never			Always
a from your colleagues?	☐	☐	☒	☐
b from your superior manager?	☐	☒	☐	☐
c from the management above your nearest superior?	☐	☒	☐	☐
d with the management?	☒	☐	☐	☐

Figure 3 Example of multiple question as part of the questionnaire

The questionnaire came to include 54 questions.
The questionnaire was handed out to all employees at the randomly chosen construction projects at Skanska Syd AB. The number involved as listed in figure 4

	Pilot project		**Normal project**		**Total**	
Category	Number	%	Number	%	Number	%
Craft man	52	75,3%	304	74,0%	356	74,2%
Salaried Employee	16	23,2%	106	25,7%	122	25,5%
Unstated profession	1	1,5%	1	0,3%	2	0,3%
Σ	**69**	**100,0%**	**411**	**100,0%**	**480**	**100,0%**

Figure 4 Compilation of the selection of the research

Interviews

The compilation of data was also carried out thorough interviews. The target group of the interviews was on principle the same as by the questionnaires.

The guiding principle in the selection of respondents was the idea that all professional categories should be represented by at least two persons. Considering the great amount of skilled workers at the randomly chosen building sites, it was decided that this category should be represented by at least 50% of all interviews.

In all, 23 interviews were carried out. 12 skilled workers were chosen at random from in turn randomly selected projects, within building and road construction.

Strategy of Analysis

The strategy has been, on the one hand, to try and obtain the largest amount of information possible from each separate data compilation instrument, on the other hand to systematically search for differences and similarities in the answers from the different sources of information.

The questionnaire answers were run through a data base programme, which offered great opportunity to deal with the material according to demands that appeared in the process of evaluation The data base was developed continually. This enhanced the possibility of processing and analysing the extensive amount of data.

In the analysis work hypotheses were formulated. The purpose of the hypotheses was to systematically establish whether the available data from the questionnaire and interviews supported or discarded these assumptions.

Members of the seminar group worked with testing the hypotheses. The seminar group was divided into three smaller groups, which each took responsibility for testing one third of the hypotheses. The work of the groups came to play a central role in the whole reporting of the analysis and results of the evaluation.

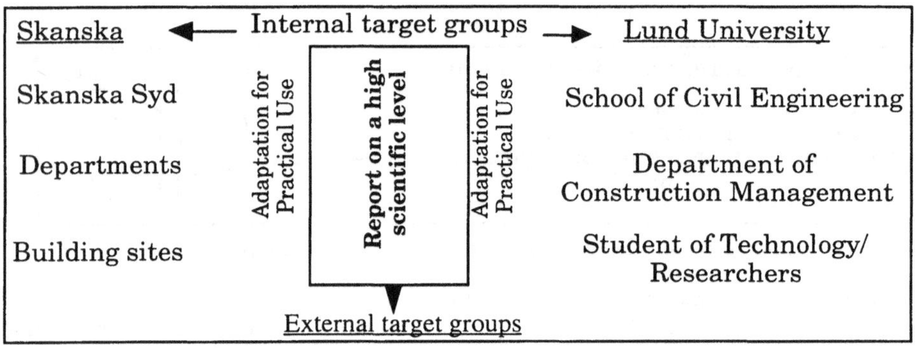

Figure 5 Target groups of the final documentation

5 Results of the Evaluation

Continued Development Work within the Company

3T/ Innovative Work

Many results in the study indicate the success of Skanska Syd's venture called 3T, and the fact that the staff within the organisation has got the message. The employees are in general sympathetic towards the changes. The questionnaire answers do however show a lack of understanding of the 3T-message within the production lines.

To increase knowledge and understanding, systematic support from the management, for instance in the form of additional training, is necessary.

Traditionally, this line of business has been marked by a limitation in the spreading of information. Our study shows that the communication between heads of departments and employees is better today than only a few years ago. There are however still traditional patterns of behaviour, which consciously or subconsciously work against a good spreading of information.

The risk of 3T/Innovative work is that the staff gradually loses interest in it and sees it as a time-limited project rather than a long-term process. It is therefore important to:

- spread the theoretical concept behind the 3T concept, in a pedagogical fashion
- make example of the theory by turning it into practical models
- systematically spread new ideas and practical solutions
- point to positive as well as negative results

Information/Communication

To illustrate the flow of information within a company, we can analyse its movements from office to work site and from building site to office. Information that comes from the office usually contains purposes, conditions and guide lines. Information from the building site, on the other hand, usually takes the form of results, experiences and solutions to problems.

It is important to demand information and to perhaps clearer formulate which kind of information one wants. Within a large information flow, there is the risk of informing on the wrong matters, which leads to the recipient still feeling uninformed or misinformed.

Working towards a more decentralised organisation increases the possibility of an effective information flow, since the number of levels the information has to pass decreases.

Lack of communication and information leads to a lack of commitment, less pleasure taken in the work as well as less understanding of and identification with set operational goals. Improved communication and information is a prerequisite for increased productivity and better experiential feed-back.

Technical Support and Experiential Feed-Back

The questionnaire shows that salaried employees as well as skilled workers wish to make more use of computers. Exactly what computers should be used for, they find it hard to define. Knowledge and proficiency when it comes to computer usage within the construction process, is little developed, which means that the co-workers find it hard to see practical advantages of computer usage.

Front competence within Skanska Syd AB must be provided, i e someone who can follow and evaluate the development in the computer area, which takes place within the construction industry as well as in society in general.

Motivation and Work Environment

In the Pilot Projects (using MbO), there is a greater awareness of the theoretical models that are the basis of the 3T process, and theories are connected to practical situations. Compared to traditional projects, the Pilot projects emphasise the practical application of the 3T-message more consciously. This shows a greater commitment, a higher level of motivation and a larger assumption of responsibility for the work.

The work environment is formally controlled by laws and regulations. The company appears humble when it comes to questions that deal with work environment, and accepts rather than questions sets of rules. Work environment questions are dealt with, in that the law is abided with, necessary measures carried out and directions followed up and checked.

There are still great differences between employees and skilled workers, for instance when it come to terms of employment, trade union affiliation, salaries and wages. This system may lead to a division in professional interests, which works against changes. This has led to increased difficulty in the development of competence and a lessened inclination towards change.

Development within the Deptartment of Construction Management

Development Projects

The development of the organisation and distribution of work within the Department can be carried out with the help of the methods that have been used in the evaluation work. Concrete needs for further training of the employees can be identified, for instance concerning methods of introducing and evaluating the quality assurance in the construction process.

The evaluation of 3T has also provided the basis for changes concerning the planning of teaching, from a traditional type of teaching to a more reflective and analysing form.

Basic Training

The evaluation project has shown that a continual survey and evaluation of all courses within the basic training is required. Such a process has been at work since 1992, but has gained increased actuality through the project results. The sections on management and leader development ought to be extended and another important goal lies in the extension of interplay with other departments within the School of Civil Engineering.

Supplementary Training

There is need for a supplementary training for active engineers and architects. In this field, experiences from this evaluation project can be very helpful. The 3T efforts at Skanska, to further develop the working methods, means of assistance, steering, and follow-up in the construction process, can provide many practical examples for practice cases and exercises.

Postgraduate Studies

The general methodical questions to do with research have been given an encompassing illumination within the project, and has shown the need for new postgraduate courses aimed at increased insight in the dynamics of the construction process.

The mode of procedure in this project will be used in a co-project with the Danish Institute for Construction Research, SBI. Researchers and contractors will co-operate in the evaluation of one or several construction projects with a focus on questions concerning productivity. The project will also serve as a postgraduate course for doctorates from both countries.

Research

We have through this project come very close to the essential questions of the contractor companies. It therefore feels urgent to continue the mutual build-up of knowledge. The forms may be continued closer co-operative projects with the industry, for instance in the development of methods and means of assistance and their testing and evaluation.

In the collected material there is further information to be dealt with in order to give a deeper insight, as to which factors support or hamper the carrying-out of innovative projects. The material can also preferably be completed through studies in other companies, in order to reach deeper insight into this group of problems.

Experiences from an Evaluation Project

At the final stage of the evaluation project, all participants were offered the opportunity to comment upon the evaluation work in written form. The following are extracts from their personal essays.

Departmental manager:

"I have learnt that a research report, which a 'normal human being' from the real world in general has the energy to read, contains a ceaseless source of information. Knowledge and facts, which, if rightly used, can achieve grand things out in the company. The perhaps largest and most important observations has however been concerned with the co-operation between university and employees within Skanska."

3T-Manager

"I believe that I will carry with me the following experiences:
-What is the truth today may not be the truth tomorrow.
-The majority of things, including 'the wonders of technology', contain human values."

Professor

"Some of the company's representatives expect to receive easily accessible, lucid and concentrated result reports, preferably with recommendations summarised in a simple and concrete fashion, that can lead straight to measurements.
Other people express expectancies that are very close to my own. We have to make the evaluation methodically planned enough, for a production of working material to be used for long-term operations within the Company as well as the Department."

Foreman

"It has been immensely stimulating to experience the successful co-working of the seminar group. Despite very different backgrounds, we have been able to speak the same language and make ourselves understood. The difference in attitudes between the representatives from The School of Civil Engineering and Skanska respectively wasn't marked until now that the report is taking shape."

Doctorate Student

"The extent of the difficulties included in the change of routines and work forms, I do not believe can be fully appreciated until one finds oneself in the middle of the changing process, either as a participant or as an observer. I have during the course of the project learnt that the most difficult thing in executing changes usually is not to make people accept new ideas so much as to make people reject the old ones."

Project Engineer

"To form a group by mixing university and company people proved interesting. The individuals had, independent of background, pretty far into the evaluation fundamentally different opinions of the relation time-quality. I do however believe that the individuals from both organisations in time learnt to not only accept but also to appreciate and acknowledge each other's best qualities."

Craft Man

" '-Would you like to participate in the evaluation of 3T, in co-operation with the School of Civil Engineering?'

'-Of course', was my answer. But I thought to myself, what contributions can I make among professors, architects, psychologists, engineers and other 'schooled people'? I was not thinking that my school consisted of 20 years of work experience at Skanska. But I have realised that what is the truth for one, is a lie to the other. And both are right, are they not?"

Secondary References

1. Thurén, T. Theory of Science for Beginners, Runa förlag, Stockholm, 1993
2. Molander, B. Philosophy of Science: A Book on Science and the Scientific Human Being. Bokförlaget Thales, Stockholm, 1988
3. Holme, M and Solvang, B. Research Methods: About qualitative and quantitative methods. Studentlitteratur, Lund, 1991
4. Repstad, P. Closeness and Distance: Qualitative Methods in Civics. Studentlitteratur, Lund, 1993
5. Borgbrant, J. Strategic Dialogue. Natur och Kultur, Stockholm, 1987
6. Borgbrant, J. Strategic Dialogue 2. Natur och Kultur, Stockholm, 1990
7. Bruzelius/Skärvad. Speed Management. Almquist&Wiksell, Ekonomiförlagen, Lund, 1992
8. Skanska AB.Annual Report, 1993
9. Skanska AB Annual Report, 1994
10. *Construction Trade Outlook 1/95*, Byggentrepenörerna
11. *A Presentation of 3T - A new Way of Thinking and Constructing*", Presentation Material, Skanska AB, 1993
12. Skanska AB. *Questions and Answers,* 1994
13. Borgbrant J, Strategic Dialogue - a framework to improve developmental process in counstuction companies, Management Quality and Economics in Building, Sydney 1991

2 The firm turning money into quality

ARE QUALITY MANAGEMENT SYSTEMS POSSIBLE?

D.W. Cheetham
School of Architecture and Building Engineering,
University of Liverpool, Liverpool, UK

Abstract
The place of quality management in the context of the evolution of management thought is reviewed. Direction and control of employee activities is essential to all management systems and the audit of quality management systems and of the individual performing tasks within it provides management with control. The audit received inadequate attention in the firms studied.

Abstrait
La place de gestion du qualite dans la contexte d'evolution de la pensee du gestion est critique. La direction et control des activites d'employe c'est essential pour tout les system du gestions. Aussi la vertification de la qualite de system du gestion et d'individual qui execute les tache du vertification qui execute les tache du vertification fournie la gestion de la compagnie avee le control. Cettee etude out a trouve que la vertification de qualite dans les compagnies etudees ne pas receve l'attention a les gestionaine.

Keywords: Management, management systems, quality, development, audit, control.

1. Introduction

Many businesses are started on the initiative of individuals seeking an opportunity for self advancement, either by higher income or profit from their activities or by perceived enhanced status. There are many such businesses in construction and with the lack of employment opportunities the number of self employed have grown. The self employed, whether consultant designer or individual tradesman, have to satisfy clients needs. The consultant seeks to convert the client's brief into drawings and specifications. The self employed tradesmen's client is the main or subcontractor who engages his services. If the client becomes dissatisfied with the standard of work produced, it is unlikely that repeat orders will follow. There is immediate feedback. As the quantity of work to be done grows beyond the ability or scope of an individual an organisation comes into existence. The relationship between two individuals in the simplest form of organisation may be that of equal partners but almost inevitably they will tend to specialise their activities. Indeed many people seek as partners persons with

The Organization and Management of Construction: Shaping theory and practice (Volume One).
Edited by D.A. Langford and A. Retik. Published in 1996 by E & FN Spon. ISBN 0 419 22230 8.

complementary talents and skills when forming partnerships. Few organisations consist entirely of partners of equal status, most acquire employees as they grow. Most contracting firms consist of the owners of the business (shareholders) and their representatives (the board of directors) and the employees. Senior employees being paid managers who organise and control the business by directing the activities of all employees towards the company objectives. When accepting paid employment a person accepts the authority of the organisation and responsibilities placed upon them. Some people have authority, the power to issue instructions, over others.

Just as the buildings created by the industry reflect the capabilities and needs of society, as society has developed and changed so have contemporary perceptions of the nature of management. This paper reviews the nature of management and the evolution of management thought, management development programmes such as management by objectives and the problems of controlling product quality by inspection. The place of quality management within the various quality systems used for buildng is described. Direction and control of employees is maintained by means of audit. It is the audit of records and individual's performance that is currently lacking teeth.

2. Evolution of management thought

Many social reformers such as Robert Owen with his Utopian plans and development of an industrial model village in New Lanark can properly be regarded as pioneers in the development of management. He sought to gain acceptance of factory discipline by dispossessed agricultural workers through the introduction of welfare schemes and improved working conditions.

Fayol [1] is credited with initiating management education by his establishment of the "Centre of Administrative Studies" and by adopting the definition "To manage is to forecast and plan, to organise and command, to co-ordinate and to control." Communication being the essential link. Though Fayol's analysis of the nature of management activity was based on his experience of a mining and metallurgical company in France during the last century these functions still serve as a basis for the study and practice of management today, both management in general and in building construction management in particular. These processes can be applied in all the departments, divisions and to all activites of the firm. Marketing, Estimating, Planning, Personnel, Production, Purchasing, Plant and so on depending upon the organisational form chosen by the management.

The body of knowledge has expanded considerably during the current century. Taylor, Gilbreth, Gantt and other pioneers of 'Scientific Management' were concerned with making manual workers more productive and therefore better paid. Most studies were based in factories mass producing goods but exceptionally Gilbreth started his career as a bricklayer and became a successful contractor before becoming a pioneer industrial engineering management consultant [2]. The two aspects of work study, method study and work measurement remain useful today. Many involved with the management of building firms find Gantt's charts useful though they may be ignorant of the origins of the bar chart programme. The techniques of work study practitioners involve recording and measuring employees movements, questioning their necessity and devising alternative better methods of working. Administrative and clerical tasks also benefitted from the application of work study techniques to the organisation and methods used in offices. Their objective being to determine the best way of performing manual tasks.

Lillian Gilbreth, wife and lifelong collaborator of Frank, was concerned with the individual, the human factor. Elton Mayo and others of the Western Electric Company of America confirmed the importance of considering human factors in their study the effects of lighting on output at the Hawthorne plant. To their surprise they found that human emotion played havoc with the results of carefully planned and controlled scientific experiments. One interpretation of their results is that if changes are imposed without consultation production will fall. If people participate in decision making production will rise despite a deterioration in physical working conditions. Concerns for personnel management, social and participative approaches to management have developed as a result of the application of the social sciences. Motivation, satisfaction and incentives have received attention from McGregor, Maslow and Herzberg. The work of all these persons is well summarised in most standard textbooks.

There are many aspects of management that have to be considered by practicing managers. A well respected author on building construction management, Calvert [3] defines management as "a social exercise, part art, part science, involving the organisation of a number of individuals in order to achieve a common purpose. Management is therefore concerned with the ways and means of getting things done. Determining corporate objectives and determining organisational structure, accepting responsibility for planning and regulating the firm by installing and operating proper procedures and ensuring the co-operation of personnel by providing the resources necessary for their work, giving them the will to work by guiding and supervising their activities."

3. Management development programmes

Improved managerial performance is seen as essential to the achievement of improved business performance. Training and development programmes have been devised to achieve this. Drucker pioneered the concept of "Management by Objectives" (M.B.O.) which has been developed by Humble and others. Humble defines MBO as 'a dynamic system which seeks to integrate the company's need to clarify and achieve its profit and growth goals with the manager's need to contribute and develop himself'. It provides a system of integrating logical business planning with management appraisal and development. MBO procedures formalise the continuous process of stating and critically reviewing an organisation's strategic and tactical plans by agreeing with each manager the Key Results Areas of his job and the performance standards he should achieve appropriate to company objectives. MBO follows a cyclic pattern of Job Analysis and preparation of Job Descriptions, identification of Key Results Areas and the determination of objectives and performance in the job, followed by a review of that performance. Key Results Areas are those activities which occupy a significant proportion of the manager's time. They are those parts of his job over which he can exercise control and is able to monitor progress. By measuring and discussing progress towards objectives at frequent performance reviews, plans for management training can be developed to help each manager build on his strengths, overcome his weaknesses and accept responsibility for self-develoment. Inherent in this is that a system of control over junior managers is established by more senior managers for at the review meeting the manager discusses his progress with his superior.

The proponents of MBO programmes believe that participation in assessment leads to understanding of the organisation's goals, to a co-ordinated application of the knowledge, skills and experience of employees, a more determined approach by each manager and greater job satisfaction for all concerned. The reality can be different. During a study of the implementation of MBO to a contracting firm described by the author [4] some fifteen years ago. The managers concerned did everything possible to frustrate the effective operation of the system. The major problem was the lack of certainty of job definition and delineation of responsibility. In particular site managers regarded the review as a challenge to their self esteem, showed aggressive and anarchic behaviour, were preoccupied with short term problems and failed to prepare for review meetings. The company training officer acted as counsellor and many site managers scorned his lack of craft and technical knowledge. They resented providing their superiors with information with which they could be controlled.

Other management development programmes such as Action Learning [5] and Action

Centred Leadership [6] provide opportunities for personal self development by mutual discussion of problems related to the work situation but outside the line management hierarchical structure. Specialist tutors and trainers give "on the job" counselling, create learning situations and seek to act as change agents.

Philosophies of organisational development recognise that managers must react not only to ever changing external situations but also to the expectations and aspirations of the people within the enterprise. Company objectives and those of its employees must be compatible if effective improvements in performance and company growth is to be achieved.

4. Procedures manuals

Contracting companies operate simultaneously on varying numbers of sites dispersed over an area. Some firms may choose to undertake a few large projects dispersed over the whole country (or even continent), others smaller projects in a region. Contracting companies must of necessity operate using management information systems which are similar on all projects in order to allow head office support departments such as wages, purchasing and plant to satisfy the production needs of the site and to allow senior management to exercise control. The system must allow effective communication between sites and offices and allow for receipt of instructions from consultants. Most firms have evolved procedures manuals which describe how administrative and supervisory functions are to be undertaken. Standard forms for reporting information encourage ease of comprehension by the recipient through familiarity and allow senior management to quickly make comparisons of achievement between various sites.

The author has previously described [7] the procedures manuals of a large national contractor. For convenience the firm separated site procedures from head and regional office procedures. Standard procedures are employed to obtain certainty that everyone knows what to do, how and when to do it. This is especially important when introducing new individuals to the company, for providing a common basis for training in offices dispersed about the country and for the providing a common basis of understanding as to how each individuals duties contribute to the overall objectives of the company. The number of persons in management and administrative roles on site varies with the size and complexity of a project. Persons having similar job titles may be undertaking different ranges of duties. For example surveyors can be found checking materials deliveries, site engineers work is sometimes limited to setting out but may include responsibility for concrete mix design, making test cubes and sampling bitumen.

Procedures manuals describe how each procedure should be carried out. Unfortunately some procedures are described in more detail than others. This provides scope for individuals to interpret them in different ways. This was found particularly to be the case with those procedures associated with the quantity surveying function. When this occurs senior management cannot be certain that financial data has been generated in similar ways using similar assumptions.

5. Controlling product quality by inspection

Ashford [8] argues that the major effect of industrialisation of production in factories was to replace the traditional skills of the self employed and self motivated craftsmen by those of the production engineer. Although spectacular increases in productivity resulted by separating planning from execution it deprived workers of the right to decide how their work should be done and at what rate. Secondly, the organisation of work into highly repetitive short-cycle activities created boring monotonous tasks for the workers. Quality suffered.

"The response of management was to establish independent internal inspection departments to restore the balance. Inevitably, since they had to be capable of distinguishing between what was satisfactory and what was not, the inspectors were drawn from the more intelligent and knowledgeable members of the work force. This had the immediate effect of diluting the skills of the production team and at the same time creating a powerful group whose existence could be justified only if it found work which had to be rejected. It is difficult to imagine any arrangement more calculated to demotivate the productive work force and create friction and bad feeling. An inspector can identify a fault only after it has been committed. He may then order the item to be scrapped or rectified in some way. Whatever the decision, waste will have occurred and harm done which cannot be undone." Ashford additionally describes an inharmonious working environment in which "at off peak periods the inspectors become under employed and occupied their time by creating bureaucratic procedures to further strengthen their stranglehold on their enemies in the production departments ...... In many cases the inspector will know the cause of the fault and how it can be prevented, but he has no incentive to pass on this knowledge to those in charge of production. They are on opposing sides and unlikely to welcome his advice."

The existence of inspection departments removes any sense of responsibility on the part of the work force for the quality of their work. The quest for improved product quality has preoccupied many both in manufacturing and in construction.

In building contractual agreements normally define the responsibilities of the various parties involved. These include inspection of the work as the work proceeds. It is worthy of note that none of the standard contracts used or agreements between clients and designers have been revised to reduce the powers of architects (or other client representatives) to inspect and approve (or condem) the standards of workmanship being achieved. The client's representative's right to take samples of materials for testing or to require the contractor to product test certificates to show that materials supplied comply with those specified remains despite the superficially wholehearted adoption of quality management by the industry.

6. Quality systems for buildings

The quality of buildings in the UK began to be questioned during the late 1970's when many reports identified major failures to achieve acceptable levels of performance from the viewpoints of both owners and users. At that time all members of the industry were having to come to terms with the long term implications of the introduction, over the previous three decades, of a large number of new materials and techniques. Economic circumstances were also creating changes in the structure of the industry toward the increased use of subcontractors and self employed labour, both of which were perceived as having adverse effects on good working practice. There was an increase in successful legal actions against designers for professional negligence. Many reports [9] have shown that faults in design and materials specification cause about as many failures as faults in site workmanship. Faults generally result from a failure to comply with recognised and published good practices.

All building designers design to standards such as Building Regulation Approved Documents and British Standard codes of design practice. They select components which have received independent third party assessment, either by product certification or product approval. *Product certification,* the oldest established, is the most common. The system provides assurance that the product being manufactured complies with specification. This system is used when the specification is well established and widely accepted. The scheme is backed up by random product testing by the certification body and the product is marked by a recognisable symbols such as the BSI Kitemark scheme. *Product approval* is based on "service in use" assessments and product testing. The aim of product approval is to give assurance, not only that the product complies with specification but also that the product and its specification are suitable for the application claimed by the manufacturer. A typical example of this type of assurance is that offered by the British Board of Agrement and its assessment certificates. The assessment process often involves developing tests for the product. A product approval scheme is for products with a degree of innovation either in their materials content or their usage.

The British Government has long been concerned that the quality of manufactured goods has to be improved for manufacturing industry to remain internationally competitive. The Mensforth Committee in 1970 expressed concern about national arrangements for the attainment of quality in the engineering industries. The costs of defective quality (scrap, re-work and defective products) were estimated by the Department of Prices and Consumer Protection to be about 10 percent of turnover. The need for quality assurance - planned systematic actions necessary to provide adequate confidence that a product or service will satisfy given requirements - was emphasised and a government strategy adopted in 1978 "to facilitate the implementation and assessment of modern quality management systems" [10] This reflected the philosophical developments, attributed to Deeming [11] and Juran [12] but developed by the Japanese, [see for instance Saski and Hutchins [13], Robson [14], that the primary responsibility for quality must lie with those during the work. The responsibility of the management of an enterprise is to establish a system for verification of work and to educate the work force in its use. These innovatory concepts, known as quality management resulted in BS 5750 Quality Systems [15] and its international equivalent ISO 9000 series.

Management system certification can cover a very wide range of products and include services such as design and contracting. A management system is assured against BS 5750. Assurance is given by the fact that the whole purpose of a management system based on BS 5750 is to ensure that the resulting product will comply with the specification. The product specification may be determined either by the producer or the purchaser.

The complication of management system certification is that BS 5750 offers three levels of quality management system - Part 1 where the manufacturer accepts responsibility for the design of the product, Part 2 where he manufactures to a set specification or Part 3 where he may only operate a process of final inspection and test The boundaries between the various schemes can and do become confused. For example the BSI in operating its Kitemark scheme now requires participants to operate their management systems to BS 5750 in as far as it affects the Kitemarked products. The Kitemark therefore gives additional assurance in that it is a combination of the two. However not all the output of the factory may be Kitemarked. Where the performance of a product depends on its installation it is not possible to prove compliance with specification at the factory. If a factory is operating to a BS 5750 system it does not mean that testing is excluded for BS 5750 requires adequate tests of the product to ensure compliance with specification. The third party assessor will be able to check that test results are adequately recorded and action is taken as a result of the tests.

7. Quality management systems

It is clear that since BS 5750 was developed for manufacturing industry a degree of interpretive licence is required to relate it to the construction industry. Most contractors can only seek accreditation to Part 2 of BS 5750 but the its limited scope, in that it does not include accepting responsibility for design and specifying materials, is not widely appreciated. Ashford [8] describes the theoretical background to BS 5750 and gives examples of its application to construction companies while Hughes and Williams [16] give detailed guidance on implementation by building firms.

The impetus for the construction industry to adopt quality management has been client led. Major industrial clients have insisted that the same quality systems and procedures developed by their own organisations are applied to the construction projects they commission. The government, through its Property Services Agency, until recently a major client, provided a major stimulus. Indeed the author knows of one small company specialising in pipe and cable laying that was obliged to introduce quality management systems and have them audited by a firm of management consultants in order to be allowed to remain on the tender list of their major client, a regional water company.

Quality management requires that the adopting organisation critically examines its procedures and methods of working. Lines of authority and communication must be clearly established. General and specific responsibilities of all members of the organisation identified and clarified. The management must critically examine and record procedures to be followed for all managerial and clerical activities and Inspection and Test Plans specific to the process being undertaken on site.

Written statements of procedures for activities enable both managers and those managed to ensure that reliable and proven methods are used. Writing down and reviewing procedures provides an opportunity for considering if there are better or more effective ways of performing a task.

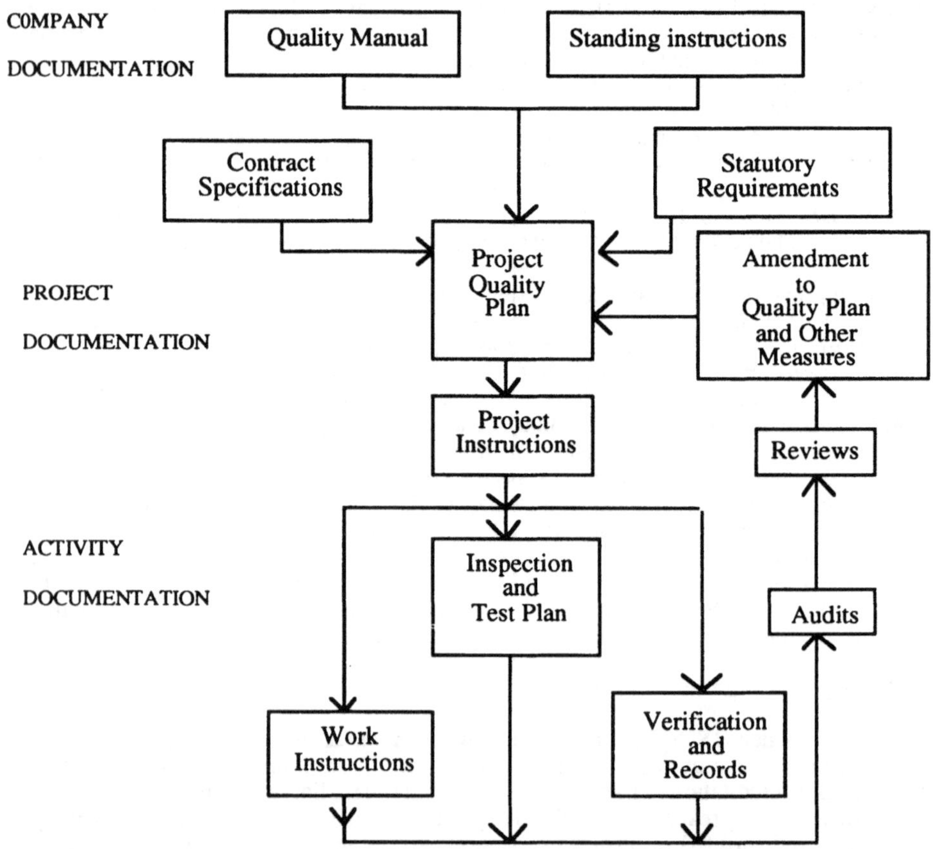

Figure 1. Relationship of documents in the quality managements system

This can increase efficiency both by reducing the number of errors and the time taken to perform operations. Records are maintained of all activities, in particular monitoring, inspections and tests, customer complaints and technical data available for use by employees. The discipline of good record keeping ought to enable problems to be traced and provide feedback for improvement.

The system should be organised so that continuous control is exercised over all activities. A company quality manual should be available for use as a training document for both staff and managers and will give assurance to senior management that a framework exists for good management control. Once laid down the systems create certainty (assurance) that everyone in the organisation knows what to do and how to do it. Records give assurance to senior management that a framework exists for good management control.

Certainty that the various processes that make up the total system are being correctly implemented is undertaken by audit. Audit, by definition is not a continuous process so it is necessary for those undertaking the process to maintain comprehensive records. These records and the auditors observations of task and processes provide the company senior management with the certainty the methods and procedures that they have determined or approved are being undertaken.

This approach to control of the activities of a company's personnel is considered by the many proponents of quality management as being innovatory and superior to more traditional methods. It empowers the workers! It is argued that the primary responsibility for achieving quality must be with those performing a task and they should be given every facility to satisfactorily complete the tasks assigned to them. The management responsibility is to establish systems for control and the verification that the work has been correctly completed and to educate and train the work force in their implementation.

The proponents of quality management argues that the introduction of quality management systems results in certainty that the final product is to the clients satisfaction. This certainty is the result of audit of all procedures. Among the benefits claimed are greater satisfaction of employees, reduction in waste materials and time due to improved efficiency, improved company reputation resulting from external recognition of achievement of a standard of management against an external bench mark, greater customer satisfaction. In this current depressed state of the market, without certification against the British Standard BS 5750 it is difficult for a construction company to remain on tender lists, to obtain work, and remain in business.

7.1 Project quality plans

Construction on site is a dynamic process and the Project Quality Plan is the key site document. Most main contractor's standing instructions require preparation of the quality plans and method statements in standard format. A general plan is prepared prior to the commencement of the building works on site. However, production of the project quality plan extends into the construction period as instructions and further drawings are received and subcontractors are appointed. Consequently the quality plans are formulated in stages. Initially at the preconstruction (mobilisation) stage and revised at intervals during construction as further information becomes available.

They contain a complete schedule of the drawings, specifications and contract documents and allocate specific responsibilities for quality related matters such as the procedures to be followed for receiving instructions and responsibility for maintaining records to named individuals. The plan are prepared to help the site managers firstly define and subsequently control the quality requirement of the projects. The preparation of quality plans requires planning and programming the works, recording proposed methods in method statements and analysis of quality requirements in the project documentation such as requirements to submit test certificates and to obtain architect's approval of samples. Hold points, where checks on site works by consultants and/or local authorities are needed are identified on work programmes. Quality plans identify which of the company standard inspection and test plans are to be used and any procedures, both administrative and site work, which are specific to the particular project. Sub-contractors quality plans are included and the documentation kept in a master file in the site offices.

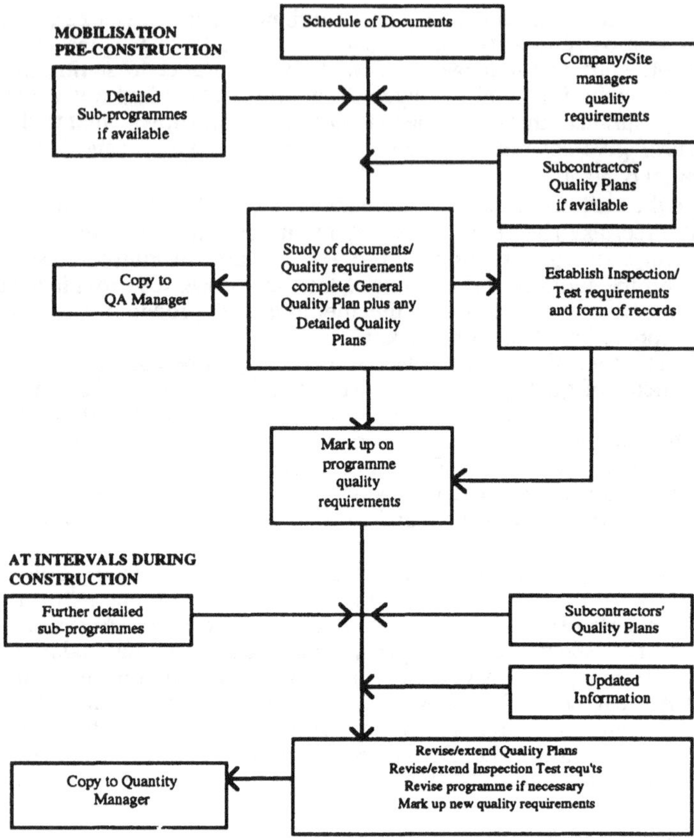

Figure 2. Preparation of projects quality plan

A study of subcontractors quality plans [17] described some very poor site workmanship practices and revealed considerable ignorance of the requirements of quality plans by subcontractors. Only the Mechanical and Electrical subcontractor produced a satisfactory quality plan. Very long chains of employee/employer were identified often ending with a labour agency. Although formal enquiries about general education, age of leaving school and appropriate trade qualifications and training were not made, the overwhelming impression was of a self taught workforce that had gained skill and experience on the job in an unstructured manner dependent entirely on the work that needed to be done. Informal discussion indicated that most had left school at the earliest opportunity and not sought formal occupational qualifications. They accepted the industry with its tradition of casual employment and were not concerned with record keeping or traceability. Key concepts of a Quality Management System. These operatives, chargehands and subcontract foremen knew they would probably not be on site if and when problems with their work might be identified.

The main contractors site staff implemented long standing Inspection and Test Plans (devised originally for directly employed labour) to overcome the deficiencies of the subcontractors. Even some of the site engineers and junior foremen doubted the likelihood of their continued employment with the firm! There was a reluctance to accept Robson's [14] concepts of "getting it right first time"; and the workmen did not regard each other as "internal customers'.

7.2 Implementing quality plans - the role of the audit

The quality management system must be thoroughly examined on a regular basis so as to assess its effectiveness and to ensure that stated objectives are being attained. This is achieved by means of audits which result in reports of problems and suggestions for possible corrective actions so as to avoid propagation of faults and failures.

An audit is a systematic, planned and documented examination to determine whether procedures, plans, documents and activities comply with requirements and are effectively implemented. The audit objective is to provide an independent review of each element within a quality system to establish:

- Whether the management system meets the requirements of the quality standard.
- That the system has been effectively implemented.
- That all documented activities and controls are being adhered to.

It is possible to categorise audits by the persons undertaking them:

- Internal: by Company on its own performance.
- External: by Company on its own suppliers.
- 2nd Party: by customers on the company.
- 3rd Party: assessment of the Company by an independent accreditation body.

There are two types of audit - system and compliance.

A System audit is used to assess both the adequacy of the total system against the requirements of the BS Quality System Standard and the extent to which it is followed in practice. This type of assessment gives comprehensive coverage and therefore is time consuming and costly to conduct. System assessments are normally carried out by purchasing organisations on their suppliers and by third party assessment bodies.

Compliance audits are more narrow in scope and are used to determine solely whether procedures and instructions are adhered to in practice. Normally compliance audits are conducted internally, although some purchasing organisations may also carry them out as part of an overall strategy of assessment activity in support of major supply contracts or projects.

The audit consists of three main stages:

- Preparation: The audit date is agreed with those concerned, and reports on previous audits studied. An audit plan is prepared giving a list of items to be examined during the audit.
- The Audit: On the agreed date the quality assurance manager meets the staff and arranges a timetable of visits for the day. Audit interviews check compliance with the various company procedures manuals. The procedures that should be used are explained where departments or sites are at variance to the procedural requirements. A closing meeting reports the results of the audit to the members of staff and identifies matters requiring further attention.
- Report and follow up: An initial audit report is then sent to solicit a reply as to the action taken to correct problems. A full audit report is then issued to senior management and copied to site if problems remain. This may be followed up by a further visit to ensure that the agreed action is being taken. A final completion report is then issued.

7.3 System audits

Two studies of the conduct of systems audits have revealed contradicting approaches. One of a large national contractor operating through regional subdivisions involved a thorough and rigorous analysis of company standing orders and procedures manuals. Over a period of two years the group quality manual was reviewed and revised, existing procedures manuals revised and re-arranged to match the sequence of BS 5750 and the position of regional quality assurance manager established. The firm eventually received independent 3rd party accreditation of its system from an internationally recognised assessing body. The other firm, a small cable and pipe laying contractor had to achieve 3rd party accreditation to remain on tender tests for its major client. The loss of this business would have considerable impact, potentially devastating, on the firm's ability to exist. The response was sensible but costly. A QA manager was recruited and required to review existing management systems and records and to develop new ones where necessary. This met with some opposition from gangers and gang foremen who were not interested in record keeping. The company directors, two brothers and a wife realised that BS 5750 had to be achieved but did not realise (or perhaps were not prepared to acknowledge) that they had a major part to play and that it was their lack of formally recorded management systems that prevented immediate third party assessment. A small independent consultancy offering accreditation, for a sum of approximately £15,000 was appointed to work with the QA manager to prepare a quality manual. Within 3 months the company quality manual was produced by recast existing procedures and developing new ones. Some such as checking on suppliers own QA procedures had not previously existed. The QA manager visited all gangs each week to ensure records were maintained, if necessary completing them himself, one month later another "consultant" from the consultancy spent a day conducting a system audit. Issued a certificate and collected a fee. All for a firm whose prime interest was digging trenches!

7.4 Compliance audits

Several compliance audits undertaken by the large national contractor have been observed over a two year period. Their conduct has been thorough and courteous but the audit depends on the regional QA manager's knowledge of all procedures described in the manuals, his ability to identify the procedures being used in a regional office department or on a particular site, and his ability to identify departures from them. Clearly it is difficult to memorise all the different procedures and requirements of an entire management information system. This was not helped by the general lack of detail in the description of many standard procedures. The consequence was that a quality assurance audit only tended to spot the obvious. This was confirmed by examination of audit reports. The conduct of the quality assurance audit was difficult because there are no procedural checklists for the person carrying out the audit. The absence of such audit checklists meant that the quality assurance manager's interpretation of standard procedures was the baseline against which compliance was assessed. Furthermore this meant that it was difficult to introduce new members of staff into the quality assurance department as the current system relied completely on the experience of the quality assurance manager.

Studies of audits conducted on sites reveal a situation distressingly similar to that revealed in the MBO study some 15 years ago, although site agents, managers and surveyors were more cooperative foremen and chargehands remained sceptical. Attempts by the QA manager to vary the audit sequence to suit the needs of individuals were interpreted as a weakness for, particularly on site, the audits were seen as an unwelcome intrusion.

Although all personnel, management, administrative, and operatives, were required to take part in the QA programme by edict from the group managing director, many took part with reluctance. It was seen as an additional chore in an already busy day. Unfortunately neither the regional QA manager or a site based QA manager had authority over the company's line management. They had to reply on co-operation and persuasion for they could not issue instructions to personnel. They did not wish to be perceived as enemies by their colleagues within the company or to be involved in unpleasantness every working day. Their task was to ensure that standard procedures were correctly followed and that anything not being done as they should were rectified. Their only sanction lay in the possible action of senior management if an adverse final audit 'completion' report was issued. The part the company's quality assurance system that investigated suppliers and subcontractors appeared to work well in checking their financial status.

7.5 Problems in practice

Many people failed to prepare for audits due to pressure of work and some managers encouraged the interruption of audit by urgent requests and telephone calls. A typical audit could only sample a small number of the standard procedures on any site or in a department. The audit process depends heavily on the Quality Assurance manager's personal knowledge and interpretation of the standard procedures and in his identification of problem areas. The audit reporting system was comprehensive but its potential educational role within the company was reduced by the limited dissemination of the audit results. Furthermore, though poor performance was criticised, good performance was not praised. Indeed only negative feedback resulted.

The standard procedures were described in varying degrees of detail in the manuals. A minority were so general, or terse, that the level of interpretation required by the user to implement them was such that neither the user nor the auditor had a basis for audit. Other better-defined procedures could be more easily checked. Even with these the audit process only checked the existence of the results rather than the stages used to produce them. In addition it was noted that the operation of some fully-defined standard procedures were varied by the user for good reason associated with the circumstances of each different project. This made the quality audit at best difficult and in some cases impossible. A well-defined procedure has clearly-defined stages and uses standard forms which require explicit entries.

A study within the joinery manufacturing subsidiary of a regional contractor revealed a more consistent level of detail of procedures and less scope for variation in interpretation by both management and operatives. The audits were again undertaken cheerfully and courteously. Although the motive, to satisfy clients requirements was worthy, the company had a longstanding reputation for good quality and innovation, particularly in laminated timber roof structures, their prices remained high. Possibly because of the cost of developing a QA manual and employing a QA manager, the company collapsed into insolvency and ceased to trade.

The site-based Quality Assurance manager limited his interest to site inspection and checks on completion of records of Inspection and Test Plans. He became unpopular with his colleagues, subcontractors and consultant designers. The relationships between him and major M&E subcontractors responsible for major work packages became so difficult and tense that his incompetence (due to lack of knowledge of M&E systems) was established and pressure put on the company management at a high level with the result that the site-based QA manager was sacked!

On site there is a constant pressure for production, for clarification, interpretation and co-ordination of design information from many sources. Procurement of resources - labour, machinery and materials - and the co-ordination of subcontract trades are the main interest of site management and trades supervisors. Although all procedures were nominally subject to the quality assurance audits, in the rapidly changing situation, a reluctance to properly implement them could occur when people failed to understand their job priorities. Resistance to change by those who did not anticipate continued employment with the firm could be concealed.

Unfortunately many people regarded the extra record-keeping associated with quality control as tedious extra paperwork. There was a reluctance to complete the forms which would enable those responsible for quality control procedures, particularly inspection and test plans, to be identified. The attitude of those responsible for supervising the work and completing the records was that records are unnecessary for they had already supervised the work! They resented the implied lack of confidence in their managerial and technical ability that the completion of records implied.

A major difficulty on site is determining when a piece of work by a trade is completed. the practices of first and second fixing and 'making good' after subsequent trades have finished their work, e.g. electrician's fixing cover plates to switches and socket outlets, give scope for foremen and chargehands to avoid signing check sheets by arguing that the wall is not yet complete, rather than signing to confirm, say, plastering is complete, or painting complete. These difficulties were compounded by the multitude of trades and subcontractors. The careless work of another trade can destroy a satisfactorily-completed piece. The focus of interest on a building changes from the component, element or work piece, wall, door, window, column etc., to the space. Should completed rooms be inspected or the internal walls between rooms? Are individual sections of services systems to be inspected and tested or complete systems?

Large areas of site operation were simply not assessed. For example site assembly of preformed components such as certain walling, suspended ceilings, platform floors, wall cladding and partitioning, by trade contractors could not be fully assessed because of lack of competence in specialist operations by those responsible for assessment. Lack of comprehensive specification on, sequence of assembly and permitted tolerances was commonplace. Some specifications and installation instructions supplied by trade contractor to the contractor's head office were not passed to site for use by the QA manager. The interpretation by those on site that quality assurance is quality control prevented a full discussion of the importance of compliance with standard procedures and completion of records.

Frequent changes to the main contractor's management and supervisory staff occurred. There were opportunities for poorly motivated individuals to use the disruption associated with personnel changes to not complete records. In the absence of a written record, the responsibility for completion of a particular piece of work on site could not be identified. Thus it could occur that records would not be filled out or be incorrectly completed to conceal mistakes. The most common problem identified during audit was incorrectly completed checklists. The monthly audit procedure, whilst being able to identify missing records or items of data, was not able to check the technical accuracy of the records.

On one site the QA manager's right to receive signed sheets to confirm that stages of work were satisfactorily completed by building services subcontractors was disputed by them. They argued that they were only required to submit a 'certificate of satisfactory completion' of the completed systems at the commissioning stages. This difference of opinion, initially identified by the heating contractor, was to lead to a major clash of personalities and the refusal, for a time, of many of the subcontractors to allow inspection of their own rather perfunctory and vague records as to who had completed particular pieces of work. On a smaller site such differences of opinion did not arise. No one sought to inspect the subcontractors' own records or to request confirmation of satisfactory completion of intermediate stages.

Almost all suppliers provided test certificates for materials when requested. These were passed to the client's representatives.

Conclusions

Quality management, like many other management systems, was initially developed as a result of studies in manufacturing industry. The building industry differs in many important respects, changing product and workplace, exposure to climate, constantly changing workforce on site dictated both by work requirements and because of turnover within the trades employed. It is a contractually complex industry.

Quality management seeks to ensure that everyone knows their job and performs tasks correctly. The checking of compliance, essential in all management systems becomes the responsibilty of internal auditor, occasionally checked by second or third party audits. The employees performng the audits spend most of their working time on these activities. They are generally well intentional people interested in improving company performance but lacked the ruthlessness and authority of line managers. Non compliance reports did not result in retribution for those not complying!

The legal and contractual framework which has evolved within the building industry ties individuals and organisations to the commitments they have undertaken regarding timing, cost, quantity and quality of their output. Standard contractual conditions still require material test certificate and samples to be provided to the architect.

The restructuring of the industry that is occurring during the recession and consequent change in culture of firms which formerly employed many trades to financially driven firms employing management staff and few trades has made implementation of Quality Management Systems more difficult.

Firms have changed in character from a command culture directed by contracts, directors with extensive industry practical experience of constructing buildings, gained over many years in relatively stable organisations, to a control culture. The control being based on contractual agreements. Unfortunately many of the trade contractors, with the exception of M&E contractors, do not have the management sophistication to respond positively.

This trend to increasing fragmentaiton of the contracting industry does not enable the lessons resulting from the introduction of quality management systems to take root. The signatories of many reports and forms are unlikely to be on site when the audits take place! This assignment of responsibilty from main contractors to others, particularly to small insubstantial firms and even individuals who failed to understand the risks assigned or their potential impact upon themselves as individuals has resulted in a litigous industry. It is to these firms which have fewest management resources that the success of implementing quality management systems is being assigned.

The quality on site is governed by the men actually doing the work. Therefore it is in the interests of the management of a subcontract building company to make all operatives full aware of their roles in the implementation of the Project Quality Plan; and ensure they are competent to carry out the works in accordance with this document.

The abilities, experience and training of those employed is very varied and the techniques appropriate for stable labour forces in manufacturing situations have needed considerable adaptation for use in building contracting.

There must be doubt as to whether or not clients realise the differences between Part 1 and Part 2 accreditation and the distinction between the roles of the contract supervising officer and that of the contractors quality management system to implement the quality standards laid down by specification in contract documents.

References

1. Fayol, H. (1916 1949 English Edition) *General and Industrial Management* Trans C. Storrs. Pitman and Sons London.
2. Steel, M. and Cheetham, D.W. (1993) Frank Bunker Gilbreth : Building Contractor and Pioneer Industrial Engineer *Construction History*. No. 9, pp. 51-69.
3. Calvert, R. E. (1986) *Introduction to Building Management*. 5th Edition. Butterworths, London.
4. Cheetham, D. W. (1980) Management by Objectives - a case history. *Building Research and Practice*. Vol. 8, No. 6, pp. 378-387.
5. Revans, R. W. (1982) *The Origins and Growth of Action Learning*. Chartwell-Bratt Ltd., Bromley.
6. Adain, J. (1986) *Effective Team building*. Gower, Aldershot.
7. Cheetham, D. W. and Carter, D. J. (1993) The challenge of assuring quality on site. *Building Research and Information*. Vol. 21, No. 2, pp. 85 - 98.
8. Ashford, J. L. (1989) *The Management of Quality in Construction*. E.&.F.N. Spon, London.

9. See for instance Freeman, I. L. (1975) Building Failure Patterns and their Implications. *Architects Journal.* Vol. 161, No. 6, pp. 303-308.
10. Anon. *A national strategy for quality - A consultative document..* Department of Prices and Consumer Protection, London.
11. Deeming, W. E. (1982) *Quality, Productivity and Competitive Positions.* MIT Centre for Advanced Engineering Study, Cambridge, Mass.
12. Juran. (1964) *Management Breakthrough.* McGraw-Hill, Manchester.
13. Saski, N. and Hutchins, D. (eds) (1984) *The Japanese Approach to Products Quality.* Pergamon, Oxford.
14. Robson, M. (1986) *The Journey to Excellence.* John Wiley & Sons, Chichester.
15. British Standards Institution (1979) *Quality Systems Parts 0-6,* London.
16. Hughes, T. and Williams, T. (1991) *Quality Assurance - a framework to build on.* B.S.P. Professional Books, Oxford.
17. Cheetham, D. W. and Lewis, J. (1993) *Implementing quality plans - the role of the subcontractor.* Advances in Construction Management Research, Proc. Ninth Annual Conference, Association of Researchers in Construction Management, pp. 124-139.

FINANCIAL FAILURE MODELS: HYPOTHESIS TESTING FOR UK CONSTRUCTION COMPANIES

L. Ruddock
TIME Research Institute, University of Salford, Salford, UK

Summary

There is a high level of financial risk in the construction industry, perhaps more so than in any other industrial sector. Previous work, by the author and others, has been undertaken to develop statistical models that can be used to predict those companies at risk. This work was mainly based on Z score models, and identified the major problem of too small a capital base.

The premise put forward in this paper is that a ratio model's predictive value can be improved by cash flow based (CFB) models. An analysis of data from the failure of several construction and property companies in the U.K. suggests the importance of cash flow in studying the causes of financial distress. In the study undertaken, the importance of cash flow is demonstrated, and additional components such as accounts receivable, capital expenditures and dividends from the cash flow funds statement are added to ratio-based models to increase their accuracy.

A vital difference between this study and its predecessors is formal hypothesis testing. Results of financial failure forecasts by three different models are compared and differences tested using statistical techniques.
Keywords: Cash flow, financial distress, risk.

Sommaire

L'industrie de la construction, peut-être plus que d'autres secteurs industriels, est exposée à un risque financier élevé. Un travail précédent, de l'auteur et

The Organization and Management of Construction: Shaping theory and practice (Volume One). Edited by D.A. Langford and A. Retik. Published in 1996 by E & FN Spon. ISBN 0 419 22230 8.

d'autres, a été entrepris afin de développer des modèles statistiques pouvant être utilisés pour prédire quelles sont les sociétés en danger. Ce travail, basé essentiellement sur les modèles "Z-score", faisait ressortir le problème important de l'insuffisance des fonds propres.

Dans la présente communication, nous avons posé en prémisse que la valeur de prédiction d'un modèle à base de ratios peut être améliorée par des modèles basés sur les cash-flows (CFB). Une analyse des données sur la défaillance de plusieurs sociétés de construction et du bâtiment au Royaume-Uni suggère l'importance de la marge brute d'autofinancement dans l'étude des causes de difficultés financières. Dans l'étude entreprise, l'importance de la marge brute d'autofinancement est démontrée et des composantes supplémentaires, telles que les comptes clients, les dépenses en capital et les dividendes à partir des tableaux des emplois et des ressources sont ajoutés aux modèles basés sur ratios afin d'accroître leur exactitude.

Une différence vitale entre cette étude et les précédentes est l'utilisation du test d'hypothèse formel. Les résultats des prévisions de défaillance financière selon trois modèles différents sont comparés et les différences sont testées au moyen de techniques statistiques.

Mots-clés: Cash flow, défaillance financière, risque.

1 Introduction

There have been several studies in recent years, concerned with research into the early detection of financial difficulty in U.K. construction companies, based on a Z score model approach.

Studies by both Langford [1] and Abidali [2] come to a general conclusion that a Z score alone provides insufficient evidence for failure prediction.

Langford applied the Z value technique to failed construction companies to test whether the technique could be used for predictive purposes with other companies in the sector, and advocated the development of Z models to assist in the assessment of construction companies.

Abidali has undertaken research directed towards the development of an operational system for identifying construction companies in danger of failure. In this research, the use of a Z score model was reinforced by a secondary method concerned with managerial performance aspects, based on the notion that both adverse financial and adverse managerial indications may be observed as a company moves towards insolvency. The study concluded that the Z score alone could not predict failure, providing only a financial indication of the solvency of a company. The model being able to indicate for example, that a company had a profile very similar to a failed company and, therefore possessed a high probability of failure.

So, whilst the usefulness of a Z score approach to the determination of a company's health may be obvious, it is also apparent that research into modifications to or expansions of the Z model may prove beneficial.

This current study tests the notion that a Z score based model's predictive value can be improved by the addition of cash-flow-based (CFB) variables; specifically, the usefulness of a cash flow identity that includes external capital flows is tested.

2 Business risk in construction

According to a major risk management consultant [3], building is the weakest industrial sector in the U.K. in terms of business failure. In mid-1993, average profit margins were as low as negative 10% and more then a half of the companies in the sector were losing money. The results of a survey by the Building Employers Confederation (see Fig 1) indicated that 40% of firms were tendering for work at a level based on negative margins.

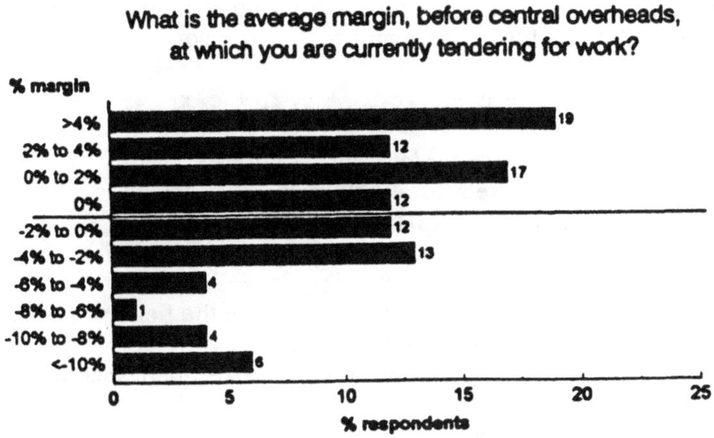

Fig.1. Average tender margins

Balance sheets have been weakened by totally inadequate levels of capital in relation to their turnover. Additionally, too high a proportion of companies' funding comes from short-term borrowing or creditors. On average, 50% of the funding of construction companies is short-term, coming from current liabilities such as bank overdrafts and money owed to subcontractors. Many construction companies believe they can manage effectively on low levels of capital. They hire much of their plant, subcontract much of the work and pay their subcontractors only when they have been paid. A larger capital base is essential to support the working capital of a business, and when new work is hard to achieve, a reservoir of capital is essential.

3 The Z score approach

A scientific approach to financial management can be used to take account of these financial inadequacies. The U.K. based Z score models described here, have been in use for several years. The first stage in building each model was to compute more than eighty ratios from the accounts of failed and solvent companies. Then, using stepwise linear discriminant analysis, the solvency model was derived by determining the best sub-set of ratios, which, when taken together and appropriately weighted, distinguished optimally between the two samples.

If a Z score model is correctly developed, its component ratios typically reflect certain key dimensions of corporate solvency and performance, such as profitability, working capital adequacy, financial risk and liquidity. Various models have been calculated using different component ratios, such as: Profit before interest and Tax/Sales; Debt/Net worth; Current liabilities/Total assets; Cash flow/Total liabilities; Debt/Quick assets; Debts/Credits.

Not surprisingly, different combinations of ratios and coefficients are required for companies operating in different sectors. By way of example, a model for the analysis of quoted U.K. building companies is shown below:

$$Z = c_o + c_1 x_1 + c_2 x_2 + c_3 x_3 + c_4 x_4$$

where

x_1 = profit before tax/current liabilities
x_2 = current assets/total liabilities
x_3 = current liabilities/total assets
x_4 = no credit interval

Table 1 shows the Z score for six listed companies in the fields of property and construction, that entered into receivership in 1990.
Every one of the companies showed a negative Z score in advance of failure.

Table 1. Analysis of receiverships

Company	Activity	Z score
Bestwood	Property developers	-4.5
City Grove	Property developers	-1.2
Federated Housing	House builders	-1.1
Stanley Miller	Building contractors	-3.3
Rockfort Group	Property developers	-1.9
Rush & Tompkins	Developers and contractors	-4.8

3.1 The risk rating

Most academic research in this field has focused exclusively on the Z score itself.

However, to improve its use as a practical tool, it had to be developed further.

Further research into companies which at some stage had a negative Z score identified three determining factors of ultimate failure.

- The degree of negativity of the Z score.
- The number of years it was negative.
- The steepness of the Z score trend.

Using these variables, an additional statistical model was developed, generating a 'risk rating' on a five point scale for all companies with a negative Z score, with 1 indicating a low probability of immediate financial distress, and a risk rating of 5, indicating a company that has only rarely survived in the past.

A study of U.K. companies in the early 1980's (in related economic conditions to the early 1990's), showed the cumulative probabilities of companies with risk ratings of 4 or 5 suffering financial distress to be 80% [4].

3.2 The PAS

Taking the Z score analysis a step further, a company's Z score can be transformed into a PAS (performance analysis score). A company's PAS is its ranked Z score in a particular year, in percentile terms. The percentage of companies with lower Z scores represents the company's PAS in that year. In this way, a simple readily interpretable performance measured on a scale 0 - 100 is derived. By deriving a PAS over a series of years, problems can be identified as the positive Z score company starts to decline and well before the company reaches the solvency threshold, giving more time for appropriate action to be taken.

Considering the case of Rush & Tompkins, Figure 2 shows its PAS trajectory and consists of three lines - the PAS for the company itself, the solvency threshold (where Z = 0), and the contracting and construction industry average. Rush & Tompkins, with a risk rating of 5, was at risk of financial distress for all of the six years of the graph, financing increasing activity levels mainly through trade credit.

While the construction industry was booming, the company was just able to survive, but as soon as the sector started to turn down, it fell into receivership.

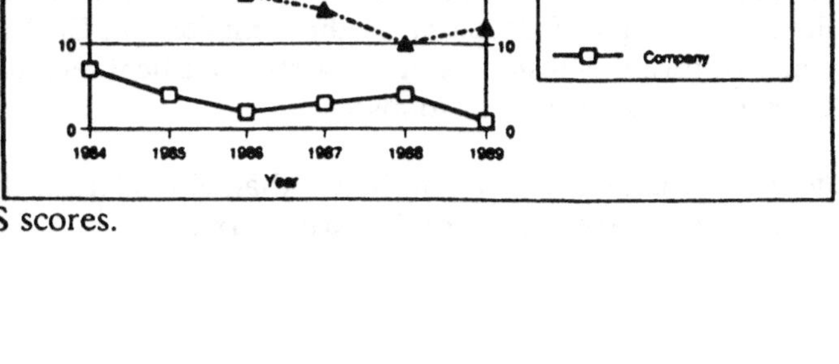

Fig.2 PAS scores.

3.3 Merits and limitations of the Z score model

The model is statistically derived, objective and unbiased. It is effectively reflecting and condensing multi-dimensional information conveyed in the accounts themselves, communicating the result in the form of a readily interpretable single index. Because of the model's multivariate methodology, it is preferable to one-at-a-time ratio analysis. It should be noted that a company rarely collapses overnight and will generally be seen to be giving out signals years in advance, once measured in this holistic way.

The analysis used in this research for the calculation of Z values is based on the multiple discriminant analysis proposed by Altman [5], assigning weightings and combining ratios to produce a composite figure to tell the story of a company's financial state.

A major short coming is the use of a sample containing equal proportions of bankrupt and non-bankrupt firms (obviously much higher then exist in the real world). Samples are selected from companies whose fate is already known and the calculated Z equations are the best fit only for companies with the same characteristics.

4 Cash flow based (CFB) models

Cash flow analysis has two justifications.

- Firstly, Gombola and Ketz [6] found that cash flow ratios contain certain information not revealed by other financial ratios.
- Secondly, an analysis of data from the collapse of several companies suggests the importance of cash flows in studying bankruptcy causes [7].

In this current study, the importance of cash flows is demonstrated. Adding components such as accounts receivable, capital expenditures and dividends from the cash flow funds statement to ratio-based models increases their accuracy.

Past studies have used different definitions of cash flows. This study tests the usefulness of a cash flow identity, initially specified by Lawson [8], which includes external capital flows. This model bases the prediction of bankruptcy on both internal earning variables and external stock market variables. Since stock market value influences the external capital a firm can raise, the probability of a firm going bankrupt would be influenced by the external capital used. According to this reasoning, a bankruptcy prediction model should include external capital flows i.e. debt as well as equity.

A critical difference between this study and previous ones, is formal hypothesis testing. Results of bankruptcy forecasting by three models (Z, CFB and Mixed) are compared and differences tested using standard statistical techniques.

The hypotheses tested are stated as follows:

- *Hypothesis 1:* CFB models perform better than Z models in the overall classification of bankrupt and non bankrupt firms.

- *Hypothesis 2:* CFB models perform better than Z models in identifying potentially bankrupt firms.
- *Hypothesis 3:* Mixed models perform better than Z and CFB models in identifying potentially bankrupt firms.

A sample of 18 firms that entered bankruptcy between 1985 and 1993 was matched with 18 firms that did not.

4.1 The cash flow identity

According to Lawson, operating cash flow applications are taxes paid, net capital expenditures and liquidity changes. Any surpluses or deficits flow to or from lenders or shareholders. Thus a firm's total cash flow statement for any year, *i* can be represented as the identity:

$$(k_i - h_i) - (A_i + R_i - Y_i) - H_i - t_i = (D_i - B_i) + (F_i - N_i - M_i)$$

where:

$(k_i - h_i)$ denotes operating cash flow in year *i* represented by cash collected from customers k_i and operating payments h_i.
$(A_i + R_i\ Y_i)$ denotes net capital investment in year *i* represented by replacement investment A_i, growth investment R_i and proceeds from assets displaced Y_i.
H_i denotes liquidity change in year *i* represented by change in cash and marketable securities and t_i denotes taxes paid in year *i*.
$(D_i - B_i)$ denotes shareholder cash flow in year *i* represented by dividends paid D_i and equity capital raised B_i in year *i*.
$(F_i - N_i - M_i)$ denotes lender cash flow in year $_i$ represented by interest payment F_i, medium or long-term debt N_i and short-term debt raised M_i.

Taken over a series of years, the equation is concerned with both economic performance and financial policy. The left side of the equation contains information pertinent to three decision areas - sales and production, capital investment and liquidity - and reveals the effective incidence of taxation on corporate cash flows. The right side provides information on dividend and debt-equity financing policies.

4.2 Sample design and model development

The study used a sub-sample of bankrupt companies and a matching sample of non bankrupt companies. A company was included if at least two consecutive years of data for the five-years preceding bankruptcy was available. This rule also applied to non bankrupt matching firms.

The sampling method can be faulted from the use of non random matched sampling but then is not entirely counter indicated.
Zmijewski [9] investigated the influence of non random, matching samples on bankruptcy studies and concluded that statistical inferences and overall classification accuracy rates are not affected.

The identity components were used as independent variables and tested for predictive ability. As already indicated, to determine the ability of the components to discriminate between bankrupt and non bankrupt firms, multiple discriminant analysis was used.

4.3 Preliminary data analysis

Basing a bankruptcy prediction model on the cash flow identity seems to be empirically justified. The use of cash flows in a model to predict bankruptcy is valid if there are significant differences between bankrupt and non bankrupt firms.

There are significant differences between operating cash flow and tax payments from the fifth to the first year prior to bankruptcy; lender cash flow is significant in the fourth and third year period before bankruptcy and the liquidity change is significant in the fifth year.

4.4 The mixed model

Three sets of models were developed for each of the five years prior to bankruptcy. One model contained only CFB variables, a second model contained only Z variables and the third contained both CFB and Z variables. (The Z variables used in the latter two models were those indicated in section 4.1 of this paper).

In order to assess the validity of the models, classification accuracy is defined operationally as the ability to classify firms in a given year using that year's data.

Table 2 shows classification accuracy for the three models.

Table 2. Comparative classification and prediction accuracy between the CFB, Z and Mixed models (in %).

Model	CFB		Z		Mixed	
Year	0	B	0	B	0	B
1	93.1	88.3	87.7	79.1	96.4	95.2
2	88.6	84.6	81.6	74.4	94.2	87.7
3	80.0	81.1	72.9	72.4	84.0	79.4
4	76.1	75.2	70.1	68.2	78.2	77.6
5	70.7	68.9	66.8	62.3	73.4	72.0

0 = Overall accuracy
B = Bankrupt firm accuracy

5 Findings and conclusion

As far as testing the hypotheses is concerned, the conclusion to be reached is that the cash flow based model does improve on the Z model's accuracy. At the $\propto$ =

0.05 level of significance, seven out of ten percentage comparisons are significant.

The results indicate that when the Mixed model values are compared to either of the other two models, the data in Table 2 show that the direction of difference is always in favour of the Mixed model.

The significant rise in predictive accuracy from the mixed model indicates that the addition of cash flow based variables contributes to lending/investing strategies and emphasis on cash flows in the monitoring of the financial health of construction companies is a positive step.

6 References

1. Langford, D., Iyagba, R. and Komba, D.M. (1993) Prediction of solvency in construction companies. *Construction Management and Economics* (11).
2. Abidali, A.F. and Harris, F. (1995) A methodology for predicting company failure in the construction industry. *Construction Management and Economics* (13).
3. Syspas (1994) *Report: Construction Companies, London.*
4. Tafler, R. (1984) Empirical models for the monitoring of U.K. corporations. *Banking and Finance,* Volume 8.
5. Altman, E.I. (1968) Financial ratios, discriminant analysis and the prediction of corporate bankruptcy. *Journal of Finance.*
6. Gombola, M.F. and Ketz, J.E. (1983) A note on cash flow and classification patterns of financial ratios. *The Accounting Review.*
7. Gombola, M.F., Haskins, M.E., Ketz, J.E. and Williams, D.D.,(1987) Cash flow in bankruptcy prediction. *Financial Management.*
8. Lawson, G.H. (1985) The measurement of performance on a cash flow basis. *Accounting and Business Research.*
9. Zmijewski, M.E. (1984) Methodological issues related to the estimation of financial distress prediction models. *Journal of Accounting Research.*

COSTS OF MANAGEMENT AND MAINTENANCE: THE ITALIAN CASE

V. Manfron
Architectural Construction Department, University Institute of Architecture of Venice, Venice, Italy

Abstract
The forecast of generalised building costs, understood as the total of all expenses necessarily incurred from the moment of conception and planning of a building to its final demolition, requires due consideration of the costs of use and deterioration (particularly management and maintenance costs) even in early phases of planning. With reference to Italy, which does not have a consolidated tradition in this field, the above mentioned costs can be analysed using a case study done by the Institute (I.T.E.A.) that administrates public housing in the Trento Province. This study analyses aforesaid costs and management and maintenance costs by taking into consideration the correlation between the evolution of construction technology and the costs in question, studying past models of building construction and relevant wear and deterioration of these products.
Keywords: Costs of Maintenance, Costs of Management, Generalised Costs, Public Housing, Quality, Reliability

Sommaire
La prévision des coûts généralisés d'un édifice, c'est-à-dire la somme de tous les coûts qu'il est nécessaire soutenir du moment de son projet à celui de sa démolition, comprend aussi la prévision des coûts d'usage et, en particulier, ceux de gestion et d'entretien. Pour ce qui concerne la réalité italienne, qui n'a aucune tradition consolidée en matière, on analyse les coûts susdits en considerant un cas étudié par l'Institut I.T.E.A. qui gère le bâtiment public de la province de Trento. L'analyse considère les coûts susmentionnés, l'influence
réciproque entre les coûts de gestion et d'entretien, le rapport entre l'évolution des technologies de construction et le coûts d'usage en question et les modéls de déterioration du bâtiment vérifies *a posteriori*.
Mots clé: coûts d'entretien, coûts de gestion, coûts généralisés, bâtiment public, qualité du bâtiment, fiabilité.

1 The planning of costs of a building product

In Italy the most widely used approach in building project and design is to consider building costs uniquely as the actual cost of construction. There exists, therefore, a general custom which considers management and maintenance expenses as an invariable in relation to decisions taken in the actual planning and design of a building; it seems almost that one does not have significant impact on the other. This practice and attitude

The Organization and Management of Construction: Shaping theory and practice (Volume One). Edited by D.A. Langford and A. Retik. Published in 1996 by E & FN Spon. ISBN 0 419 22230 8.

does not find comparison in other industrialised nations and is found in Italy as a result of the peculiar structure of its building market (lack of co-ordination among the public corporate bodies that manage large property estates).

This has rendered management and maintenance cost studies of building products unserviceable, and so a delay in national research on this subject has resulted. Such delay makes it difficult to initiate a policy of accurate economic forecasts of all actual costs in the expected life cycle of buildings. Therefore, the total costs of a building from its conception to the end of its life cycle are not calculated when the building is on the drawing board. This situation is in direct conflict with the necessity for the Italian construction sector to conform with other countries within the European Community. The singularity and individuality of each separate market makes the importing of data and procedures from other countries relatively useless. For this reason, recently introduced national legislation which requires that management and maintenance costs be included in the economic forecast of any building project, risks failure in its attempt to control expenses by calculating these costs.

A research project which has been going on for the last ten years at the University Institute of Architecture of Venice, has studied public housing with reference to the northeast part of Italy, Veneto, Trentino and Alto Adige, Friuli and Venezia Giulia, with prevailing, but not exclusive, attention paid to residential, partially or wholly public financed, housing and its management and maintenance.

This type of analysis of management expenses has been made possible through the existence of data banks, although quite elementary, found at specifically interested institutes. The statistics are, unfortunately, relative only to that part of the above mentioned costs which has been assumed by the owners of the estates, and thus, expenses incurred by individual home owners are not included. To study maintenance costs, expenses incurred by this corporation from 1983 to 1992 were compared. These expenses have been documented because of the recent application of a national law (Number 457/78) which provides state financing of such interventions.

2 Theoretical framework

This data is aimed at appraising:

1. In which manner the selection of design influences management and maintenance costs
2. In which way the technological evolution of construction influences management and maintenance costs
3. The means in which a subsequent analysis of maintenance projects can give, not only an idea of general costs, but also an idea of the reliability of the constructions, by studying the type of work done on different types of buildings. In this way, the durability, reliability and quality of the constructions can be analysed over a certain period of time.

3 The generalised cost of the building product

In calculating general costs or costs during the expected life span of the building which we are referring to, one must consider "cost of construction" as the sum of all costs sustained during the entire time the building is used-that is, from the moment it is conceived as a project to its final demolition. The following correlation can be considered valid:

$$Cg = Cbui + Cman + Cmai + Cue \quad (1)$$

that is, generalised costs (Cg) being the sum of building costs (Cbui), management costs (Cman), maintenance costs (Cmai) and the costs of unpredictable events (Cue). From this correlation and in this study, building costs (considered as the total of briefing, design and construction costs) [1] and the costs of unpredictable events, will be omitted. We classify unpredictable events as the probability of a certain event and the expense of obviating its consequences: the probability of building collapse is [2] one in a million over a year but minor building accidents frequently occur in the order of one in a hundred or even one in ten.

Considering the other two components of our sum, that is management and maintenance costs, it is necessary to refer to the research which Ciribini [3], Dell'Isola e Kirk [4], Marshall [5], Di Battista [6], Di Giulio [7] and Arbizzani [8] have done on the subject. For management costs, the following formula can be considered valid:

$$Cman = Cre + Ccn + Ccl + Che + Ccg + Cse + Cfi \quad (2)$$

Here management costs are the sum of replacement costs (Cre), conversion costs (Ccn), cleaning costs (Ccl), heating costs (Che), cooling costs (Ccg), security costs (Cse) and financial costs (Cfi). This type of generalised formula, to which various studies tend to add more and more expressions, must be properly adapted, as will be shown, to the particular type of construction being examined and to the objectives of its financiers and its administrators.

Maintenance costs can be expressed as follows:

$$Cmai = Cmn + Cml + Crh \quad (3)$$

with the three expressions of the formula being post natal maintenance cost or maintenance costs in the early life of the building (Cmn), maintenance costs caused by the ageing of the building (Cml) and the rehabilitation costs (Crh) necessarily incurred at the end of a building's life cycle when restoring the building becomes necessary to render it reusable. The introduction of the first voice (Cmn) and its distinction from traditional maintenance costs (Cml), which are directly tied to the ageing of a building, is most necessary when one realises that the major part of maintenance problems occur in the early years of the building's live span.

This fact, which distinguishes building and construction production from that of other industrialised production, (as represented in Fig. 1) is not only an Italian phenomenon, but can be found in other countries as well. One can consider that Sycodés [9], for example, indicated how, in the French market, 75% of all problems in construction occur in the first six years of the life of a building-a phenomenon not found in ordinary industrial products which tend to wear out with ageing.

In Fig. 1, the "a" curve represents the reliability in a certain time span of a typical industrial product. One can see that initial reliability is high but decreases as the product grows older. Curve "b", on the other hand, refers to a building or construction product; three distinct phases can be noted. The first part represents a low reliability of the product until a general stabilisation of regime is reached (the second part). In the final phase, the product is more and more unreliable; the third part of the curve represents this final senescence of the building.

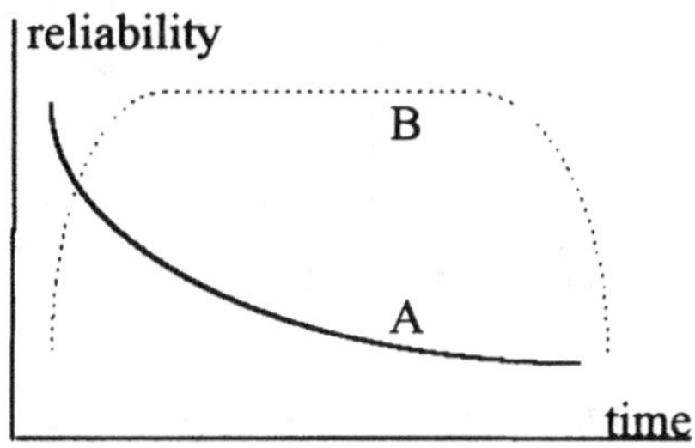

Fig. 1. The reliability of building products as compared with industrial products.

These numerous failures during the early years of a building are a result [10] of the impossibility of experimentation on a prototype of said construction. This possibility of experimentation and testing is fundamental in the industrial world. Every new building, although built with materials of certifiable quality, has, in fact, an extremely complex comportment which is very difficult to anticipate in the planning and development stages, because no valid experimentation on a prototype is possible for obvious economic reasons.

4 Cost of management

The case which will be considered is made up of properties which constitute the estate of the (ITEA) Institute that manages public housing in the Trento Province. This estate [11] is made up of 749 buildings with 9,493 flats or residences. We are speaking of buildings of various ages (as we can see in Fig. 2) built from the end of the First World War up to 1990. The type of construction to be considered includes relatively small units-four, six or twelve flats, built in the smaller communities of the Province and larger buildings that house thirty or forty flats which were built in larger more important communities in the seventies and eighties. Building technology used up to the 1960s, was composed of traditional construction techniques, load-bearing walls, floors and roofs in layered wood, vertical divisions in brick and window and door frames in wood. With the 1970s and the introduction of much larger buildings, reinforced concrete was adopted in the building field (reinforced concrete frames or concrete bearing walls, reinforced concrete slabs or hollow type slabs) and the use of special technology or industrialised casting (tunnel, demi-tunnel, banches et tables, banches et predalles) was also developed. Our sampling, taken into consideration for the above sited characteristics, can be judged sufficiently valid as a means to study the Italian national building market.

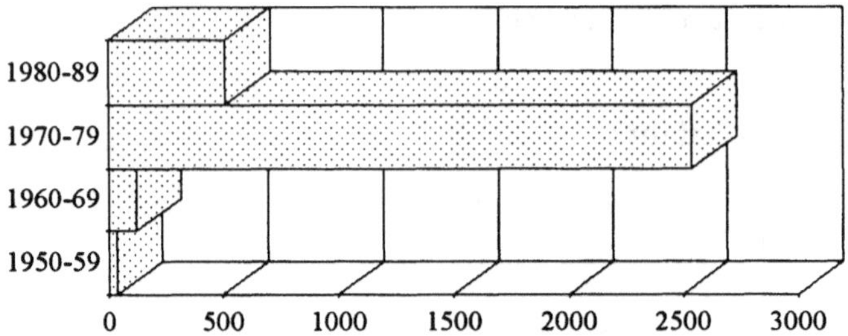

Fig. 2. Buildings for which management costs classified by year of construction have been analysed. Note that more than 2/3 (71%).have been constructed after 1970.

The management costs which have been analysed are those sustained and administrated by estate owners for the years 1991, 1992 and 1993; expenses paid by single tenants (such as electricity, water, telephone, etc.) have been omitted. It should be underlined that heating costs paid individually by each tenant has become a common practice in Italy with the introduction of methane gas for heating to each individual flat.

In detail the statistics which we have on hand regard: cleaning of communal zones, lighting of these areas, lift expenses, as well as heating and hot water supply to each single residence. The aforesaid study of maintenance costs (2) should be, therefore, adapted to our case study in the following way and with these symbols: (2 bis) costs of cleaning the communal areas (Ccl), lighting them (Clh), lift costs (Clf) and heating (Che) and hot water (Chw) costs.

$$Cman = Ccl + Clh + Clf + Che + Chw \qquad \text{(2 bis)}$$

Our sample for these costs is made up of 3,202 residences, that is 1/3 of the cases under study; a small part of these buildings were built in the fifties and sixties, 1.06% and 3.72% respectively, but for the most part, they were built in the seventies, 79.02%, and eighties, 16.02%. We must take into consideration, therefore, that these buildings are generally large scale constructions of thirty or forty flats and "reinforced concrete structures" was the building technology used for construction.

To analyse the costs which we are looking at, we must consider the percentage of each one in the yearly consideration of building costs (Cbui), the first addend in our equation (1). In Fig. 3, the aforesaid incidence is represented for the various time periods of the construction of a building.

Variation from the values indicated in Fig. 3, is considerable. One finds, in fact, that there is a range from 1.29% to 2.00% each year. It must be further pointed out that administration costs for buildings built in the seventies are 55% higher than for buildings built in the fifties. Thus, one can assume that the different building technology used in various periods of time has significant influence on management costs. This renders the following consideration important:

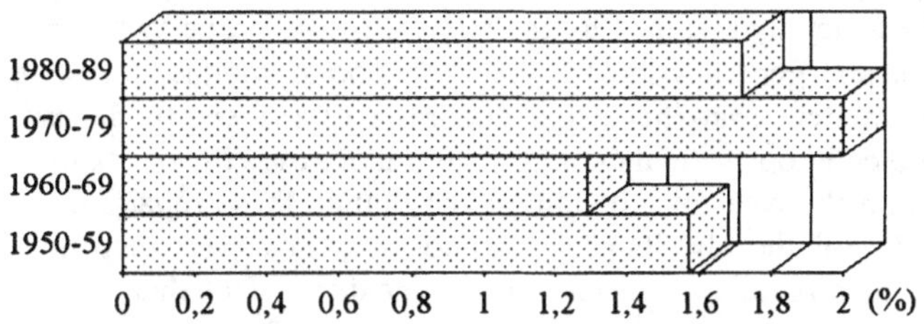

Fig. 3. Yearly incidences of management costs on building costs classified according to the age of a building.

1. The management costs under consideration are made up of 74% for heating costs (Che), followed by hot water costs (Chw = 16.18%), lift expenses (Clf = 7.83%), cleaning (Ccl = 0.26%) and lastly lighting (Cli = 1.73%) of communal areas. It immediately meets the eye that heating costs are by far the most important and the most subject to fluctuation in the global consideration of total management costs.
2. The progression of statistics in Fig. 3, should be taken into consideration remembering that in the seventies, traditional construction technology, that is, masonry bearing walls with excellent thermal insulation properties were substituted with reinforced concrete structures. This solution, while permitting the building of thinner and lighter external walls, does not give the same thermal insulation properties as the older method and, therefore, heating costs naturally increase as a direct result.
3. The noted reduction of heating expenses in the seventies and eighties (16%) is due to the better thermal insulation properties of building materials imposed by a national political policy (Law 373/76) adopted during the famous oil crisis after the noted Kippur War, a policy that fixes energy consumption for heating a building within given limits and is aimed at reducing national energy consumption on all levels.

5 Costs of maintenance

As in our study of management costs, only certain components of maintenance costs are available-thus, generalised cost is defined as expressed in the aforementioned equation (3). These missing statistics are a direct consequence of the general practice adopted by estate administrators in public housing of carrying out only emergency repairs and modifications while ignoring a more enlightened policy of general programmed maintenance. The above equation (3) should, then, be rewritten for our purposes as:

$$Cmai = Cmfl + Cta + Cpa + Cda + Crh \quad (3\ bis)$$

In equation 3 bis, the first addend represents the cost of repairing failures (Cmfl) reported by tenants. The next three represent the cost of adapting thermal insulating materials (Cta), technological installations (Cpa) and living unit modifications (Cda). Our fifth and final voice (Crh) represents the costs of rehabilitating empty properties for re-use; these costs are generally assumed by the owners of the property before a flat is rented to a new tenant and generally include work on floors, window frames etc. and other installations no matter what the actual condition of the empty flat was when evacuated. This last voice in our formula could, then, reflect the cost of restoration and not of general repairs on the wear of a building even though these repairs do probably prolong the life cycle of the dwelling. They do not, however, depend on the deterioration or wear of a building but only on the incidental fact that a lodging has been vacated by a former tenant and must be prepared for a future resident. We can point out that these five expressions in our equation can generally represent the whole maintenance operation that is carried out by national managers of public housing, and it is for this reason that we have cited a general lateness by these corporations in planning a proper economic policy of programmed maintenance of said properties.

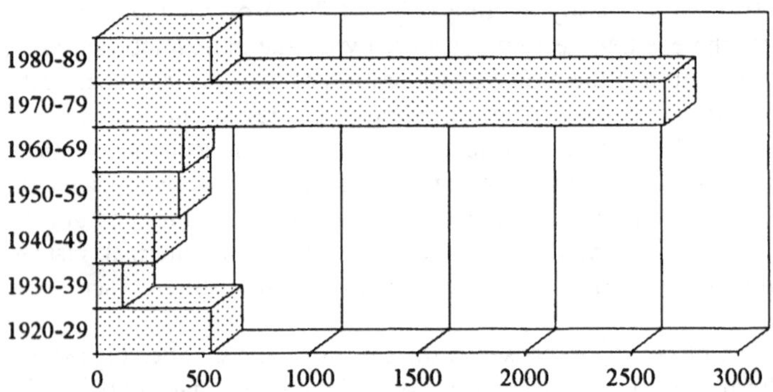

Fig. 4. Buildings in which maintenance costs have been classified according to the year of construction.

This time our sampling is greater than that used for the analysis of management costs: we are considering 4,933 residences, that is, 52% of the building estates under study. Different age groups are represented in Fig. 4, and one can note, even though we are referring to a longer period of time than before when we were studying management costs, that, also in this case, about 2/3 or 65% of the residences were built after 1970 and thus, with the use of modern technology.

Such costs can be divided into three almost equal parts: repairing of failures (Cmfl) 36.88% of the total, modification of thermal insulation of installations in common areas and in individual flats (Cta + Cpa + Cda) 30.22% and finally the restoration of empty flats to prepare them for new tenants (Crh) which is 32.90% of our total.

Detailed consideration of these statistics and an accurate analysis of the single expressions under maintenance costs are necessary. In the first voice (Cmfl), costs of repairing failures, we can note that constructions built in the 1920's (5.67% of total estates) weigh upon total reparation costs for 26.59%. On the other hand, buildings of the 1980s cannot be considered homogeneous, because maintenance was done on these buildings immediately, that is, practically at the same time they were being built (in 1983 to 1992). In the other age classifications, there is a direct relationship between size of the categories and amount of expenses. Thus we can presume that, leaving out the oldest constructions, there is very little connection between age of the building and resulting problems in need of repair.

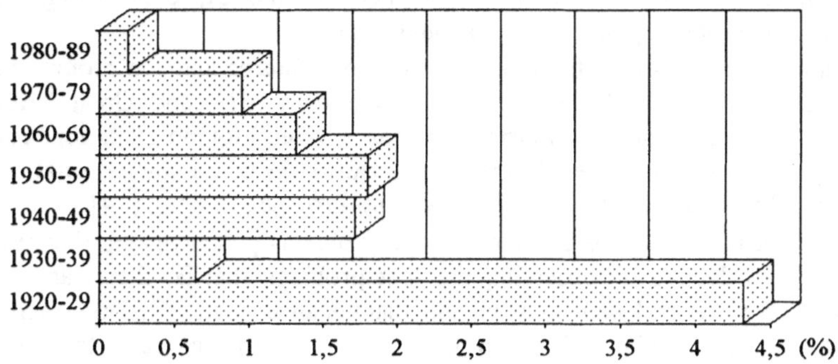

Fig. 5. The components of maintenance costs divided into age groups of buildings considered in our study.

If we look at adaptation and restoration costs, we can immediately note that the cost of adapting thermal insulation to new standards (Cta) regards only buildings built from the 1970's. In the same way, as was previously seen in maintenance costs, the greatest factor influencing these costs is the change in construction technology, that is, the passage from buildings with masonry bearing walls to buildings in reinforced concrete that makes increased costs inevitable. When considering the adaptation of installations (Cpa), we note that 88% of total expenses fall in the same age group. The same is true for restoration and adaptation costs of single residences that make up 77% of the total. In relation to this, we should remember that buildings from the 1970's and on, make up 67% of the total estates in the hands of large scale proprietors. There is, then, a direct relationship between the period of a building's construction, the relative technology used and the costs of restoration and adaptation. New technology seems, surprisingly, less trustworthy than older traditional construction technology.

Lastly, considering rehabilitation and restoration costs (Crh), it appears that older buildings are the ones for which this type of intervention is most burdensome. Buildings of the 1920s, in fact, which weigh upon total estate value for 5.67%-weigh upon the voice of total rehabilitation costs for 42.62%. Looking at the successive age classifications, the relationship between the incidence of the number of residences and amount of expenses seems to balance itself out. Buildings of the sixties that are 7.95% of total estate value, weigh upon restoration and rehabilitation costs for 12.83%, - those of the fifties, 9.43% for 23.83% - those of the forties, 4.41% for 15.35%, and those of the thirties, 1.4% for 4.96%. Thus, we can deduce that an increase in the age of a building necessarily increases the rehabilitation and restoration costs which must be sustained.

6 Conclusion

The over all management costs analysed weighs annually for 1.92% on the total costs of production (understood as the total of briefing, design and construction costs) of a building. As we have seen, the principle part of this voice is heating costs which contributes as much as 74%. Of lesser importance are the expressions regarding maintenance costs that make up 0.4% of the annual total.

One can note the remarkable difference between our figures and the programmed maintenance costs estimated [12] in Sweden and Germany, 3%, and Great Britain, 2.5%. Our statistic is substantially confirmed in France when looking at its repair and maintenance costs [13] which generally make up about 0.6% of the annual total, as it does in the Venice Province (0.5%) and in the Bolzano Province (0.8%), [14].

In our final consideration, we can look at the problems posed in this theoretical framework. We can deduce by observing the first problem, that is, the relationship between design and plans and management and maintenance costs, that management costs (see Fig. 3) of a building built in the seventies using modern technology (reinforced concrete casting, tunnel, demi-tunnel, banches et tables, banches et predalles) are over 55% higher than those of traditional buildings with masonry bearing walls and that this cost is, above all, heating.

There is not, however, the same correlation regarding maintenance costs (see Fig. 5), in so much as there seems to be a direct relationship in each different age classification between the amount of expenses and the number of flats taken into consideration. The technical evolution does not seem to have had undue influence on differences in the maintenance costs analysed. An exception to this premise can be found, however, analysing restoration and adaptation costs (Cta, Cpa, Cda) of buildings built from the seventies and onward.

In our case study, the technological construction evolution has resulted, and this is our second problem, in a worsening of the costs of deterioration and wear, thanks to the increase of management costs. Traditional technology seems, thus, more economical in the long run when considering management costs, than the new technology adopted in the seventies. This statistic appears to contradict recent strategies used in other sectors of industrial production where the ultimate object is to create a product in which the costs of deterioration are as low as possible.

The third problem-models of decay and deterioration-maintenance costs that must be assumed by owners (we must underline that we are referring particularly to the problem of breakage and eventual repairs), it must be noted that the major part of these expenses concern external elements, that is, walls, roofs, galleries and other external works, that absorb as much as 64% of the necessary cost of maintenance. The second expense factor regards installations, hydraulic, heating, electric, lifts etc. by which 22% of total costs are incurred. Longevity is threatened by those technical elements that suffer from atmospheric phenomenon-rain, for example. Above all, it is the various technological systems, heat, electricity etc., necessary for the life and use of a building, which are at greatest risk in construction, and it is to these elements that it is indispensable that greater attention and care be paid, especially in the design and planning state.

7 References

1. Manfron V., *A Later Analisys of the Models of Degradation of Buildings*, in *Proceedings of the International Symposium Dealing with Defects in Building*, Varenna, 1994
2. Manfron V., Verso la qualità: un appproccio statistico-strategico alla definizione della qualità, in *Modulo* n. 7/8, Be-Ma, Milano, 1985
3. Ciribini, G., Durabilità e problemi manutentivi negli interventi di recupero, *Recuperare* n. 6, Peg, Milano, 1983.
4. Dell'Isola, A.J., Kirk, S. J., *Life Cycle Costing Data*, MacGraw Hill, New York, 1983.
5. Marshall, H., Standards Practices for Measuring Life Cycle Costs of Building and Building System, *Astme 917*, Washington D.C., 1983.
6. Di Battista, V., Manutenzione, *Recuperare* n. 32, Peg, Milano, 1991.
7. Di Giulio, R., *Qualità edilizia programmata*, Hoepli, Milano, 1991.
8. Arbizzani, E., *Manutenzione e gestione degli edifici complessi*, Hoepli, Milano, 1991.
9. Sycodés, *Guide de l'expert*, E.G., Paris, 1986.
10. Manfron V., *Qualità e affidabilità in edilizia*, Angeli, Milano, 1995.
11. Pompermaier, F., *I fattori componenti il costo generalizzato*, thesis, Universitary Institute of Architecture of Venice, A.A. 1993-1994, Prof. Manfron, V., tutor.
12. Hammarlund, Y., Josephson, P. E., Sources of Quality Failures in Building, in *Management, Quality and Economics in Building*, E & FN Spon, London, 1992.
13. Toubon, J. C., Le politiche di recupero in Francia, *Recuperare,* n. 3, Peg, Milano, 1983.
14. Manfron, V., Le cornacchie del Middle-West: comunicazione e successo tecnologico, *Performance*, n. 25, Ance, Rome. 1993

A PILOT SURVEY OF THE SUPPLIER/SUB-CONTRACTOR PAYMENT SYSTEM

R. Kenley

Faculty of Architecture, Building and Planning,
University of Melbourne, Melbourne, Australia

B.M. Wood

School of Architecture and Building, Deakin University, Australia

Abstract

The analysis of the cumulative working capital which is available to a building company from operations has focussed in the past on the net cash flow of individual projects. There has been a search for an ideal reference curve, an examination of applied systematic delays and analysis of the individual source curves for specific projects to derive net cash flow. However there has been little exploration of the cumulative effect of a significant number of projects, and the resultant net cash flow, or working capital generated from operations, available at the organisational level.

This paper explores some of the relationships between the supplier/sub-contractor payment system - the timing of client, sub and head-contract payments - and the resultant cumulative working capital at the organisational level. A pilot survey is used to develop an understanding of contractor financial policy regarding the timing of contractual payments, credit terms and credit control. It is shown that there is a wide variance between firms in their approach to cash flow management, and that this has implications for organisational financial policy.

Precis

Dans le passé, l'analyse du capital de roulement accumulé dont une compagnie de construction dispose en raison d'opérations précédentes, s'est concentrée sur le mouvement d'argent comptant venant de projets individuels. On a essayé de découvrir une courbe de référence idéale, on a examiné les retards systématiques appliqués et on a analysé des courbes individualles concernant certains projets spécifiques et la façon

The Organization and Management of Construction: Shaping theory and practice (Volume One).
Edited by D.A. Langford and A. Retik. Published in 1996 by E & FN Spon. ISBN 0 419 22230 8.

dont ils obtenaient leurs des fond de roulement. Cependant on a négligé l'exploration de l'effet accummulé d'un nombre significatif de projets et du capital de roulement en découlant, disponible au niveau opérationel. Cet exposé explore certains des rapports entre le systéme de paiment en ce qui concerne le client, le contrat principal et secondaire - et le capital de roulement qui en résulte au niveau organisationel.

Nous nous servons d'une étude pilote pour essayer de comprendre la politique financiére des entrepreneurs en ce qui concerne la date des paiements contractuels, les clauses et le contrôl du crédit. On peut voir qu'il y a une énorme diversité entre les firmes dans leur attitude envers la gestion d'argent comptant et que ceci a des implications pour la politicue financiére organisationelle..

Keywords

Cash flow, Working Capital, Trade Credit, Payment Terms.

Introduction

The management of the flow of funds on individual projects has been investigated over the past 25 years, and several mathematical models have been developed for analysing the net cash position on a given project. There has been a search for an ideal reference curve [1], an examination of applied systematic delays [2] [3] [4] [5] [6] and examination of the individual source cash flow curves for specific projects [7] [8]. However, there has been little exploration of the cumulative effect of a significant number of projects, and the resultant net cash flow, or working capital generated from operations, available at the organisational level.

Models of cumulative multi-project organisational cash flow [1] [3], have assumed that estimated receipt-disbursement time-lag factors [3] can be used to calculate monthly cash flow figures, and that these once calculated can be overlaid cumulatively. This approach is supported by Archer [9] who analysed the literature and concluded that monthly transaction or activity summaries could be summed to arrive at what was termed a "stock of cash" needed for the management of the firm.

While there is agreement that overlaying individual project net cash flows will result in the net organisational contribution, there is no such agreement on the method for modelling the net cash flow source curves. Kenley & Wilson [7] argued that there is not a standard curve and that all projects will be unique. Other models have used some variation on the idea that a standard delay applied to either an inward or outward flow, or in fact two delays applied to a work-in-progress curve, would provide a net cash flow curve.

If the assumption that a standard delay may be applied is to be utilised, then it is necessary to increase understanding of the types and extent of delay involved in the management of cash flow by construction companies. These delays relate back to the work-in-progress curve and apply to both the flows of cash in from the client, and the flows of cash out to the sub contractors.

Cash in-flow delays relate to the delay between the work being completed on the site, and the payment by the client. This delay typically relates to the delay in claiming (typically at the end of the month) and the subsequent delay while the client values the work and then processes the payment. These delays may be assumed to relate to the contractual terms applicable and could vary between one week from the end of the month in which the work was carried out, to one month from certification which might be 2 weeks from the end of the month in which the work was undertaken (a total of approximately 6 weeks). These delays relate to individual circumstances, client and certification team practices, contracts, etc and are beyond the scope of this paper.

In contrast cash out-flow delays relate to the financial policy decisions made by individual contractors. Rather than define weighted average delays based on individual projects [5], it is believed that individual contractors will have delays which apply based on their own financial management policy, including the supplier/sub-contractor payment system.

Accordingly it is necessary to examine individual contractor firms to assess their cash flow management policies and the implications which this might have on a delay based analysis; and which in turn these might have on the overall financial management of the firm.

It is contended that builders have different interpretations and management of trade credit which has significant implications for organisational working capital.

In order to assess this contention, a pilot study was established to analyse individual contractors understanding of cash flow management and their policies. It is intended that this pilot study be followed by a full analysis of major construction firms within Australia. This paper presents the results of this pilot study.

Method

With the objective of determining where future research in the area of contractor credit control could be directed a pilot study of eleven (11) contractors was conducted. The contractor sample was chosen to include, medium to large employers of subcontractors, from both country and metropolitan Victoria. The method of data collection was by way of structured interview technique, employing both closed and descriptive questions.

The questions seek to investigate the following financial policy issues which relate to cash flow management

- the management of retention
- the management of the timing of payments
- adherence to contractual conditions

Results

In order to assess the management of retention, contractors were requested to provide their average annual turnover for the past three financial years and whether it was company policy to allow subcontractors to provide bank guarantees in lieu of cash retentions and proportionally by value, what percentage of subcontractors work would be included in each category. They were also requested to indicate their policy toward the provision of security to the client.

When asked whether it was company policy to provide bank guarantees when allowed under the contract, 80% responded in the affirmative. The results in Table 1 indicate that contractors, in approximately 60% of cases, withhold cash retention from subcontractors. When questioned whether interest earned on subcontractor cash retention was returned, 75% responded in the negative.

Table 1. Percentage of subcontractors work in cash or bank guarantee by turnover

Contractor	1	2	3	4	5	6	7	8	9	10	11	**Average**
Cash Retention %	70	60	50	50	0	60	50	70	0	0	30	**40%**
Bank Guarantee %	30	40	50	50	100	40	50	30	100	100	70	**60%**
Turnover $m	300	75	30	300	30	30	300	300	4	8	2	**Adjusted Average**
Cash Retention	15	3	1	11	0	1	11	15	0	0	0	**58%**
Bank Guarantee	7	2	1	11	2	1	11	7	0	1	0	**42%**

The general policy then appears to indicate, as would be expected, that contractors utilise bank guarantees to avoid cash retentions being held by clients, but receive both cash and subsequently interest payments from subcontractors. Thus, considering weighted average delays only, and not referring to financial management policies would give an erroneous financial picture of overall trade credit management. Consider, that by mid-contract on a ten million dollar contract at 5% subcontractor retention, the contractor may have at its disposal $500,000 plus interest accruing. By disregarding this cash input which has no corresponding cash outlay (as the contractor has provided bank guarantees) the true company financial position would be distorted.

There is, of course, no guarantee that the cash retention/bank guarantee scenario would be the same on every project. This finding is supported by the pilot survey and is consistent with the Kenley & Wilson [7] case for non-standard delays. Therefore, unique projects may have differing cash retention/bank guarantee policies. Any model developed must, once applied, account for differing contract conditions, trade emphasis and client certification procedures.

In order to answer the management of trade credit, or terms of payment, contractors were asked for their standard terms of payment for different work trades. In order to ensure that terms have equivalency, they were also asked to describe what they understood "30 day terms" to mean. Understanding the definition and use of such terms is critical when developing any model which addresses the management of trade credit, such as weighted mean delays.

The construction industry has for years claimed the general credit terms are "30 days". This pilot has attempted to try and define the term 30 day payment, for, when considering weighted delays for payments in and out this concept is critical in any calculation.

When asked for their general terms of payment for subcontractors and suppliers, the answers were unanimous in stating 30 days (although minor exceptions were made by most for labour only sub-contracts particularly form-worker, reinforcement fixer, bricklayer and carpenter). However, when asked what they understood 30 day terms to mean, the answers were totally diverse. Of the 10 respondents who replied to the question the following definitions were provided:

1. Payment 30 days from valuation of certification of work completed
2. Payment 30 days from invoice date
3. Supplier's/sub-contractor's claims in by the 25th of the month, payment on the 30th of the following month
4. Supplier's/sub-contractor's claims at end of the month, payment 30 days later
5. Payment 30 days from date of invoice
6. 30th day of month following receipt of invoice
7. Payment within 30 days of invoice being presented
8. Payment 30 days from valuation - back to back, ie: client pays 30 days after head contract claim, suppliers/sub-contractors paid after head contract payment
9. Payment 30 days from end of month that claim presented
10. Payment made by the end of the next month goods/ services invoiced

It is interesting to compare these individual definitions (all provided by the finance director or managing director) with accepted accounting definitions. Warren & Fess [10] define "net 30 day terms" to mean 30 days from the date of invoice. Almost all of the definitions adopted by the building firms examined are beyond the accounting definition.

These replies indicate that any model which uses delays based on a single definition of "30 day terms" should be treated very warily. 30 days takes on various meanings according to the above definitions, such as being 30 days from a particular date (eg 25th of month or end of month) or from valuation date, or from date of invoice, or from when invoice presented, or 30th of the month following invoice, or 30 days from end of month claim is presented! Therefore, if we take a scenario which has work completed and accepted on the 5th of the June 1995 and invoice dated 12th of June and presented two days later using the above responses would give the following ranges for the term 30 day payment. These delays are presented in Table 2. Also presented there are the equivalent delay figures for the instance when the invoice is not presented until the 1st of July.

Table 2. Delay from completion of work to payment

Respondent (Contractor)	date of payment (to supplier/ sub-contractor)	Actual delay with invoice received 14/6	Actual delay with invoice received 1/7
1	5th July	30	56
2	12th July	37	86
3	31 July	56	86
4	31 July	56	47
5	12 July	37	86
6	31 July	56	37
7	3rd July	27	86
8	7th July	31	86
9	31 July	56	37
10	31 July	56	30
Accounting	12 July	37	37

The variance in definition of 30 day payment results in a range of the actual number of days from work completed to payment of 27 - 56 days. The comparison with the accounting definition can be viewed in figure 1. It is clear that using the accounting definition of net 30 day payment and applying weighted delays without taking into account the actual company policy would not be representative of the industry. Further to this it can be seen that the situation can be exaggerated, particularly when receipt of an invoice is delayed.

Figure 1: Chart of payment delay compared with the accounting definition

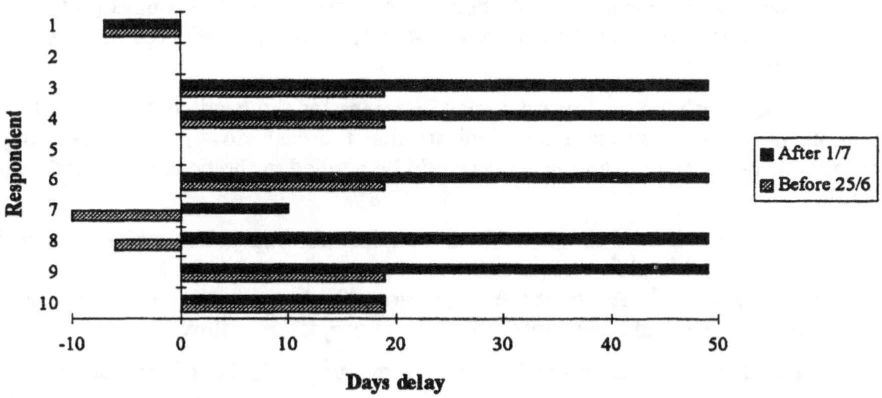

The supplier who is three days late with supplying invoice can actually be paid 90 days after completing the work. If we take for example respondent's 3 and 9 definition of thirty days a claim made three days after the 25th or the end of the month would result in a payment being made to these suppliers on 30th August or a period of 86 days after work complete. If the contractor had claimed from the client at the end of June, as would most probably be the actual situation, and therefore received payment at the end of July, then the weighted delay for the contractor in respect of their cash flow would be recorded as positive; again with accruing interest charges.

A premise of much research to date has been that contractors strictly follow the contract terms with regard to payments. It is assumed payments to suppliers and subcontractors are made on dates specified in the contract usually claimed by the 25th of the month and payment made 30th day of following month. However, the results of the pilot study indicate that payments are frequently withheld as a carrot and stick method of controlling suppliers and subcontractors. Asked whether payments to suppliers are ever delayed due to reasons other than contractual ie. slow progress five(5) or 45% replied, yes. Similarly, when asked whether subcontractor payments were ever withheld for reasons other than contractual ie. as a penalty for being behind program, disagreed back-charges etc., eight(8) or 73% replied yes.

Thus company policy could have potentially significant implications on the company operational cash flow, especially if it is found to be commonplace throughout the course of the contract.

Conclusion

It is clear that contractor financial policy may have hitherto unknown implications on the development of mathematical models for the analysing of the net cash on any given project. This in turn impacts on any overlaying of individual project net cash results required to give an overall company working capital profile.

The accounting definition of "net 30 day terms" being payment 30 days from date of invoice was only being employed by two of the eleven contractors. The remainder have a subcontractor payment distribution range between 23 and 49 days (and longer in certain circumstances). This would obviously impact on any weighted delay analysis of project cash flow.

The contractor's policy toward bank guarantees for the head contractor and cash retention for the subcontractor may indicate that the cash flow may be receiving a considerable boost and weighting delays could be applied in the positive as well as the negative.

Respondents indicated that it was commonplace for contractual obligations to be overlooked if the withholding of payment served as an incentive for the subcontractor to increase rate of work. Again, if the suspension of subcontractor payments is in any way prevalent, this could have a marked effect on project cash flow.

The significance of these results is that previously held beliefs of conformity in the application of payment terms, retention policy, and adherence to contract conditions have been brought into question. If the wide variance to accepted norm is, in fact the norm, then models of working capital derived from building projects may need to be reviewed. Future research must make allowance for this variation in the mathematical models developed.

This pilot study has demonstrated a requirement for more detailed research in the area of the building firm's financial policies and its subsequent impact on working capital generated from individual projects.

Acknowledgments

The researchers would like to thank the contractors who participated in the pilot study for their time and co-operation. The work described herein has been financially supported by Deakin University.

References

1. Nazem, S.M. (1968), 'Planning contractor's capital', *Building Technology and Management*, **6**(10), 256-260.

2. Hardy, J.V. (1970), Cash flow forecasting for the construction industry, M.Sc. Report; Dept. of Civil Engineering, Loughborough University of Technology

3. Peterman, G.G. (1973), 'A way to forecast cash flow', *World Construction*, October: 17-22.

4. Trimble, E.G. (1972), 'Taking the tedium from cash flow forecasting', *Construction News*, **9**, March.

5. McCaffer, R. (1979), 'Cash flow forecasting', *Quantity Surveying*, August: 22-26.

6. Ashley, David B. and Tiecholz, Paul M. (1977), 'Pre-estimate cash flow analysis', *Journal of the Construction Division*, American Society of Civil Engineers, Proc. Paper 13213, **103**(CO3), 369-379. See also discussion by Gates, M. And Scarpa, A. (1978), **104**(CO1), 111-113, and closure by Ashley *et al.* (1978), **104**(CO4), 554.

7. Kenley, R. & Wilson, O.D. (1989), 'A Construction Project Net Cash Flow Model', *Construction Management and Economics* 7, 3-18.

8. Kaka, Ammar and Price, A.D.F. (1991), 'Relationship between value and duration of construction projects', *Construction Management and Economics*, **9**, 383-400.

9. Archer, S.H. (1966), 'A model for the determination of firm cash balances', *The Journal of Financial and Quantitative Analysis*, 1-14.

10. Warren, C.S. & Fess, P.E. (1989), *Principles of financial and managerial accounting*, 2nd Edition, South Western Publishing Co., Cincinnati.

MANAGING PROJECT QUALITY

J.M. Mars
Stichting Bouwresearch, Rotterdam, The Netherlands

According to the NEN-ISO 9000 standards series, each company must prepare a "quality plan" for the individual tasks which it performs. In building, people often talk about a "project quality plan". A project quality plan covers in essence the tasks which the company performs in a project, who bears responsibility for them, how the company manages its own project processes, and what steps it has taken to ensure that the required quality is actually delivered.

Usually in a building project, various planning and executive companies work together to achieve a joint end-product. To be able to supply a good quality end product, all participating companies must work well together. That means that their own individual project quality plans must be harmonised with each other. This can be achieved through an "integral project quality plan" (IPQP).

A model is made for this. The IPQP model is a "blank" model, which must be filled in for each project. It will be completed differently for each project; this is dependent on all sorts of project-specific circumstances. The IPQP model is then nothing more nor less than a starting point, a pattern with the tools and accessories to be brought in to suit the specific project.

In practice not all building partners need have a quality system, nor have their system based on a fixed methodology. Additionally, each project will again in practice turn out differently, because of all kinds of circumstances. However, the IPQP model is not just suitable for projects where all participants operate a standard quality system, or where projects neatly fit into the "ideal" procedure. In principle the model is universally adaptable, even in projects where no individual participant operates a quality system which has been put down in writing.

The Organization and Management of Construction: Shaping theory and practice (Volume One).
Edited by D.A. Langford and A. Retik. Published in 1996 by E & FN Spon. ISBN 0 419 22230 8.

Suivant la norme ISO série 9000, toutes les sociétés doivent établir un "plan de qualité" pour les travaux individuels qu'elles réalisent. Dans le domaine de la construction, on parle souvent de "plan de qualité de projet". Dans un plan qualité de projet, sont mentionnés - brièvement résumés - les travaux que la société exécute dans le cadre d'un projet, les personnes qui en ont la responsabilité, la façon dont la société maîtrise les processus de projet qui lui sont propres et les mesures qu'elle a prises pour veiller à ce que la qualité demandée soit réellement fournie.

Au niveau d'un projet de construction, diverses sociétés d'études et de réalisations collaborent la plupart du temps pour réaliser un seul produit fini commun. Afin de pouvoir livrer un bon produit fini de qualité, toutes les sociétés participantes doivent mettre au point une bonne collaboration. Cela signifie que les plans de qualité de projet distincts doivent s'harmoniser les uns aux autres. Ce qui est possible au moyen d'un "plan de qualité de projet intégral" (PQPI).

Un modèle a été réalisé à cet effet. Le modèle PQPI est un modèle lacunaire qui doit être rempli pour chaque projet. Il sera complété différemment pour chacun des projets ; cela dépend des différentes circonstances liées au projet lui-même. Le modèle PQPI n'est donc ni plus ni moins qu'un point de départ, un exemple proposant une aide et des conseils, qu'il faut remplir en fonction d'un projet déterminé.

Dans la pratique, tous les partenaires participant à des travaux de construction ne disposeront pas d'un système de qualité ou n'auront pas fondé leur système sur une méthodologie déterminée. De plus, dans la pratique, chaque projet se déroulera chaque fois de façon différente en fonction de plusieurs circonstances. Cependant, le modèle PQPI n'est pas seulement applicable à des projets pour lesquels tous les participants disposent d'un système de qualité standard et où les projets suivent correctement le processus "idéal". En principe, le modèle est universellement applicable, même dans les projets pour lesquels aucun participant ne dispose d'un système de qualité consigné par écrit.

Introduction

In Holland an increasing number of construction companies, architectural firms, consulting engineering firms, suppliers and building managers are developing internal quality systems in accordance with the NEN-ISO 9000 series. The (large) construction contractors have taken the lead, but in the last two years various architectural and engineering firms have also made a start.
One of the reasons for this is that the (international) market is increasingly asking firms to prove that they can properly manage and control quality in projects. In this context many firms are pursuing certification.

The quality plan

According to the NEN-ISO 9000 series every company must produce a "project quality plan" for individual orders. This is therefore the translation of such a company's general quality system for a specific order or a specific project.

The project quality plan states which tasks are carried out by the firm, who is responsible for this, where the different authorities are situated, how the firm manages its own project processes and those measures it has taken to ensure that the quality required is actually delivered.

However, different design, construction and supply firms work together in a construction project to produce one collective end product. All of the participating parties must therefore cooperate well if they are to be capable of delivering a good quality end product, a building that satisfies the client.
This means that quality care among all the construction partners, with their individual project quality plans, must be well synchronised; an Integral Project Quality Plan (IPQP) can be used for this.
An "IPQP Model" has been developed for this and this can be used as a manual for producing project orientated quality plans. The "Model" incorporates a number of general procedures that can be filled in depending on the project circumstances.

The Model is **primarily** intended for participants who are held responsible by their clients for the management of construction projects in the different phases of the construction process.
The initiative for producing an IPQP must come from the client or the managing contractor.
This party must appoint a manager who is responsible for producing the IPQP for each phase of the construction project. The form of organisation chosen for construction will decide which of the participants is mainly responsible in a specific phase of the construction process.

The QSM framework

The Model is based on the Construction Quality System Model (QSM framework). In Holland, various participants in the construction sector have derived their quality care models from the QSM framework. The most important feature of the QSM framework is the structure for building a quality system in the form of a matrix. The functional steps of the business process are set out horizontally in the matrix; the aspects which have to be managed in each process are set out vertically.

QSM-Framework *structure*	0. GENERAL	1. INITIATIVE	2. DESIGN	3. ELABORATION	4. PREPARATION/ IMPLEMENTATION	5. IMPLEMEN-TATION	6. USE
0 goals/principles							
1 organisation							
2 communication							
3 requirements							
4 resources							
5 purchasing/third parties							
6 time							
7 finances							
8 realisation							
9 experiences							

It is derived from the Norwegian Bygg-forsk system for the construction industry.

In cell 0-0 of the matrix, "general - goals and principles", a firm can describe the general quality policy and the managing strategy for accomplishing the policy objectives. The firm can formulate derived objectives for each phase in the business process in the remaining cells of row 0.
In column 0, "General", the firm can further describe how it generally goes about the management, quality control and quality assurance of its business processes based on the quality policy and strategy from cell 0-0.
This can be worked out in detail, as required, in the other fields of the matrix (cells 1-1 to 6-9), and may be backed up by all kinds of tools such as check-lists, standard forms, data files and calculation methods.
In practice this means that all procedures and tools within the quality system are given codes that correspond to the relevant cells in the matrix.

This may be seen as a chest of drawers, every drawer of which can be used separately without losing its cohesion with the whole.

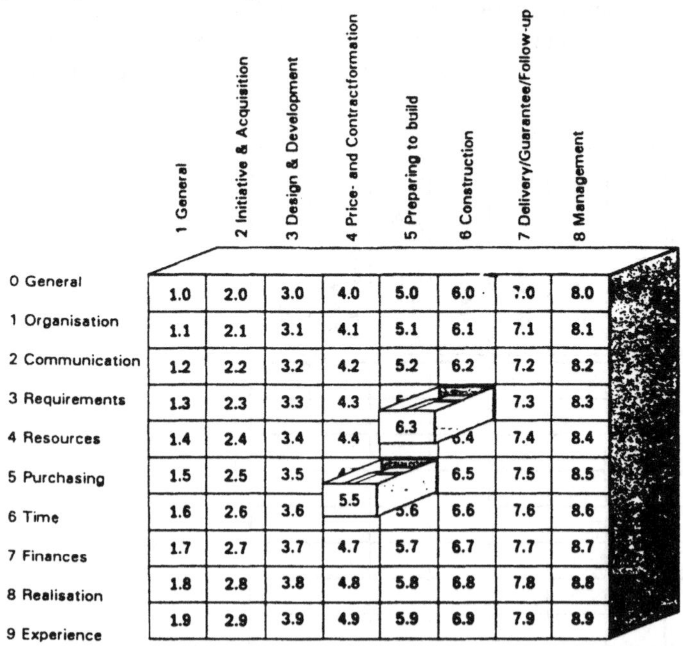

Good conditions for integral quality care in a construction project are created as a result of the layout, which is jointly chosen by the construction partici-pants.

The Model

The IPQP model is built up in the same way. It can therefore be used as a guideline when producing project orientated IPQPs. The IPQP Model contains an indication of what information could or should be incorporated in an integral project quality plan - depending on the specific project situation - and an indication of how such a plan can be drawn up.

The information in the IPQP mainly concerns the intrinsic and organisational synchronisation of the tasks of the construction partners involved. The managing. It is aimed at answering questions from the client relating to the form of organisation chosen for construction. An IPQP incorporates the agreements and practical tools which are necessary for this synchronisation. When considering "tools", one can think of checklists, standards, contract schedules, planning systems and other management tools.
Unfortunately, it is still true in practice in the construction sector that not all construction partners have a quality system. Others have not based their system on the QSM framework. Many installation firms for example have synchronised their quality system with dozens of sector-specific assurance criteria and are even certified in compliance with these criteria. Other firms have synchronised their quality system directly with NEN-ISO 9000.

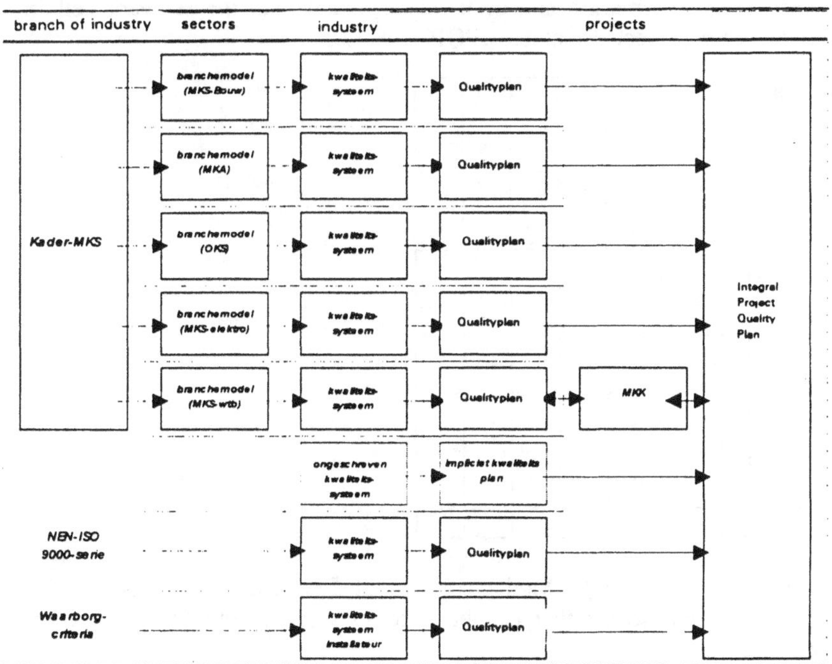

However, the IPQP Model is not just applicable in projects where all the participants have a quality system based on the QSM Framework. In principle it may be applied universally.

The Model also provides the possibility of adapting the phasing of the project to the circumstances and principles of the project concerned. Firms that have a quality system à la QSM Framework will be able to provide input to an IPQP relatively simply and - conversely - will be able to adapt their own project quality plan relatively simply to the conditions imposed on this by an IPQP.

The structure of the IPQP Model

The IPQP Model contains directions for producing the project orientated IPQPs. How this is actually worked out will be different for each project; this depends on all kinds of project-bound circumstances.
The matrix columns, or phases in the construction process, and the "General" column form the main sections of the Model.

The vertical layout of the matrix consists of "management aspects", which are process and product aspects that have to be managed in every project. The management aspects make up the paragraphs of the IPQP Model; each cell in the matrix represents an IPQP Model paragraph. There is a description for each paragraph of what information can or must be included in the relevant cell in the framework of a project orientated IPQP

IPQP-Model *structure*	0. GENERAL	1. INITIATIVE	2. DESIGN	3. ELABORATION	4. PREPARING TO BUILD	5. CONSTRUCTION	6. USE
0 goals/principles							
1 organisation							
2 communication							
3 requirements							
4 resources							
5 purchasing/third parties							
6 time							
7 finances							
8 realisation							
9 experiences							

The information for the IPQP Model consists of "work orders". Appendices, standards, reports and so on that can be used in working out the project are referred to here.

The first cell of column "0 - general" contains a general project description.

The nature of the project; who is the client, who are the (envisaged) users, what accommodation problems does the project have to solve? General policy principles can also be formulated, including those in the areas of environment and working conditions.
The goals and principles for the different phases can be described in the remaining cells on row "0". This is done by describing start documents and results (decision documents) for each phase. The results from one phase form the basis for the "start documents" for the next phase.
In each project the project partners must agree on what information has to be included in a start document, what is taken into a phase, what basic documents will be supplied in the different phases and what information must at least be contained in the decision documents.
The project organisation is described in cell 0.1, "General - organisation); the form of organisation chosen for construction along with the associated main responsibilit(y/ies) for each phase and the apportionment of tasks, responsibilities and authorities.
A conclusive apportionment of tasks is important for the project partners' contract agreements; reference may be made to the task apportionment charts in the contracts. I will come back to this later.

Consultation structures and principles for information exchange in a project are described in row "2", "Communications".
This can be done in a general sense in cell 0.2 "General - Communication"; the principles can be worked out for each phase in the remaining cells of row 2, including the information carriers to be used for this.
Row 3 contains a description of how every participant can always be assured of having the most up to date versions of internal and external requirements and conditions applicable to the project. These can be listed, specified by phase, in the remaining cells of row 3.

Row 4 "Resources" indicates for each phase the common tools (including calibrated inspection and approval resources) which will be jointly used to achieve the objectives.

The procedures and tools for the selection of partners who are to be added during the process are described in row 5, "Purchasing / Third parties".
Procedures are also included for the purchase of jointly acquired goods and materials.

Row 6, "Time", is the overall planning and detailed planning of each phase. A place must be given not only to the plans themselves, but also to the set up and supervision procedures.

The information for managing investment costs is described in row 7 "Finances".

The operational procedures for quality management in each phase are described in row 8 "Fulfilment". The procedures must ensure that project management always has insight into progress of the process so that this can be driven intrinsically. There must for example be a specification of who inspects what, and when and where.
Finally, in row 9 "Experiences", there is a description of how experiences, test results and so on have to be collected and fed back both within the project and for the benefit of subsequent projects. In principle the process must be evaluated at the end of every phase, in relation to both quality achieved and process progress. The evaluation may lead to adjustments in the following phase.

Now I am going to give a few examples of how this is filled in.

Column 2: design

2.0 Goals and principles

- Using the procedure in paragraph 0.0 establish what the **results to be achieved from the design phase** have to be:
  - check the important decisions that have to be taken in the design phase and, based on this, decide on possible sub-phasing;
  - check what **start information** is necessary to be able to take well founded decisions (check results of previous phase);
  - establish what minimum information has to be in the dossier at the end of the (sub-)phase(s) to enable well founded decisions on progress of the project to be taken. Use the task instruction here.

2.1 Organisation

- Check whether changes are necessary in allocation of the **final responsibility for the entire project.** Determine who will be acting as **mainly responsible partner for the design phase** and who will be acting as the **party responsible for drawing up and implementing the IPQP**.
  Work out the **organisation structure** for this phase in more detail in an **organisation chart**.
- Apportion the **product development tasks, responsibilities and authorities** in accordance with paragraph 0.1.

Elements that have to be discussed here are:

- list of tasks which must be performed to enable realisation of the **decision documents / phase results** or to enable the dossier to be filled for each sub-phase;
- the **composition of the team** for this phase;
- appointment of a **responsible partner for each decision document**;
- draw up **task apportionment plans**, which include indications of the product development tasks and (derived) responsibilities and authorities of the participants involved for production of the decision documents;
- check that the apportionment of tasks has been intrinsically concluded.

- Where necessary, supplement paragraph 0.1 of the (project-bound) IPQP with the most up to date information.

Column 5; construction

5.3 Communication

- Establish in detail the **consultative structure** for the construction phase in accordance with paragraph 0.2. Aspects that have to be discussed here are:
  - types of consultation, participants in each type of consultation, (standard) agendas;
  - full meeting schedule;
  - arrangements for convening, chairing and minute taking;
  - arrangements for the distribution of minutes.
- Implement the procedure for **document management** and **information exchange** for the fulfilment phase (see paragraph 0.2). Establish the routing and handling of vital (decision) documents in a **"document schedule"**.

5.3 Requirements

- Implement the procedure for **requirements management** (paragraph 0.3) which apply to the project process in the fulfilment phase (relevant legislation and regulation, product and process requirements):
  - list the requirements relevant to this phase;
  - ensure that the latest known versions are available in the right places at the right times.

5.4 Resources

- Manage the **tool and/or product resources to be supplied jointly by the process partners** for the fulfilment phase in accordance with the agreed procedure (see paragraph 0.4), i.e.:

- list the resources to be used jointly in this phase;
- ensure that these resources are always available at the right time, in the correct quantities, at the right place and in good condition.

- Check whether individual partners manage their own digital and measurement systems properly and whether they are working with accredited software.

Tasks Model

Base documents for each phase have to be present in the quality plan in order to enable the right decisions to be taken. A task inventory is used for this.
Project managers can use these as a checklist when listing the tasks in a project. This involves tasks which would normally be carried out in every construction project with "ideal" progress of the process.
In practice every project is different and the processes and tasks will therefore differ on many points from the "ideal" process each time.

The task inventory is derived from the "decision documents" in the construction process. These decision documents are the interim results for each phase of the construction process. The tasks are linked to decision documents.

IPQP			2. design						
	0. general	1. initiative	2.1 structural design	2.2 pre-design	2.3 final design	3. preparing to build	4. preconstruction	5. construction	6. use
0 goals/principles									
1 organisation					■				
2 communication									
3 requirements									
4 resources									
5 third parties									
6 planning									
7 cost control									
8 realisation									
9 experiences									

Tasks in the framework of a construction project may be separated into "intrinsic professional tasks" or "product tasks" and "coordination and steering tasks" or "process tasks".

Here, this mainly involves the product tasks: the tasks that must be executed in order to obtain a good (interim) product; the decision document, based on which a go/no go decision is possible. This involves filling in row 1, "Organisation", for each phase.
The coordination and steering tasks will come up for discussion themselves when working out the remaining matrix lines.

A "task" is understood to be:
a cluster of associated activities, aimed at achieving a well defined result, which is generally executed by one participant.
I will also give an example of this.

Final Design Task Inventory

- Working out the functional and architectural structure and the layout of the land in detail (final building block borders, final spacial plan, location and size of "material box plan" building sections, outer wall layouts, principle details, material use and colour of image-governing elements);
- establishing the dimensions of and drawing foundation and load bearing structures, developing principle details for the foundations and load bearing structures;
- establishing the dimensions of and drawing the water and electrical installations and ducting packages, specifying the capacity of the installations;
- developing the Final Design base on the pre-design results with integration of the detailed functional, structural and technical installation calculations;
- consult with municipal services;
- initial consultation with the building inspectorate;
- consult with the local fire service;
- consult with the utility companies;
- calculate construction costs (in elements and sub-elements) and investment costs, inspections and so on;
- estimate operational costs and incomes, inspections and so on;
- coordination of partners' health and safety obligations in the design phase, based on the Working Conditions Act - Construction Process Decree; maintain/update the Safety & Health plan and the Safety & Health dossier;
- possibly: produce environmental effects report;
- test final design against the current Programme of Requirements; accept the Final Design;
- submit the provisional building application;
- decide on next step.

Finally

Good communication between the partners from the initiation phase right through to the management phase of a construction project is important in managing a good quality product. It seems that in practice, good communication is still a difficult problem to solve in construction and because of this it is one of the greatest factors in the costs of failure.

The Integral Project Quality Plan Model (IPQP) provides ample opportunities to improve and optimise mutual communication. It gives an indication of what participants have to communicate and when, what they have to communicate about and where the relevant information is to be found. The agreements that are made are recorded and controlled. Everyone can draw on these. Because of this Quality Management is guaranteed.

EVALUATION OF THE QUALITY ASSURANCE SCHEME IMPLEMENTED IN PUBLIC HOUSING PROJECTS OF HONG KONG

C.M. Tam
Department of Building and Construction,
City University of Hong Kong, Hong Kong
A.P.C. Chan
School of Building and Planning, University of South Australia,
Adelaide, Australia

Abstract
Quality assurance has attracted much attention in the last few years. The release of the ISO 9000 in 1987 further fuelled the 'quality debate' in Hong Kong and all over the world. The construction industry in Hong Kong is no exception. In April 1990, the Housing Authority mandated the requirement that main contractors must take ISO 9000 certification before 31 March 1993. Contractors then made a dash for ISO 9000 registration. In less than two years' time, more than 50 contractors have successfully acquired certification and many of them are still in the queue. The industry has expended large amounts of money and resources in setting up and running the system. In fact, the study shows that the setting up cost of the system ranged from HK$1 - 3 million with a mean running cost of around 0.2% of contract values. Given the above, reasonable questions might be, "What are the benefits in return? Has the quality of buildings really improved?"

This paper aims to study the benefits claimed in exercising the system. The study has exposed a number of difficulties encountered by contractors in meeting the ISO 9000 requirements, due partly to the unique characteristics and attributes of the construction industry in Hong Kong.
Keywords: Quality Assurance, Main Contractors, ISO-9000, Quality System, Quality.

Sommaire
L' assurance de qualité a attiré beaucoup d' attention récemment. La création d' ISO 9000 en 1987 a relancé la discussion sur le "débat autour de la qualité" à Hong Kong et ailleurs. L' industrie de construction à Hong Kong n'y échappe pas. En avril 1990, le Ministère du Logement a obligé les entrepreneurs principaux de souscrire à la riglementation d' ISO 9000 avant le 31 Mars 1993. En consequence, les entrepreneurs

The Organization and Management of Construction: Shaping theory and practice (Volume One).
Edited by D.A. Langford and A. Retik. Published in 1996 by E & FN Spon. ISBN 0 419 22230 8.

se sout précipités pour s'enregistrer à ISO 9000. En moins de deux ans, plus de 50 entrepreneurs out obtenu le certificat arec succis et un grand nombre d' entre eux font toujeurs la quene. L' industrie a consacré grandes sommes d' argent et de ressources pour créer et gérer le système. En fait, l' étude indique que le coût pour le fondement du système était compris entre HK$1 et 3 millions avec un coût moyen des frais de gestion d' environ 0.2% des valeurs de contrat. Compie -tenu de ces données, on pourrait poser les questions suivantes, " Quels sout les avantages en échange? Est-ce que la qualité des immeubles s'est vraiment améliorée? "

Cet article vise à étudier les avantages tirés par la pratique de ce système. L' étude a revelé un certain nombre de difficultés rencontrées par les entrepreneurs pour satisfaire aux critères d' ISO 9000, dûes partillement aux caractéristiques uniques et aux specificités de l' industrie de construction à Hong Kong.

1 Introduction

The increasingly competitive and quality discerning international markets indicate a compelling need for all industries in Hong Kong to upgrade the quality of their 'products'. In March 1990, the Hong Kong Government launched the 'Quality Awareness Campaign' which stressed the importance of 'Quality'. Thereafter, 'quality' became the centre of debates in Hong Kong.

The ISO 9000 series of quality management and assurance standards was published by the International Organisation for Standardisation (ISO) based in Geneva in 1987. In the following three years, as recorded in 1990, over 30 countries [3] adopted the ISO 9000 Certification Scheme. In the same year, the Hong Kong Quality Assurance Agency (HKQAA) started administering the certification exercise.

The first ISO 9000 certificate in Hong Kong was awarded on 16 April 1991 to a semi-conductor manufacturer [4]. The first ISO 9002 certified contractor was designated in June 1992. Up to May 1994, over 200 companies have been awarded ISO 9000 certificates. Table 1 shows the distribution of organisations certified up to the end of April 1994 [1] .

Table 1: Types of Business Obtained ISO 9000 Accreditation as at April 1994

Type of Business	Number of Organisation
Construction Contractors	55
Concrete Production Plants	47
Construction Related Engineering Firms	12
Telecommunication	15
Electronic/ Computer Production Firms	36
Manufacturing Firms: Toys, Office Appliance, Bicycle, etc.	7
Others	14

It will be noted that the construction industry sector takes up a significant proportion. This is attributable to promotion of the scheme by the Housing Authority; implementation of quality assurance system and ISO 9000 registration through the Hong Kong Quality Assurance Agency has been made mandatory by the Hong Kong Housing Authority for main contractors tendering for public housing projects in Hong Kong after mid 1992. This makes Hong Kong unique in the world, having the largest percentage of building contractors registered in the ISO 9000 scheme, over 50% of the accreditated ISO 9000 certification companies falling into the construction sector.

2 Background of Quality Management in the Construction Industry

Due to scarcity of land and very rapid increases in population, the Hong Kong government was concerned about quantity rather than quality of housing in the 60's and 70's. This resulted in the poor quality public housing and concomitant high maintenance and remedial costs incurred in the last decade. In pace with the economic progress and the worldwide quality awareness, the aspirations of Hong Kong people have increased and the whole of society has become more quality aware. In recent years, the Housing Authority introduced the Performance Assessment Scoring System (PASS) and ISO 9000 certification requirement in order to overcome its poor quality image. The Housing Authority required that before 31 March 1993, all construction contractors must be ISO 9000 certified. Although there was a leeway period from April 1990 to March 1993, most contractors rushed to obtain the accreditation before March 1993. Most people believed that the accreditation exercise would be beneficial to the quality of the industry.

It has been two years since the introduction of the Quality Assurance Accreditation Scheme in the construction industry. The writer has conducted a comprehensive survey to investigate the scheme.

3 The Study

The study was conducted in 1994, and comprised two parts: mailed questionnaire survey and structured interviews.

3.1 Mailed Questionnaire Survey

The questions were designed to assess the benefits of the quality assurance (QA) scheme that were partly envisaged by the author and partly claimed by previous researchers [5]. Each question was split into two parts; the first being the expected benefits before running the QA scheme, the second concerning the actual benefits experienced after its implementation.

There were 42 replies from 136 dispatches sent by facsimile. The targeted organisations were obtained from the Buyer's Guide, a register of the HKQAA ISO 9000 certified companies [2]. Companies are those who have obtained the ISO 9000 certification. The response rate was 31%, an acceptable figure in comparison with similar surveys in Hong Kong. The results of the questionnaire survey are summarised in Table 2. The answers were then verified by 'Prior Probability'. Answers favoured by around 50% respondents or less are obtained by no better than chance. Results which were lying well clear of 50% are significant. The following section discusses a few interesting findings.

1. Site conditions: Most people expected an improvement in site control and conditions. Table 2 shows that most people were actually enjoying the advantage. The high percentages obtained indicate that there was an actual improvement in site conditions. This may be the result of improvements in proper material delivery, handling, storage and packaging systems on site.
2. Overall quality awareness: 95% of respondents agreed that the QA programme can induce quality awareness potentially and 70% actually encountered the expected outcome. This is the only significant item among the various quality issues proving to have a positive effect on quality.
3. Reduction in defects, rejection and remedial works: 85% of respondents expected that there would be a reduction in defects, the rejection of works, and remedial works. However, only 40% of the respondents actually encountered these benefits. The low percentage (below 50%) cannot statistically verify the actual reduction in defects. This shows that the quality of 'products' is not actually improved with the implementation of QA system in an organisation. One of the reasons was the huge amount of paper work, deterring people from actually carrying out the necessary inspections before putting their names on the checking forms.
4. Reduction in quantities of defects during the defect liability period: Once again the findings agree with the above result that few people experienced such a benefit as shown in Table 2 although 85% of respondents expected such an outcome.
5. Improved filing and documentation system: 95% and 85% of respondents expected and actually experienced such a benefit. The high percentages indicate that the documentation system has been improved.
6. Delineation of Responsibility: 95% and 80% of respondents expected and actually experienced such a benefit as shown in Table 2. The high percentages indicate the improvement in this respect due to the implementation of QA system.
7. Motivation: Just over half of the respondents expected an improvement in staff morale. This infers that people are not quite sure whether the system would motivate their staff. However, none of them did observed the benefit in reality. This means that the QA system may even generate a negative effect on staff morale.
8. Improved documentation leading to less reliance on memory: 95% and 85% of respondents expected and actually experienced such a benefit. The high percentages indicate that there was some improvements in this respect.
9. Speedier information update: 85% and 70% of respondents expected and actually experienced such a benefit. The high percentages indicate that there was actual improvement in this respect.
10. Controlling: 89% and 71% of respondents expected and actually experienced such a benefit respectively. The high percentages indicate the likelihood of improvement in this respect.
11. Setting up costs: Small contractors estimated the costs to be around HK$1 million, while large contractors quoted HK$3 million.

12. Running costs: The costs quoted were quite variable; ranging from 0% to 1 % of the contract values. The mean lies at 0.2%.

Table 2: The Overall Results of the Questionnaire Survey

FEATURES	EXPECTED		ACTUAL	
	YES	NO	YES	NO
General Site Discipline				
Improvement in site conditions such as cleanliness and tidiness	95%	5%	70%	30%
Improvement in site safety and security	75%	25%	50%	50%
Reduction in wastage and re-work	75%	25%	45%	55%
Project Management				
Reduction in project duration due to less remedial works	55%	45%	10%	90%
Improvement in competitiveness and marketability	90%	10%	75%	25%
Internal Management				
Improvement in internal management discipline	85%	15%	55%	45%
Quality of Works				
Overall improvement in quality awareness	95%	5%	70%	30%
Reduction in defective works	85%	15%	40%	60%
Reduction in defects during the defect liability period	85%	15%	45%	55%
Planning				
Overall improvement in planning	88%	12%	73%	27%
Organising				
Improvement in site supervision and management	95%	5%	60%	40%
Improved filing and documentation system	95%	5%	85%	15%
Clearer delineation of responsibility	95%	5%	80%	20%
Motivation				
Improvement in staff morale	55%	45%	0%	100%
Communication				
Improvement in site communication	75%	25%	50%	50%
Improvement in communication with architects and consultants	55%	45%	25%	75%
Easier agreement on performance criteria with clients	50%	50%	20%	80%
Improved documentation leading to less reliance on memory	95%	5%	85%	15%
Speedier information update	85%	15%	70%	30%
Controlling				
Improvement in site control due to formalised checking procedures	89%	11%	71%	29%

3.2 Structured Interview

The interviewees included site supervisors, project managers and quality assurance managers from eleven large and small building contractors who have obtained the ISO 9000 certification; some employing more than 500 staff and with annual turnover greater than $500 million. The interview results have highlighted problems encountered by ISO 9000 certified contractors and the problems are summarised as follows:

1. Control of Subcontractors: The labour supply structure in the industry is characterised by the presence of multi-layers of subcontractors. It is quite impossible to impose the requirements of ISO 9002 on these subcontractors as few of them have any management system or quality system; some are not even registered as a company. The QA managers must assume 100% management control on these subcontractors. However, as the skill of labour is not tested or proven, and labour mobility is high, this makes the tasks very difficult and costly.
2. Control of Suppliers: Most materials are purchased through agents who named themselves 'suppliers' in the industry. These agents normally deal with many sub-agents who order goods from manufacturers all over the world; such as the PRC, the Philippines, etc. where no quality system is in existence and the arm of the main contractor hardly reaches.
3. Inspecting, Measuring, and Testing Equipment: Owing to the vast workforce involved and the large number of trades in a job site, calibration, identification and traceability of all the measuring and testing equipment are very difficult to manage. For example, it is impossible to control measuring tapes used if everybody brings his own tape. Moreover, most small contractors usually hire plant rather than owning, and calibration and control of a third party's plant could be very difficult.
4. Changes in Design and Drawings: Changes in design and drawings is very common in the industry. This results in many variations and additional works hampering the documentation system and communication efficiency.
5. Site Management and Control: ISO 9000 involves setting up a proper documentation system resulting in a great deal of paper work and control forms. It is, perhaps easier for younger staff but very difficult to educate older and more experienced site staff. These result in site staff bypassing the checking procedures by just signing the forms without doing the actual checking. For small contractors who have limited additional resources, the lack of additional staff to support the QA system may jeopardise it.
6. Multi-quality system: In cases where most nominated subcontractors and suppliers may have their own quality systems, and different procedures, then it may be difficult to reach a compromise on the system used. Thus the amount of paper work may be doubled or even tripled.
7. Considerable Amount of Paper Work: In order to ensure 100% quality and compliance with specification, at least 10% checking to every process of work is required. Considering the large number of different trades working simultaneously in hundreds of different locations on a large site, there are thousands of check forms to be handled daily and each form may need to be signed and verified by sometimes three or more persons. This considerable

amount of paper work often annoys site management staff, who then may cut corners in handling the paper work.

8. Acceptance Criteria: Different people may have different acceptance criteria, especially as regards qualitative works such as finishing trades. Subcontractor themselves may accept works which are not acceptable by the main contractors. This always creates disputes and confrontation.
9. Document Control: Often, the quality system stipulates a document control procedure that all obsolete documents must be promptly removed from all points of issue. That is extremely difficult in the industry because:
   - the existence of the multi-layers of subcontractors makes the chain of issue very long and hardly reachable.
   - many drawings may be affected by a single change.
   - it is still very common for architects and engineers to give verbal instructions, sometimes without written confirmation.
10. Material control: The handling, storage, packaging and delivery methods stipulated in the quality system may sometimes be very difficult to follow due to site congestion and restricted site conditions as encountered in most private jobs in Hong Kong.
11. Contract Review: Some interviewees asked sarcastically, "Would any one believe that a contractor would submit his non-conformance report to the Architect?"
12. Competitiveness and Marketability: For contractors who work both for the public housing and private jobs, they may be deterred from applying quality systems in private jobs in order to cut the administrative costs and reduce abortive works. Nevertheless, the Hong Kong Quality Assurance Agency requires contractors to implement the quality assurance system in all construction projects under the same company name once it becomes certified. As the adoption of quality systems is more costly, it will no doubt hamper contractors' competitiveness in other business sectors. In the meantime, private clients and architects may not be ready for a quality assurance system and therefore, may not welcome and even refuse to adopt it.
13. Client's Control: One of the major principles of QA is that all parties to the building process must play an active role and their respective contributions may be integrated [6]. However, with the current contractual agreements, clients and architects retain the ultimate authority in approving subcontractors, materials used, and construction process. Hence contractors need to follow instructions in carrying out their works. Thus there is little room for contractors to contribute innovative, cost saving and quality enhancing suggestions.
14. Pre-screening of Tenderers by HKQAA Certification: The contractual requirement for contractors to be certified with ISO 9000 by HKQAA takes away the authority of clients to choose the right contractor for their works. The authority is conveyed to an independent outside agency whose only interest is the contractors' compliance with the 18 criteria of ISO 9002 and may not necessary make any references to contractor's past performance on building projects. Some interviewees asked, "Is the Housing Authority really concerned with the contractor's non-compliance with contract review, internal audits, or statistical techniques?"

15. Full Top Management Support: In order to secure the accreditation and window-dress the quality manual, the quality policy always stressed the need for full support of the quality system from top management. However, some interviewees mentioned that profitability is more important to them; sometimes in the sacrifice of quality, rather than following the quality management system. Most people admitted that the QA system is designed to meet the contractual requirement rather than trying to improve the company's internal efficiency and external image.

4 Conclusion

The questionnaire survey unveils that site conditions such as tidiness and cleanliness, were indeed improved but safety, security, wastage and rework remained unchanged after the implementation of the QA system. Item 10 of the interview highlighted the difficulties in following the material control procedures due to site congestion and restricted site conditions which are the attributes of most private building sites in Hong Kong. The project duration was unaffected, but the competitiveness of contractors was improved after adopting the QA system. Contractors did not see any improvement in their internal management affairs.

Perhaps this may be due to the large amount of paper work generated which would of course not be welcomed by site staff and thus cause dissension that induced frustration and had negative effects on staff morale.

As regards quality of works, only the quality awareness of organisations was improved while the actual defects, rejection and reworks were not reduced. The frequent change in design and the difficulties in controlling the multi-tier subcontractors and suppliers hampered the quality improvement effort.

Documentation systems and information updates were improved but otherwise communication was generally no better. This is due to the existence of multi-quality systems and the difficulty in quantifying the acceptance and performance criteria.

The major benefits of the QA system were the improvement in site planning and procedural control, site supervision, documentation and delineation of responsibility. However, the benefits may not outweigh the costs of the QA system which include a setting up cost of HK$1 to 3 million and a mean running cost of 0.2% of the contract values.

If the advantages of the QA system are to be fully achieved, especially the improvement in quality, wholehearted commitment from top management of contracting firms and more effort input by clients to encourage the system are required. Although the quality of works has not been improved two years after the implementation of QA systems, the rise in quality awareness both in the public and construction professionals may eventually culminate in the fruitful results.

References

1. Hong Kong Quality Assurance Agency, HKQAA (Up to 13 May 1994, issue 26), ISO 9000 Certification (Buyer's guide) - Register of HKQAA certified companies according to ISO 9000: 1987. Hong Kong Government Printer.
2. HKQAA (September 1994, issue 29)., ISO 9000 Certification (Buyer's guide) - Register of HKQAA certified companies according to ISO 9000: 1987. Hong Kong Government Printer.
3. Shui On Construction Review (1991), Focus on QAA.
4. Industry Department: Quality Quest, Issue No.9, March 1994.
5. BSI Quality Assurance "International Quality Assurance Management System Standard. " Milton Keynes: BSI Quality Assurance 1990. (Videorecording)
6. The Chartered Institute of Building, CIOB, 1990, Quality Assurance in the Building Process.
7. Hong Kong Quality Assurance Agency, HKQAA, (1992), An Executive's Guide to the Use of the International Standard for Quality Systems. Hong Kong Government Printer.

INDEX

www.ingramcontent.com/pod-product-compliance
Lightning Source LLC
LaVergne TN
LVHW011105300926
849160LV00008B/280

9780367579579